AF344437

R. Schlingensiepen · W. Brysch · K.-H. Schlingensiepen (Editors)

Antisense – From Technology to Therapy

R. Schlingensiepen
W. Brysch · K.-H. Schlingensiepen (Editors)

Antisense – From Technology to Therapy

Lab Manual and Textbook

With 101 illustrations

EX LIBRIS <ROCHE>
Volume 6

Blackwell Science

Blackwell Science Ltd
Editorial Offices:
Osney Mead, GB-Oxford, OX2 0EL
25 John Street, London WC1N 2BL
23 Ainslie Place, Edinburgh EH3 6AJ
238 Main Street, Cambridge,
Massachussetts 02142, USA
54 University Street, Carton,
Victoria 3053, Australia

Other Editorial Offices:

Blackwell Wissenschafts-Verlag GmbH
Kurfürstendamm 57, 10707 Berlin, Germany *and*
Zehetnergasse 6, 1140 Vienna, Austria

Blackwell Arnette SA
224, boulevard Saint-Germain
75007 Paris, France

Editors' addresses:

Dr. Reimar Schlingensiepen
Max-Planck-Institut für
Biophysikalische Chemie
Abt. Neurobiologie
Am Fassberg 11
D-37077 Göttingen

Dr. Wolfgang Brysch
Biognostik GmbH
Carl-Giesecke-Straße 3
D-37079 Göttingen

DISTRIBUTORS

Marston Book Services Ltd
PO Box 87, Oxford OX2 0DT
(Orders: Tel: 01865 791155
          Fax: 01865 791927
          Telex: 837515)

North America
Blackwell Science, Inc.
Commerce Place, 350 Main Street
Malden, MA 02148
(Orders: Tel: 617-388-8250
          Fax: 617-388-8255)

Australia
Blackwell Science Pty Ltd
54 University Street, Carlton, Victoria 3053
(Order: Tel: 03 347-5552)

Dr. Karl-Hermann Schlingensiepen
Biognostik GmbH
Carl-Giesecke-Straße 3
D-37079 Göttingen

© 1997 Blackwell Wissenschaft Berlin · Vienna

Set by deutsch-türkischer fotosatz, Germany
Printed and bound by Druckhaus Schöneweide, Berlin, Germany

A catalogue record for this title is available from the British Library

ISBN 0-86542-669-4
Printed in Germany

Table of Contents

Preface

Antisense technology has been moving at an unforeseeably rapid pace. More than 15 clinical trials with antisense oligonucleotides were initiated within the last two years, and technology is rapidly generating new drug candidates in such diverse fields as oncology, immunology, cardiology and infectious disease. Antisense is a platform technology with innovative potential in basic as well as applied science.

Gene function analyses using the antisense technique in both *in vitro* and *in vivo* experiments have yielded new insights into the function of genes, ranging from fundamental processes such as cell proliferation and differentiation to such complex processes as animal behaviour.

This book was developed due to the initiative of Dr. Stefan Manth from Hoffmann-La Roche, who suggested to the editors that the exponentially growing field of antisense technology demanded a practical "hands on" book with detailed protocols from experienced scientists in the field. Taking a keen interest in the innovative potential of antisense technology, especially from the perspective of the pharmaceutical industry, Dr. Manth actively participated in the late stages, i.e. the hot phase of preparation of the 1st and 2nd "International Conference on Antisense Nucleic Acids" in Garmisch-Partenkirchen in 1993 and 1995. This subsequently prompted him to suggest editing the book to the organizers.

This volume is designed to give newcomers to the antisense field an easy introduction to the technology and, hopefully, to help them avoid potential pitfalls and time-consuming errors. Experienced scientists will find numerous suggestions and ideas on how to design more complex *in vitro* and *in vivo* experiments.

Practical protocols for antisense experiments in crucial biological and biomedical subjects are the core of this book. Reviews of the literature in certain key fields provide an overview of current antisense work and allow easy access to further reading in these areas. Due to the rapid growth of the field, providing a complete review of antisense literature in the individual fields is already beyond the scope of a book designed to serve as a guide to practical application of the antisense technique in basic research, drug development and experimental medicine.

We are indebted to Dr. Stefan Manth for his continued moral support during the preparation of this book. We thank Georg-Ferdinand Schlingensiepen for his advice and competent help in many aspects of the preparation of this work. Furthermore, the enthusiasm of Dr. Axel Bedürftig from Blackwell Science to take on this project and his competent and practical support throughout the project were very helpful. Special thanks go to Dr. Tina Schubert from Blackwell for her expertise, competence and excellent assistance. Her helpful cooperation in all phases of the preparation of this volume is gratefully appreciated.

Last but not least, without the continued help, support and endurance of our wives, Sousan, Irene and Margarethe this work would not have been possible.

Göttingen, July 1996
Reimar Schlingensiepen, Wolfgang Brysch, Karl-Hermann Schlingensiepen

List of Principal Contributors

Dr. Sudhir Agrawal
Vice Pres. of Discovery
Principal Scientist
Hybridon Inc.
One Innovation Drive
USA-Worcester, MA 01605

Dr. Lauren E. Black
Division of Antiviral Drug Products
Office of Drug Evaluation II, FDA
5600 Fishers Lane,
USA-Rockville, MD 20857

Prof. Dr. Ulrich Bogdahn
Neurologische Universitätsklinik
im Bezirkskrankenhaus
Universitätsstraße 84
D-93053 Regensburg

Dr. Wolfgang Brysch
Biognostik GmbH
Carl-Giesecke-Straße 3
D-37079 Göttingen

Prof. Dr. Joachim W. Engels
Universität Frankfurt
Institut für Organische Chemie
Marie-Curie-Str. 11
D-60439 Frankfurt am Main

Prof. Alan M. Gewirtz
University of Pennsylvania
Dep. of Pathology
Room 230, J. Morgan Building
36th Street and Hamilton Walk
USA-Philadelphia, PA 19104-6082

Dr. Markus Heilig
University of Göteborg
Dep. of Psychiatry and
Neurochemistry
Mölndal Hospital
S-43180 Mölndal

Dr. Piotr Jachimczak
Neurologische Universitätsklinik
im Bezirkskrankenhaus
Universitätsstraße 84
D-93053 Regensburg

Ingrid Klinger
Max-Planck-Institut für
Biophysikalische Chemie
Abt. Neurobiologie
Am Fassberg 11
D-37077 Göttingen

**PD Dr. Wolf-Bernhard
Offensperger**
Medizinische Universitätsklinik
Dept. Innere Medizin II
Hugstetterstrasse 55
D-79106 Freiburg

Dr. Karl-Hermann Schlingensiepen
Max-Planck-Institut für
Biophysikalische Chemie
Abt. Neurobiologie
Am Fassberg 11
D-37077 Göttingen

Dr. Reimar Schlingensiepen
Max-Planck-Institut für
Biophysikalische Chemie
Abt. Neurobiologie
Am Fassberg 11
D-37077 Göttingen

Markus Schweitzer
Universität Frankfurt
Institut für Organische Chemie
Marie-Curie-Str. 11
D-60439 Frankfurt am Main

Dr. Nanda D. Sinha
Perkin-Elmer Corporation
ABI Division
Applied Biosystems Division
850 Lincoln Center Drive
USA-Foster City, CA 94404
Present address:
Boston BioSystems Inc.
75A Wiggins Avenue
USA-Bedford, MA 017130

Prof. Paul C. Zamecnik
Worcester Foundation for
Experimental Biology
222, Maple Avenue
USA-Shrewsbury, MA 01545

I. Technical Aspects of Antisense Oligonucleotides

1 Introduction

Antisense Oligodeoxynucleotides – Highly Specific Tools for Basic Research and Pharmacotherapy

Reimar Schlingensiepen and Karl-Hermann Schlingensiepen
Max-Planck-Institut für Biophysikalische Chemie, Göttingen, Germany

1. **Introduction**
2. **Design of Antisense Agents**
3. **Toxicity**
4. **Specificity**
5. **Delivery Systems**
6. **Development of Antisense Therapeutics**
7. **Conclusions**

1 Introduction

Selective inhibition of gene expression allows the study of physiologic and pathophysiologic cellular processes on a molecular basis. The exciting concept of using antisense oligodeoxynucleotides (oligos) to block expression of a single gene was originally introduced by Zamecnik and Stephenson [81] [94]. In 1978 they published their results on blocking the replication of Rous Sarcoma Virus in infected chicken fibroblasts by adding a synthetically assembled oligo directed against a specific sequence of the viral genome. At that time, performance of such an experiment was limited by a lack of readily available reagents. Chemical synthesis of oligos was still a cumbersome process which only a handful of experienced laboratories were able to perform and only very few genes had been sequenced. Automation of oligo synthesis and mass production of synthesis reagents were an important step towards broader availability. Still, oligos had to be protected against rapid degradation in biological fluids. Thus, a further important step towards the applicability of synthetic antisense oligos was the introduction of DNA analogs that increase the stability by several orders of magnitude [55] [30].

Today, three principal fields of application are emerging: The investigation of gene function by loss-of-function analysis and the development of antisense drugs for therapeutic applications. Current research focuses on the mode of action of oligos, and their specificity, stability, cellular uptake, organ distribution and pharmacologic action in cell culture and *in vivo*. These aspects of the development and application of antisense oligos will be discussed.

1.1 What is "Antisense"?

To maintain its function a cell must express thousands of genes simultaneously. The specific information of each gene is encoded in the DNA by the four nucleotide bases adenine (A), guanine (G), cytosine (C) and thymine (T). These bases are strung along the DNA strand in a defined sequence that is specific for each gene. To be expressed, a gene is transcribed from the DNA into messenger RNA (mRNA) and subsequently translated into the corresponding protein (Fig. 1). Antisense oligos are designed to "switch off" gene expression by interfering specifically with the translation of the encoded protein at the mRNA level (Fig. 2, gene A, blue). The oligo hybridizes (i.e. binds) to that part of the mRNA that contains the complementary sequence by Watson-Crick base pairing [53]. The base cytosine forms hydrogen bonds exclusively with guanosine (C-G), and adenine with thymine or uracil (A-T or A-U[1]) (Fig. 3). Non-homologous or mismatched sequences of other genes do not form stable enough hybrids for the anti-

---

[1] Both DNA and RNA consist of nucleic acids. The bases are the same with the exception that RNA contains uracil instead of thymidine which is exclusively incorporated into the DNA.

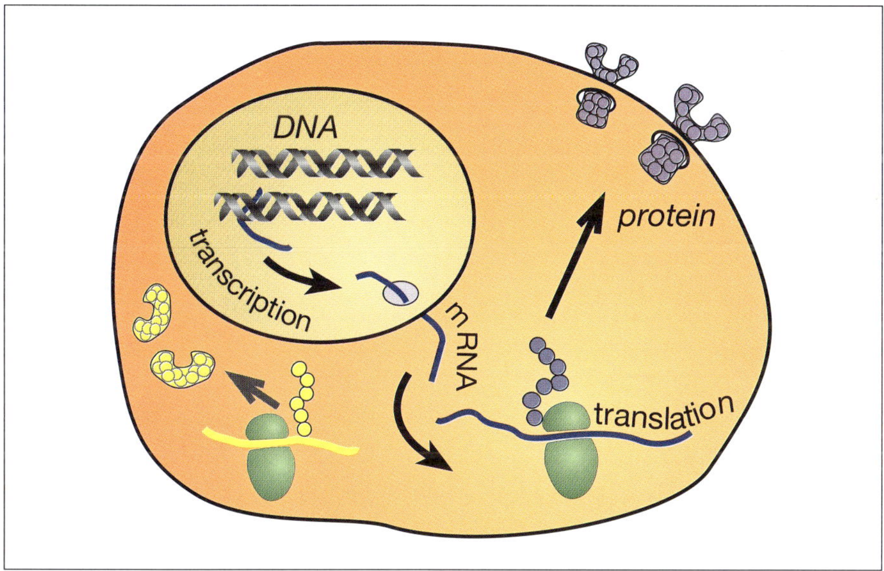

Figure 1: Process of normal gene expression. The information for a gene product is encoded in the DNA. For the correspondent protein to be expressed, the gene is first transcribed into the intermediary messenger RNA (mRNA). The mRNA leaves the cell nucleus and serves as a template for protein translation. Ribosomes (green) read the encoded information and string together the appropriate amino acids to form the encoded protein. This principle is the same for all kinds of proteins (here represented by a blue cell surface receptor and a yellow intracellular enzyme).

sense oligo to inhibit protein translation (Fig. 2, gene B, yellow). Thus, it is possible to target a unique sequence present in a single gene. Ideally, the cells becomes deficient only of the specific protein, while other proteins are regularly transcribed[2]. The properties of the targeted gene may be deduced from the resulting "loss of function" of the deficient cell.

By definition, the mRNA is the "sense" strand and any complementary sequence is "antisense" to it. Since the antisense orientation is a prerequisite for sequence-specific inhibition of translation, this term is attributed to the technique as a whole[3]. Of course the antisense mechanism has not been "invented" by man but by nature. Bacteria are well known to produce "antisense" RNA to regulate gene expression, for example, by binding to the complementary sequence in the promoter region of a bacterial gene [54]. Experimentally, antisense inhibition may be achieved with results comparable to those occurring naturally in bacteria.

Generally, any sequenced gene may be targeted by antisense oligos. Even

---

[2] In practice related genes often contain homologous sequences. Thus, any targeted sequence must be checked for homologies to other genes in a computer data base.

[3] This must not be confused with probes used for the in vitro detection of specific mRNA by the Northern or dot blot techniques, where probes are also in an antisense orientation.

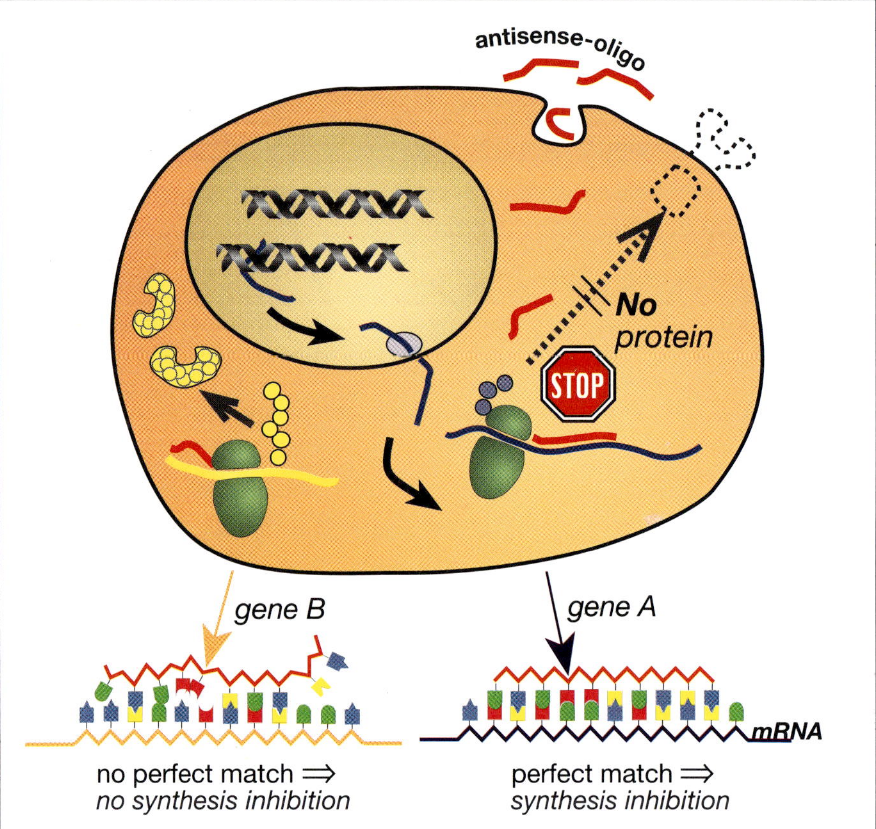

Figure 2: Antisense oligos are designed to interfere with gene expression. The oligos (red) enter the cell and bind (hybridize) to a messenger RNA. The antisense oligo contains a sequence that allows only to bind to the mRNA containing the complementary sequence (see bottom, gene A, blue). Other genes must not contain the same sequence in order that the oligo does not match (see bottom, gene B, yellow). The duplex between gene A and the oligo is strong enough to abort translation of the encoded cell surface receptor, while the imperfect match to the yellow gene is not strong enough to interfere with translation of the intracellular enzyme (yellow).

highly homologous members of a gene family may be inhibited selectively (see also section 4.). This offers a hitherto unknown potential for the investigation of different members of gene families, for example, glutamate receptors or dopamine receptors [84] [87].

1.2 Mediation of the Antisense Effect

At present, 4 different mechanisms of antisense inhibition of gene expression are being discussed:

1. Oligos directed against the coding region of the mRNA are thought to arrest protein elongation by a steric block of the ribosome (Fig. 4A). Furthermore, oligos directed to any region of the mRNA can cause conforma-

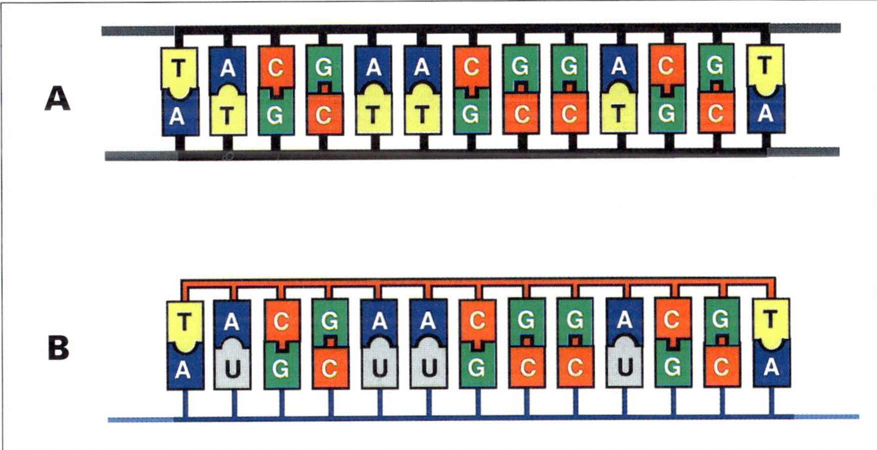

Figure 3: A: Watson-Crick base pairing. The nucleic acids of DNA and RNA can form so-called hydrogen bonds. For this reason the DNA usually exists as a duplex of two complementary strands. Adenine (A) binds to its counterpart thymine (T) whereas Cytosine (C) binds to Guanine (G). **B**: Antisense oligos may also form stable hybrids with the complementary mRNA by Watson-Crick base pairing. In the RNA molecule thymine is replaced by uracil.

tional changes which interfere with protein translation [45]. Evidence for a specific translation arrest has been provided by several studies showing that the antisense oligos block synthesis of a single protein without reducing the level of the correspondent mRNA [8] [11] [71].

2. Initiation of protein translation is inhibited when the oligo is targeted to the promoter region or around the initiation codon (Fig. 4B). It has been shown experimentally that the constitution of the ribosomal complex and initiation of protein translation may be inhibited by specific antisense oligos [48].

3. Before translation takes place in the cytoplasm, the RNA is processed in the nucleus by splicing into its final bioactive form. Antisense oligos directed to pre-mRNA may inhibit polyadenylation or transport into the cytoplasm [47]. Oligos directed to the splice junctions of the pre-mRNA interfere with mRNA maturation. [49].

4. In addition to the above mentioned mechanisms the enzyme RNase H has been described to cleave the mRNA strand while leaving the DNA oligo intact (see Fig. 4C) [26]. However, the significance of RNase H for the mediation of the antisense effect has not yet been fully elucidated. Most assays have been performed in *in vitro* systems. In cell-free reticulocyte or wheat germ lysates RNase H cleaves the mRNA specifically in the presence of unmodified oligos and phosphorothioate oligos (S-ODN) but not methylphosphonate oligos (Me-ODN) [56]. *In vivo,* the importance of RNase H for the antisense effect remains to be proven for most cell types. High RNase H activity naturally occurs in tissues of very high proliferative activity like embryonic tissue or Oocytes [28]. However, in other cell-types the

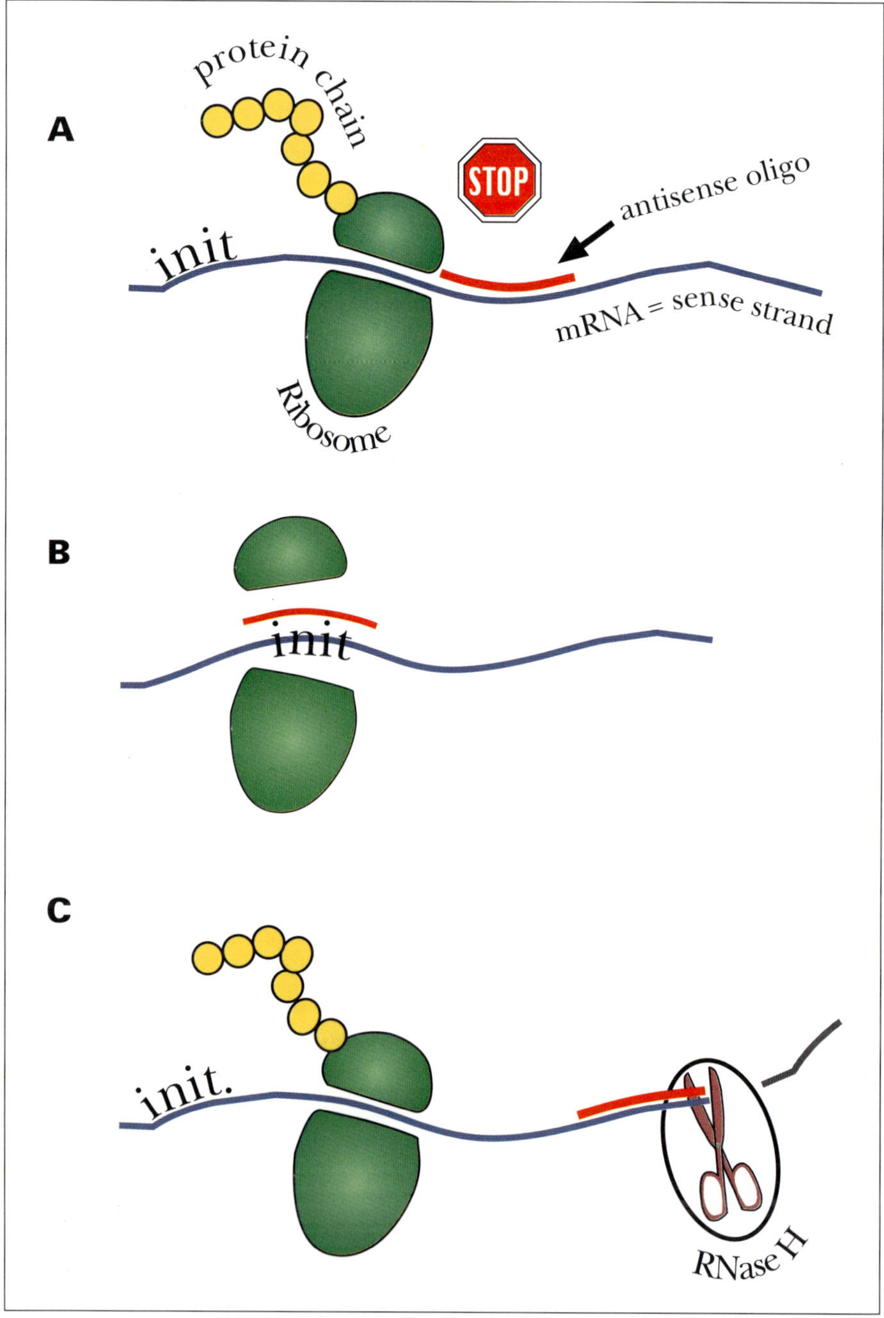

Figure 4: Different mechanisms of protein synthesis suppression. **A:** Oligos targeted to the coding region of the mRNA interfere with protein chain elongation. **B:** Targeting the initiation area of the mRNA (init.) prevents constitution of the ribosomal complex. Ribosomes are necessary to read the encoded information and to translate it into the protein. **C:** The mRNA may become a substrate for the enzyme RNase H, which recognizes the duplex of the oligo and the mRNA. RNase H cleaves the mRNA strand while leaving the oligo intact.

role of the enzyme appears to be minor. It has been demonstrated, in various non-embryonic cultured cells, that down regulation of protein translation is not enhanced by degradation of the respective mRNA even after exogenous addition of RNase H [57]. This may even be regarded as being advantageous since specificity of the antisense compound may be reduced by RNase H since RNase H cleaves the RNA strand if only five to eight contiguous base pairs of an RNA-DNA duplex form [89].

1.3 Other Approaches for Modulation of Gene Expression

Specific gene suppression may be achieved by agents other than antisense oligos. These techniques shall be briefly discussed:

Antigene approach: In the so-called antigene approach oligos are directed to the DNA double strand instead of the mRNA [60] [50]. The specificity for this binding is not Watson-Crick pairing but triplex Hoogsteen base pairing, where a third base interacts with an already formed pair of double stranded genomic DNA. Theoretically, there are advantages to this approach, mainly that normally no more than two copies of the gene are present in the cell. In practice, the concentration of antigene agents needed for effective inhibition are in micromolar ranges equivalent to antisense agents [65] [40]. Restrictions are that stable triplex formation is only achieved with 20 to 40 bases long polypurine strands, at well defined pH. Divalent or trivalent cations are needed to prevent dissociation of the three negatively charged strands. Minor changes of one of these parameters may already abolish the triplex formation. [20] [70]. Thus antigene agents are most suitable for investigations in a cell free system [18].

Ribozymes: Another compelling approach has been made possible by mimicking the naturally occurring ribozymes which contain a catalytic domain in addition to a complementary sequence. They inactivate mRNA by specifically cleaving it into two strands [15] [39]. Since ribonucleic acids are prone to rapid degradation in biological fluids, backbone modifications have been introduced to improve stability. However, development into pharmacological agents is hampered by the fact that RNA oligomer synthesis is still too expensive and that ribozymes need to be longer (from 30 up to 140 bases) than a corresponding antisense oligo due to their catalytic domain [69] [7]. In addition, to become catalytically active, ribozymes often require the presence of divalent cations and a defined pH [63]. Once the problems of stability, synthesis and *in vivo* application have been solved, ribozymes may become a new generation of oligomer therapeutics.

Gene Transfer: This approach involves transecting cells with a vector containing a cloned gene which is to be expressed at high levels. If the cell's phenotype/behavior changes in the infected cell, but not a sham transfected cell, one can tentatively attribute the change in phenotype to the newly expressed gene. Function may therefore be inferred. This approach, while also elegant and informative, has a number of drawbacks. First, it involves gene manipulation with all its hazards and it doesn't offer the advantages of a conventional drug like reversibility and individual dosing. Generally it is not certain that the changes observed are directly related to the gene's normal function as any effect observed may be an indirect one. The experiment is not really "physiologic" since overexpression of the gene may exaggerate its normal function or impart new functions by leading to an excess of the encoded protein [37]. Finally, this approach is often limited to leukemic cell lines since normal cells are often much more problematic to work with in terms of both transection and stable expression of the vector.

2   Design of Antisense Agents

2.1 Optimal Length of Antisense Agents

It has been calculated that the minimum length of an oligo has to be 12 bases to recognize a single specific sequence in the genome. This assumption is based on the estimated number of base pairs in the human genome and a random distribution of bases. It is more difficult to define a maximum length for an antisense compound. A 30 mer oligo may serve as a good probe for *in vitro* detection assays like Southern or Northern blot since both temperature and ion strength may be adjusted. In living cells, both, temperature and the ionic composition are tightly balanced. Thus, the oligo has to be adapted to the given conditions and not vice versa. Furthermore, it has been shown that antisense molecules of more than 30 bases trigger a genetic cascade of cellular antiviral response [52]. Depending on the chemistry used and base composition 14 to 25 mer oligos offer optimal specific hybridization properties. Increasing oligo length may lead to reduced specificity but increased toxicity by inhibiting different genes. In fact, if a 15 mer is sufficient for optimal translation inhibition, a 20 mer would be composed of 6 different 15 mers. In the worst case these 6 different 15 mers might hybridize to 6 different genes. It has also been demonstrated experimentally that an increase of length over a certain point may lead to a decreased specificity [41].

2.2 Backbone Modifications

Unmodified phosphodiester oligos (ODN) (Fig. 5A) offer the advantage of excellent solubility in hydrous solutions and good hybridization characteristics which are attributed to their negatively charged achiral backbone. Furthermore, automated synthesis is straightforward. However, their use is limited to selected applications due to their susceptibility to degradation by exo- and endonucleases. Indeed, in fetal calf serum the half-life does not exceed 5 min [75] and low nuclease conditions are required to prolong half-life. In cell-culture this may be partially achieved by using serum-free media. *In vivo* antisense effects were observed in short term assays in the

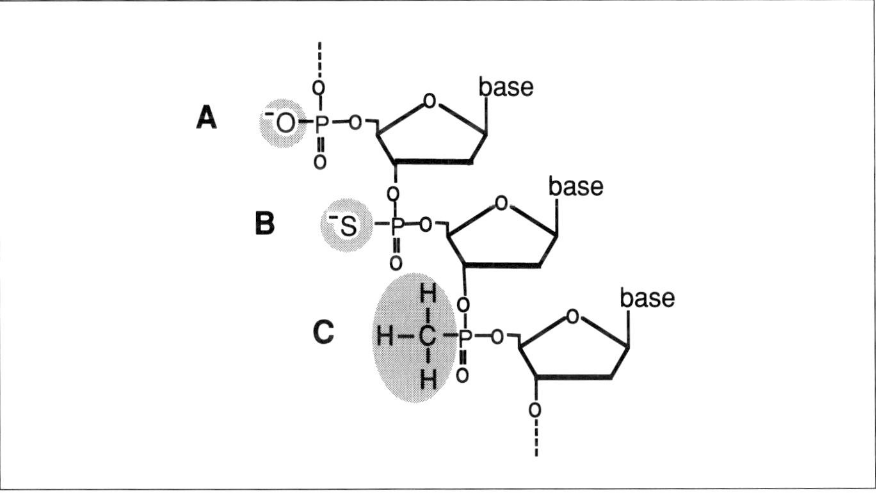

Figure 5: Structure of natural and modified ODN. **A**: unmodified phosphodiester; **B**: Phosphorothioate; **C**: Methylphosphonate.

brain where nuclease activity is particularly low [92] [86], although unmodified ODN do degrade in nervous tissue within hours, increasing the intracellular mononucleotide pool. Increased amounts of mononucleotides like dGMP or dCTP have been shown to have pronounced cytotoxic effects [77, 78].

S-ODN: In phosphorothioates one of the non-bridging oxygens of the phosphate backbone is substituted by a sulfur atom (see fig 5B) [30]. This may be achieved during automated synthesis e.g. through post-coupling oxidation with Beaucage sulfurizing reagent (see chapter 2). The product retains characteristics of unmodified oligos such as the net charge and aqueous solubility, and in addition a significantly increased stability *in vitro* and *in vivo*. S-ODN have a half-life greater than 18 h in pure fetal calf serum and more than a week in cell culture medium containing 10 % FCS [75]. However, the presence of a single phosphodiester within an S-ODN decreases stability significantly [42], pointing to the importance of a highly homogenous backbone [12].

Although they are charged macromolecules S-ODN are taken up by cells and accumulate in the cytoplasm (see also chapter 6: cellular uptake). Specific antisense activity of S-ODN has been proven repeatedly in different assays *in vitro* and *in vivo* [29] [58] [59] [11] [76] [72] [12]. Due to these properties they have become the most extensively studied and best characterized antisense agents in cell-culture, animals and man to date. They are the first antisense agents to have entered clinical trials (see chapter 9, Virology and chapter 13, Hematology).

Some properties of S-ODN remain controversial:

1. They exhibit a higher protein binding than unmodified compounds

which may reduce their immediate bio-availability. They bind mainly to albumin and alpha-2–macroglobulin with an affinity comparable to a number of drugs like aspirin or penicillin [25]. On the other hand, serum protein binding provides a repository facilitating a slow rate of total body clearance and more constant plasma levels. Furthermore, binding to cell surface proteins may explain the augmented uptake by endocytosis [10]. Too strong a protein binding that may have an impact on enzyme function can be significantly reduced by designing short oligos of less than 20 bases.

2. Chemically synthesized phosphorothioates are diastereoisomers since the sulfur atom may replace either of the two non-bridging oxygens. As a result the melting temperature and consequently the binding affinity to the complementary RNA target is reduced as compared to an equivalent unmodified ODN. In fact enzymatically synthesized all Rp-S-ODN show an improved binding affinity [79]. On the other hand improved hybridization characteristics are coupled with a reduced resistance to nucleases [82]. More importantly specific binding affinity can be increased by selecting an appropriate base composition. Thus, S-ODN diastereomers with a balanced sequence may be regarded as having an optimal profile combining a long half-life with good hybridization characteristics and sufficient cellular uptake. This is consistent with empirical data showing the highly specific action of properly designed S-ODN [29] [59] [74].

Me-ODN: In methylphosphonate oligos the negative charge bearing-oxygen of the phosphodiester bond is replaced by a neutral methyl group (see Fig. 5C) [55]. Me-ODN are highly resistant to nuclease degradation with a half-life of approx. 10 h in serum containing media [83]. Due to their lipophilic nature they enter cells by passive diffusion or pinocytosis (see also chapter 6, Cell culture). However, their use as antisense compounds *in vivo* is hampered by lower water solubility and comparatively poor duplex formation. Steric hindrance by the methyl group has been described as the main cause for inefficient duplex formation. [85]. Consequently 20 to 100 fold higher concentrations are needed to achieve specific inhibitory effects comparable to S-ODN [51] [5].

Other backbone modifications: A large variety of new backbone modifications like hydroxymethyl phosphonates, methyl phosphonothioates, peptide nucleic acids, phosphoramidates, carbamates or aminoalkyl derivatives have been synthesized (for a review see [1]). Attempts are made a) to increase nuclease resistance b) to provide an achiral isosteric backbone to improve the affinity to the target RNA c) to keep the backbone uncharged to facilitate cellular uptake by passive diffusion. Many of these agents have shown activity *in vitro*, but due to the complexity of the intracellular environment it is difficult to predict at such an early stage of investigation whether protein binding, inhibition of enzymes, complex formation with metal ions and solubility will allow these compounds to become useful agents [1]. For example, peptide nucleic acids (PNA) are extremely stable towards nuclease degradation and form strong duplexes *in vitro*. But the major drawback are their reduced solubility in water and a lack of polarity due to their achiral backbone [31]. Although investigated for years, antisense activity, cellular uptake or *in vivo* activity of PNA analogs have not yet been well characterized [1].

Chimeric ODN: To combine the lower protein affinity and the higher duplex formation of unmodified ODN with the increased nuclease resistance of modified oligos, natural phosphodiesters have been protected with a 3´ or a 5´ cap of methylphosphonates or phosphorothioates [83]. These chimeric oligos exhibit increased stability towards exonucleases but not endonucleases. Thus degradation in biological fluids is more rapid than that of fully substituted oligos [42]. Other chimeric oligos have been proposed more recently combining different analogs, exhibiting either an increased nuclease resistance, improved duplex formation or facilitated cellular uptake. Preliminary data on oligos with a phosphorothioate core and methylphosphonate caps at both ends report favorable characteristics of *in vivo* stability and distribution (reviewed in [1]). Further studies on toxicity and antisense activity in different cell systems and animals will be needed to assess their therapeutic potential.

2.3 Target Selection

Theoretical considerations play an important role for the selection of the optimal target site. It has been proposed that oligos are most effective inhibitors when targeted to the promoter region or to the translation initiation site to prevent translation initiation (see also 1.2) [27]. Targeting the initiation area appears to be most straightforward and has proven to be effective for a number of genes [84] [8] [29]. Still, targeting this area bears important restrictions. First, the initiation site often shows high homology for related and non-related genes. For example, an S-ODN directed against the initiation site of the G_S alpha-subunit would be 100 % complementary to ferritin. In most cases the selection of an oligo with a length-adapted base composition is often impossible when targeting the narrow initiation site.

Thus in many cases regions other than the initiation site are preferable. In fact, oligos targeted to different parts of the coding region of various genes were shown to be at least as effective as those directed to the initiation area. Although typically the coding region is targeted [96] [44] [72] [14] [7], sequences in non-translating regions have also shown to be good targets. For example, the 5' cap region of the c-Ha-*ras* mRNA was a significantly better target than the initiation codon region [27]. S-ODN targeted to different areas of the human nucleolar antigen p120 mRNA revealed best efficacy of an oligo targeted to the 3' non-translated region. Ten other sequences including the initiation region were less active [64].

Apart from choosing the best area of the gene, target selection is complicated by the interaction of proteins with certain base sequences. Antisense compounds may contain motifs which interfere directly with protein function. For example, the AACGTT palindrome is a potent inducer of interferon production [90], whereas GT rich oligos exhibit an inhibitory action on interferon gamma [66]. Four consecutive guanosines (G-4 quartet) also exhibit non-specific antiproliferative action independent of the overall sequence of the oligo [91]. The significance of different motifs has to be investigated *in vivo* to assess their importance at therapeutic concentrations.

In addition, target selection depends on the chemical modifications and on base composition. Different backbone chemistries greatly influence the stacking energy of the compound. For example, an S-ODN of a certain length and base composition may hybridize specifically, whereas an unmodified ODN exhibits a higher stacking energy leading to a higher rate of mismatch duplex formations. The unmodified ODN may be equally specific if length and base composition are adapted to its higher binding capacity. Vice versa, the S-ODN would exhibit too low a stacking energy to allow any efficient translation inhibition if the optimal unmodified ODN sequence is taken. Data on duplex formation and specificity *in vitro* are of very limited value for the design of bioactive compounds, if pH, salt concentration and ion composition are not equivalent to the intracellular environment. Undoubtedly, a chemistry-adapted balanced base composition in combination with database assisted exclusion of cross homologies and the avoidance or preference of certain motifs and selection of stable RNA loops [32] [9] are of fundamental importance for target selection [13].

3   Toxicity

Work from many laboratories including ours has shown that insufficient purity is one of the main reasons for toxicity even of otherwise properly designed oligos[4] [13] [12] [62] [68]. Toxic agents like acetonitrile, trichloroacetic acid or methylene chloride are used for chemical synthesis. Purification by reversed phase HPLC is often the only post synthesis method, where detritylated oligo fragments, although not all toxic by-products, are adequately removed. We found that properly designed S-ODN exhibited few non-specific effects at therapeutic doses if organic and non-organic contaminants were removed through different purification steps including anion exchange HPLC and cross flow dialysis [73] [13] [11] [12]. Application of consecutively purified S-ODN to cultured SK-Br 3 mammary carcinoma cells during log phase growth illustrate the importance of toxic organic and inorganic byproducts (Fig. 6A, B). Furthermore, at concentra-

---

[4] Further important factors for non-specific toxicity have been discussed above: protein binding and toxic sequence motifs.

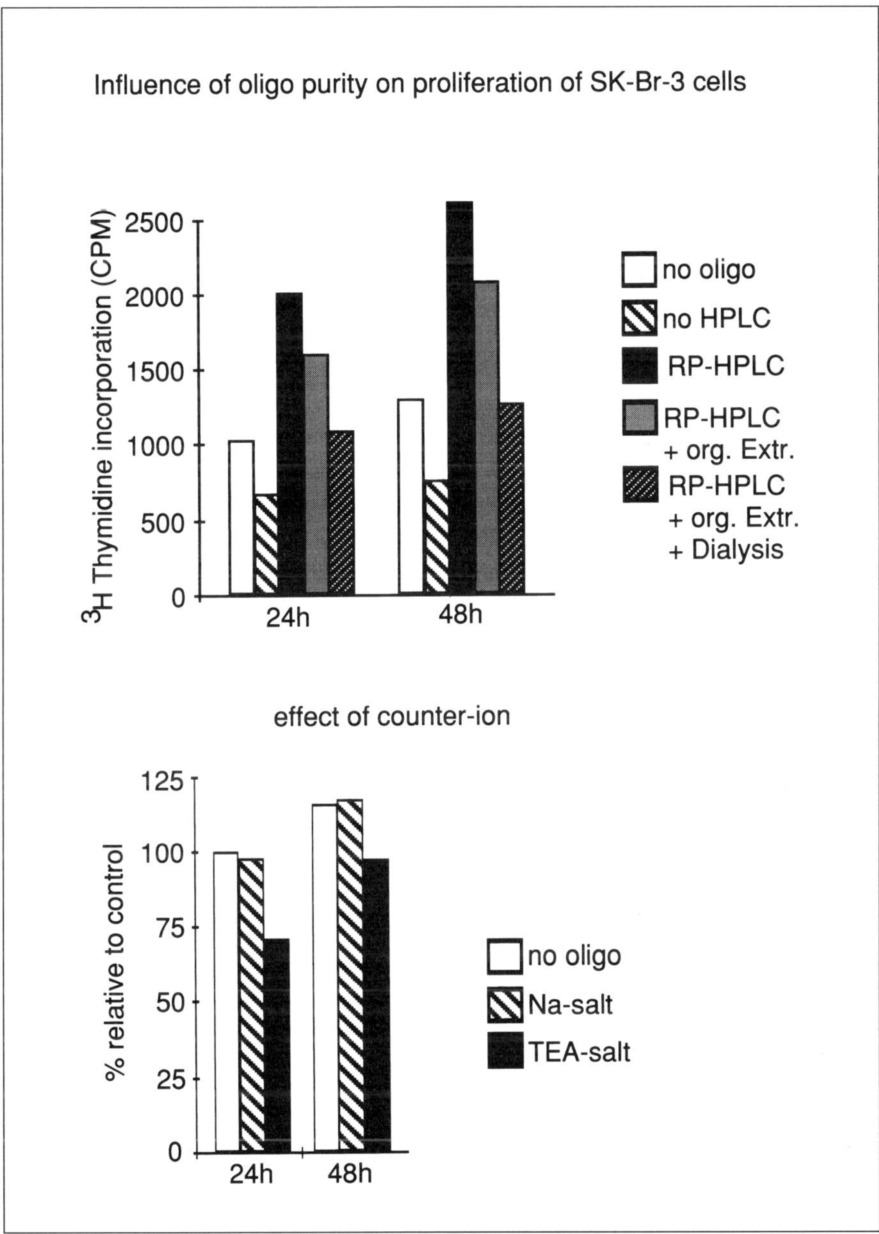

Figure 6: Comparison of non-specific toxicity of consecutively purified random S-ODN compounds on proliferation of SK-Br-3 cells. A. Effects of purifications grade on SK-Br-3 cell proliferation. White bars: Untreated control cells; Hatched bars: Treatment with S-ODN that were just deprotected and lyophilized; Black bars: Application of S-ODN, subjected to RP-HPLC; Gray bars: S-ODN purification with RP-HPLC followed by organic extraction; Dark hatched bars: S-ODN purification with RP-HPLC + organic extraction + dialysis. S-ODN concentration: 2µM in all experiments. B. Effect of counter-ion on SK-Br-3 cell proliferation: White bars: No S-ODN; Hatched bars: S-ODN-sodium-salt; Black bars: S-ODN-ammonium salt.

tions approximately 10–100 times higher than the effective dose, even highly purified and properly designed oligos cause non-specific toxicity *in vitro*. This has been attributed to chemical toxicity, an increased non-specific binding to related RNA sequences and an increased interaction with proteins [23], [13]. *In vivo*, cardiovascular problems have been described after intravenous bolus injection of phosphorothioates in monkeys, where both cardiac output and the mean arterial blood pressure dropped significantly [33], [21]. These side effects can be overcome by slow intravenous infusion or by choosing other parenteral routes of administration [33] (see also 6.2). The fact that *in vivo* breakdown products are rapidly eliminated through excretion may explain why oligos are well tolerated if infused continuously or repeatedly over several days [2] [4]. In the CNS continuous infusion of S-ODN into the cerebrospinal fluid at 1.5 nmol/hr for at least 1 week caused no obvious systemic or neurological toxicity despite maintenance of high levels in the CSF and extensive penetration into the nervous tissue [88]. Injection of different S-ODN into pregnant mice at different gestational stages revealed neither acute toxicity leading to abortion nor teratogenicity [34]. Phosphorothioates tested for mutagenicity were found to be non-mutagenic at different concentrations [24].

4 Specificity

The most important feature of antisense oligos is their ability to hybridize in a highly sequence-specific manner to their corresponding mRNA without binding to related or non-related mRNAs. As has been outlined above, specificity may only be achieved reliably when factors like length of the oligo, its base composition and absence of homology to non-targeted genes, purity and backbone homogeneity are carefully considered. Many non-sequence specific effects may be attributed to neglecting one or more of these factors. Empirical data are necessary to confirm the specificity of the compound. Comparison of antisense oligos with control oligos is necessary to rule out non-antisense effects. Ideally, specificity is proven by direct comparison of related genes. For example, in our lab antisense oligos were targeted against the structurally highly homologous but functionally different genes c-*jun* and *jun*B [72]. Specificity of both oligos was demonstrated by Western blot analysis (Fig. 7). Four demands have to be fulfilled: 1. Antisense oligo A (here c-*jun*) must efficiently suppress expression of its target gene. 2. No suppression of the structurally related gene B (here junB) must occur 3. The reverse must be true for agent B. 4. Neither of the targets must be affected significantly by control oligos. Validation of specificity is a prerequisite for reliable interpretation of functional tests.

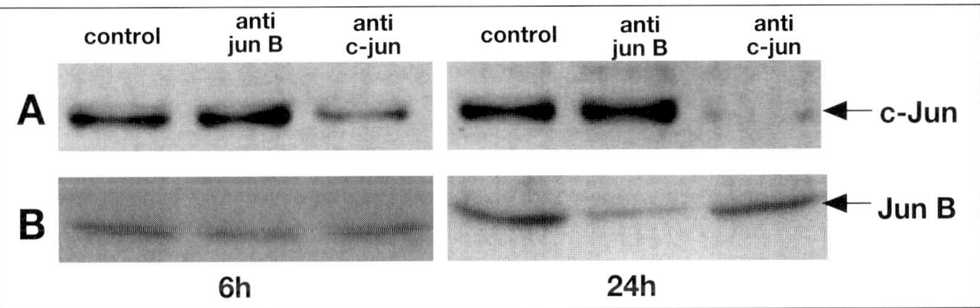

Figure 7: Western blot analysis of structurally highly related gene products to verify the specificity of antisense oligos. NIH 3T3 mouse fibroblasts and SK-Br-3 mammary carcinoma cells (upper and lower row, respectively) were incubated with 3 different oligos for 6h (lanes 1–3) and 24 h (lanes 4–6). Lanes 1 and 4: control S-ODN; Lanes 2 and 5: anti-*jun*-B S-ODN; Lane 3 and 6: anti-c-*jun* S-ODN. 10 µg of total protein were electrophoretically separated. Jun-B protein levels were reduced by anti-*jun*-B S-ODN (B, lane 5) to 26% compared to control-S-ODN treated cells (B, lane 4) but not by the anti-c-*jun* oligo; Treatment with anti-c-*jun* S-ODN (A, lane 6) reduced c-Jun protein levels to 4.8% of the amount of c-Jun protein in cells treated with control-S-ODN (A, lane 4) at 24 h, whereas the JunB protein was not reduced. These data show high specificity of antisense oligo for the targeted gene with no apparent suppression of the related gene.

5 Delivery Systems

Enhancement of cellular uptake and organ targeting *in vivo* and *in vitro* is a major topic of ongoing antisense research. Macromolecular materials in which the oligo is dissolved, entrapped or encapsulated or to which the oligo is adsorbed or attached are currently under investigation.

Cationic carriers: Cationic lipids have been used to mask the polyanionic backbone of the unmodified ODN or S-ODN. It has been reported that co-incubation of S-ODN with cationic lipids (e.g. lipofectin / DOTMA) increases cellular uptake several fold and increases the inhibitory effect of the oligo in cell culture [6] [19]. On the other hand, cationic lipids like lipofectin cause non-specific cytotoxic effects if applied alone or together with antisense oligos [93]. Only recently new generation compounds with significantly less cytotoxic effects have been developed. These new compounds may become interesting tools for antisense delivery. Nevertheless, the benefit of cationic lipids for *in vivo* applications remains unclear since improved uptake and antisense inhibition in cell culture are not necessarily observed in laboratory animals [29].

(Immuno-)liposomes: Liposomes may be used to encapsulate the oligo. The advantage lies in a further protection of the oligo from extracelluar nucleases. An enhanced cellular uptake has been described in cell culture (for review see [46]). Furthermore, liposomes may be linked to specific antibodies to achieve specific tissue targeting (immuno-liposomes) [95]. To date, the major drawbacks for the therapeutic use of immuno-liposomes are the high cost of the comparably complex production process and the limited availability of suitable antibodies.

Polymeric nanoparticles: Oligos may be delivered *in vivo* after adsorption to macromolecules like lactose- or polyalkylcyanoacrylate-polymers. Such formulations are stable, may improve the uptake of oligos and provide protection from nuclease degradation [16]. Their usefulness for delivery of antisense compounds and mediation of an improved antisense effect *in vivo* will have to be investigated. Future studies will need to reveal whether the altered delivery and biodistribution is advantageous and whether non-specific toxicity may be reduced in such a mixed compound. This area of research will most probably be of great value for the development of different galenic preparations.

6 Development of Antisense Therapeutics

The antisense technique combines the specificity of a genetic approach with the reversibility and temporal control of a classical pharmacological approach [61]. There is hardly any class of drugs for which the term "rational drug design" is more appropriate. Early on, antisense research focused mainly on cancer and viral infections because these areas offer the most compelling therapeutic prospects. In diseases like AIDS, viral hepatitis, malignant solid tumors and many hematological disorders no curative treatments are available today. Since antisense oligos act on the causative genes or viruses development of antisense therapeutics is highly attractive. Still, investigations in cell culture and in mammals point to highly interesting targets for antisense therapy in other fields like neurology (see chapter 8), infectious disease (see chapters 9–11), immunology (see chapter 7) or cardiology [58] [76] to name but a few examples. In the following, important aspects for the development of antisense therapeutics are discussed.

6.1 Pharmacokinetics

Comparison of different antisense oligo compounds shows that pharmacokinetics depend on the chemistry and to a lesser extent on length or sequence. To date, unmodified oligos, methylphosphonates and phosphorothioates have been studied in different animals (reviewed in [2]):

ODN are degraded rapidly in the plasma with a half-life of about 5 min if injected intravenously into monkeys [3] or rats [43]. After 15 min the entire dose is degraded so that organ-uptake of the intact compound is virtually impossible.

Half-life of *Me-ODN* in mice plasma is 6 min for body distribution and 17 min for elimination [17]. 70 % of the total administered dose is excreted in the urine within 2 hours. Although Me-ODN are only slowly metabolized, their rapid elimination hampers *in vivo* administration.

Pharmacokinetics of S-ODN was investigated in different animals, including mice [3], rats [22] dogs and monkeys [2]. Clinical studies were carried out in humans (see chapters 9 and 13). These studies reveal comparable results for oligos of different length and sequence. They may be summarized as follows:

Intravenously administered S-ODN are cleared from plasma biphasically with an alpha phase of about 50 min and a beta phase of 40 hours. Initial short half-life is due to distribution into most major organs, with the brain having the lowest concentration. S-ODN are then eliminated from the body slowly, primarily with urine. Up to 30 % is excreted within 24 hours and 70 % within 10 days after single administration. Up to 10 % is excreted in the faeces. Constant plasma concentrations are obtainable by once daily injections. S-ODN are very stable in all tissues except liver and spleen where 50 % of any particular S-ODN is degraded within 2 days. Breakdown products appear to be eliminated preferentially since almost no intact oligo is found in the urine whilst plasma contains predominantly undegraded oligo.

Generally S-ODN distribute broadly to all peripheral tissues. The highest percentage accumulates in liver, kidney, bone marrow, skin and skeletal muscle. The oligos accumulate most rapidly in the liver (about 20% of the administered dose within the first 2 hours) but get eliminated and redistributed more rapidly than from other tissues [22].

6.2 Route of Administration

Pharmacokinetics of S-ODN are largely independent from the parenteral route chosen, whether administered intravenously, subcutaneously or intraperitoneally [2]. Subcutaneous administration facilitates research in small laboratory animals but more importantly it is a convenient route for therapy. Long-term systemic treatment of chronic disease would benefit greatly from this mode of application avoiding prolonged stays in the hospital. Due to their size and charged backbone, S-ODN do not cross the blood-brain barrier and thus are virtually not taken up by the CNS after systemic administration. On the other hand, they have proven to be taken up efficiently by nervous tissue and to exert specific inhibition of targeted genes, if applied to the cerebro-spinal fluid (see chapter 8, Neurology). Treatment of CNS disease may require intrathecal injections or implantation of an intraventricular catheter. On the other hand, inflammatory or malignant lesions are known to disturb the blood-brain barrier allowing for systemic application [67]. An example of local delivery of S-ODN is their application to injured arterial intima during angioplasty in rats [76]. Also, topical application to the skin is possible as has been shown for the treatment of viral papilloma [36]. Altogether, these data show that in general, oligos may be applied successfully via different routes either systemically or locally.

6.3 Antisense Oligonucleotides as Informational Drugs

The antisense approach differs from classical pharmaceutical approaches to drug development. Classical compounds target the effector molecule, i.e. receptors, enzymes or transporter proteins. Since the structure of each molecule is different, the shape, chemistry and size of a classical drug must be adapted (Fig. 8A). The conventional approach to the development of pharmaceuticals requires on the average the testing of up to 10.000 substances, leading to one or more lead compounds for further development [38]. Antisense offers an appealing simplification of drug design. The basic chemical structure of the molecule is the same for different compounds; a homogenous backbone and the four bases A, G, C, T. What differs between different drugs is the sequence of the bases, and thus the encoded information (Fig. 8B). The term "informational drug" has been coined to describe this unique property [35]. Development of different antisense therapeutics may be greatly facilitated by the fact that different compounds have virtually identical pharmacokinetic and toxologic properties [80].

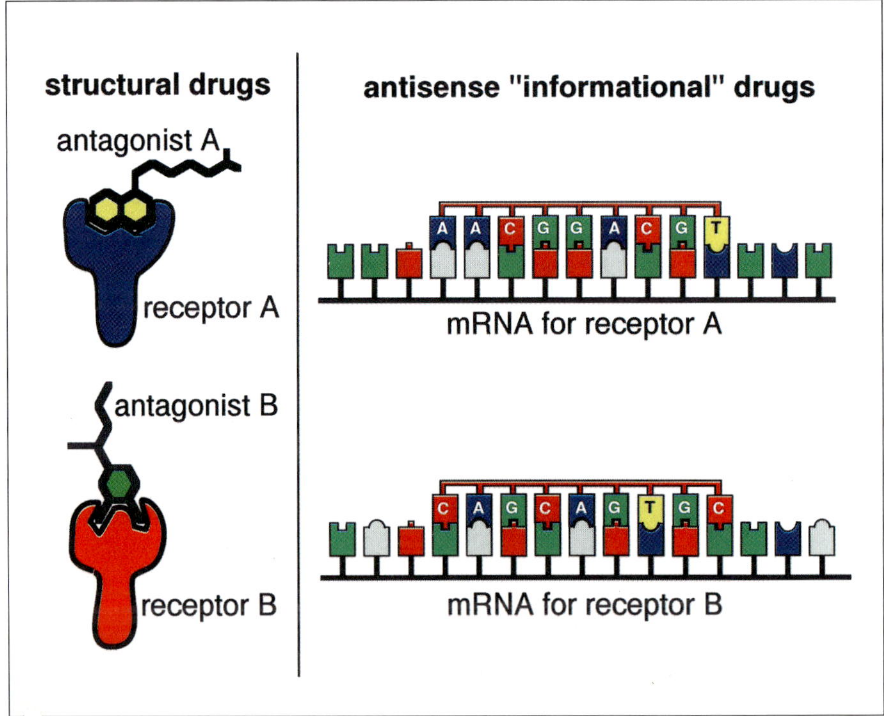

Figure 8: Comparison of classical drugs and antisense drugs. Structural drugs (left) have to be adapted to structure and function of the target molecule. Different compounds may differ considerably in structure, size, chemical and physical properties. In contrast antisense compounds are composed of the same chemical components. Specificity for the targeted gene is encoded in the nucleotide sequence, i.e. in the information of the molecule more than in its chemical or physical properties.

6.4 Production

High synthesis standards are required to produce antisense compounds. S-ODN may be produced on large scale automated synthesizers (see chapter 2). Use of standard chemicals and rapid synthesis cycles are another important reason for the widespread usage of this modification. Further research will have to focus on the development of new large scale synthesizers and possibly new chemical procedures like fluid phase synthesis. Cost effective large scale synthesis of other modified compounds are not yet fully developed [56]. However, once reliable data are obtained for novel modifications it will be of interest to develop comparably cost-effective synthesis procedures. Generally, cost-effective production is a central point for developing antisense agents as therapeutic drugs. At present, most DNA oligomers are produced in milligram amounts. Routine synthesis of kilogram amounts of an antisense oligo must be performed at lower prices than today to become attractive for pharmaceutical companies. The reduction of prices over the past ten years indicates that this goal will soon be reached. Mass production of the chemical building blocks together with improved production processes is likely to push the price of bulk oligos far below today's levels.

7 Conclusions

Antisense oligos have proven to be effective and highly specific inhibitors of gene expression. Increasing knowledge about the biological role of genes makes the antisense technique a rational and intriguing approach to study gene function and for drug development. Antisense oligos combine the advantage of conventional synthetic drugs, with the vast therapeutic opportunities that molecular genetics have opened up over the past two decades. While retaining important properties of small molecular drugs, oligos can act directly on the cause of many diseases like an aberrantly expressed gene or a viral gene.

References

1.   AGRAWAL, S. and R. P. IYER: Modified oligonucleotides as therapeutic and diagnostic agents. Curr. Opin. Biotechnol. 6(1) (1995) 12.
2.   AGRAWAL, S. , J. TEMSAMANI, W. GALBRAITH, and J. TANG: Pharmacokinetics of antisense oligonucleotides. Clin. Pharmacokinet. 28(1) (1995) 7.
3.   AGRAWAL, S. , J. TEMSAMANI, and J. Y. TANG: Pharmacokinetics, biodistribution, and stability of oligodeoxynucleotide phosphorothioates in mice. Proc. Natl. Acad. Sci. USA 88(17) (1991) 7595.
4.   AGRAWAL, S. , X. S. ZHANG, Z. H. LU, H. ZHAO, J. M. TAMBURIN, Y. M. YAN, H. Y. CAI, et al.: Absorption, tissue distribution and in vivo stability in rats of a hybrid

antisense oligonucleotide following oral administration. Biochemical Pharmacology 50(4) (1995) 571.

5. AGRIS, C. H., K. R. BLAKE, P. S. MILLER, M. P. REDDY, and P. O. Ts'o: Inhibition of vesicular stomatitis virus protein synthesis and infection by sequence-specific oligodeoxyribonucleoside methylphosphonates. Biochemistry 25(20) (1986) 6268.

6. BENNETT, C. F., M. Y. CHIANG, H. CHAN, J. E. SHOEMAKER, and C. K. MIRABELLI: Cationic lipids enhance cellular uptake and activity of phosphorothioate antisense oligonucleotides. Mol. Pharmacol. 41(6) (1992) 1023.

7. BERTRAM, J., K. PALFNER, M. KILLIAN, W. BRYSCH, K. H. SCHLINGENSIEPEN, W. HIDDEMANN, and M. KNEBA: Reversal of multiple drug resistance in vitro by phosphorothioate oligonucleotides and ribozymes. Anti Cancer Drugs 6(1) (1995) 124.

8. BOIZIAU, C., R. KURFURST, C. CAZENAVE, V. ROIG, N. T. THUONG, and J. J. TOULME: Inhibition of translation initiation by antisense oligonucleotides via an RNase-H independent mechanism. Nucl. Acids Res. 19(5) (1991) 1113.

9. BRESLAUER, K. J., R. FRANK, H. BLOCKER, and L. A. MARKY: Predicting DNA duplex stability from the base sequence. Proc. Natl. Acad. Sci. USA 83(11) (1986) 3746.

10. BROWN, D. A., S. H. KANG, S. M. GRYAZNOV, L. DEDIONISIO, O. HEIDENREICH, S. SULLIVAN, X. XU, et al.: Effect of phosphorothioate modification of oligodeoxynucleotides on specific protein binding. J. Biol. Chem. 269(43) (1994) 26801.

11. BRYSCH, W., E. MAGAL, J. C. LOUIS, M. KUNST, I. KLINGER, R. SCHLINGENSIEPEN, and K. H. SCHLINGENSIEPEN: Inhibition of p185c-erbB-2 proto-oncogene expression by antisense oligodeoxynucleotides down-regulates p185–associated tyrosine-kinase activity and strongly inhibits mammary tumor-cell proliferation. Cancer Gene Therapy 1(2) (1994) 99.

12. BRYSCH, W., A. RIFAI, W. TISCHMEYER, and K. H. SCHLINGENSIEPEN: Rationale drug design, pharmacokinetics, intracerebral application and organ uptake of phosphorothioate oligodeoxynucleotides. In: S. Agrawal, ed. Antisense Therapeutics. Humana Press Totowa, NJ (in press).

13. BRYSCH, W. and K. H. SCHLINGENSIEPEN: Design and application of antisense oligonucleotides in cell culture, in vivo, and as therapeutic agents. [Review]. Cellular & Molecular Neurobiology 14(5) (1994) 557.

14. CAZENAVE, C., C. A. STEIN, N. LOREAU, N. T. THUONG, L. M. NECKERS, C. SUBASINGHE, C. HELENE, et al.: Comparative inhibition of rabbit globin mRNA translation by modified antisense oligodeoxynucleotides. Nucl. Acids Res. 17(11) (1989) 4255.

15. CECH, T. R. and B. L. BASS: Biological catalysis by RNA. Annu. Rev. Biochem. 55(599) (1986) 599.

16. CHAVANY, C., D. T. LE, P. COUVREUR, F. PUISIEUX, and C. HELENE: Polyalkylcyanoacrylate nanoparticles as polymeric carriers for antisense oligonucleotides. Pharm. Res. 9(4) (1992) 441.

17. CHEN, T. L., P. S. MILLER, P. O. Ts'o, O. M. COLVIN, and T. C. T. C. T. CHEM: Disposition and metabolism of oligodeoxynucleoside methylphosphonate following a single i.v. injection in mice [published erratum appears in Drug Metab. Dispos. Biol. Fate. Chem. 1991 Nov-Dec;19(6):1165]. Drug Metab. Dispos. Biol. Fate. Chem. 18(5) (1990) 815.

18. CHUBB, J. M. and M. E. HOGAN: Human therapeutics based on triple helix technology. Trends Biotechnol. 10(4) (1992) 132.

19. COLIGE, A., B. P. SOKOLOV, P. NUGENT, R. BASERGA, and D. J. PROCKOP: Use of an antisense oligonucleotide to inhibit expression of a mutated human procollagen gene (COL1A1) in transfected mouse 3T3 cells. Biochemistry 32(1) (1993) 7.

20. COONEY, M., G. CZERNUSZEWICZ, E. H. POSTEL, S. J. FLINT, and M. E. HOGAN: Site-specific oligonucleotide binding represses transcription of the human c-myc gene in vitro. Science 241(4864) (1988) 456.

21. CORNISH, K. G., P. IVERSEN, L. SMITH, M. ARNESON, and E. BAYEVER: Cardiovascular effects of a phosphorothioate oligonucleotide with sequence antisense to p53 in the conscious rhesus monkey. Pharmacol. Communications 3(3) (1993) 239.

22. COSSUM, P. A., H. SASMOR, D. DELLINGER, L. TRUONG, L. CUMMINS, S. R. OWENS, P. M. MARKHAM, *et al.*: Disposition of the 14C-labelled phosphorothioate oligonucleotide ISIS 2105 after intravenous administration to rats. Journal of Pharmacology & Experimental Therapeutics 267(3) (1993) 1181.

23. CROOKE, R. M.: In vitro toxicology and pharmacokinetics of antisense oligonucleotides. Anticancer Drug Des 6(6) (1991) 609.

24. CROOKE, S. T.: COMMENTARY: Regulatory issues affecting oligonucleotides Antisense Res. Dev. 3 (1993) 301.

25. CROOKE, S. T.: Progress in antisense therapeutics. Hematologic Pathology 9(2) (1995) 59.

26. Crouch, R. J. and M. L. Dirksen: Ribonucleases H. In: R.J. Roberts, ed. Nucleases. Cold Spring Habor Laboratory, Cold Spring Habor (1982) 211.

27. DAAKA, Y. and E. WICKSTROM: Target dependence of antisense oligodeoxynucleotide inhibition of c-Ha-ras p21 expression and focus formation in T24-transformed NIH3T3 cells. Oncogene Research 5(4) (1990) 267.

28. DAGLE, J. M., D. L. WEEKS, and J. A. WALDER: Pathways of degradation and mechanism of action of antisense oligonucleotides in Xenopus laevis embryos. Antisense Res. Dev. 1(1) (1991) 11.

29. DEAN, N. M. and R. McKAY: Inhibition of protein kinase C-alpha expression in mice after systemic administration of phosphorothioate antisense oligodeoxynucleotides. Proc. Natl. Acad. Sci. USA 91(24) (1994) 11762.

30. ECKSTEIN, F.: Phosphorothioate analogues of nucleotides – Tools for the investigation of biochemical processes. Angewandte Chem. 22 (1983) 423.

31. EGHOLM, M., O. BUCHARDT, P. E. NIELSEN, and R. H. BERG: Peptic nucleic acids (PNA). Oligonucleotide analogues with an anchiral peptide backbone J. Am. Chem. Soc. 114 (1992) 1895.

32. FREIER, S. M., R. KIERZEK, J. A. JAEGER, N. SUGIMOTO, M. H. CARUTHERS, T. NEILSON, and D. H. TURNER: Improved free-energy parameters for predictions of RNA duplex stability. Proc. Natl. Acad. Sci. USA 83(24) (1986) 9373.

33. GALBRAITH, W. M., W. C. HOBSON, P. C. GICLAS, P. J. SCHECHTER, and S. AGRAWAL: Complement activation and hemodynamic changes following intravenous administration of phosphorothioate oligonucleotides in the monkey. Antisense Res. Dev. 4(3) (1994) 201.

34. GAUDETTE, M. F., G. HAMPIKIAN, V. METELEV, S. AGRAWAL, and W. R. CRAIN: Effect on embryos of injection of phosphorothioate-modified oligonucleotides into pregnant mice. Antisense Res. Dev. 3(4) (1993) 391.

35. GHOSH, M. K. and J. S. COHEN: Oligodeoxynucleotides as antisense inhibitors of gene expression. Prog. Nucleic Acid Res. Mol. Biol. 42(79) (1992) 79.

36. GILLARDON, F., I. MOLL, and E. UHLMANN: Inhibition of c-Fos expression in the UV-irradiated epidermis by topical application of antisense oligodeoxynucleotides suppresses activation of proliferating cell nuclear antigen. Carcinogenesis 16(8) (1995) 1853.

37. GREEN, M. R.: When the products of oncogenes and anti-oncogenes meet. Cell 56(1) (1989) 1.

38. GRIFFIN, J. P., J. O'GRADY, and F. O. WELLS: The textbook of Pharmaceutical Medicine. 1993, Belfast: The Queen's University of Belfast.

39. HASELOFF, J. and W. L. GERLACH: Simple RNA enzymes with new and highly specific endoribonuclease activities. Nature 334(6183) (1988) 585.

40. HELM, C. W., K. SHRESTHA, S. THOMAS, H. M. SHINGLETON, and D. M. MILLER: A unique c-myc-targeted triplex-forming oligonucleotide inhibits the growth of ovarian and cervical carcinomas in vitro. Gynecol. Oncol. 49(3) (1993) 339.

41. HERSCHLAG, D.: Implications of ribozyme kinetics for targeting the cleavage of specific RNA molecules in vivo: more isn't always better. Proc. Natl. Acad. Sci. USA 88(16) (1991) 6921.

42. HOKE, G. D., K. DRAPER, S. M. FREIER, C. GONZALEZ, V. B. DRIVER, M. C. ZOUNES, and D. J. ECKER: Effects of phosphorothioate capping on antisense oligonucleotide stability, hybridization and antiviral efficacy versus herpes simplex virus infection. Nucl. Acids Res. 19(20) (1991) 5743.

43. INAGAKI, M., K. TOGAWA, B. I. CARR, K. GHOSH, and J. S. COHEN: Antisense oligonucleotides: inhibition of liver cell proliferation and in vivo disposition. Transplant Proc. 24(6) (1992) 2971.

44. JACHIMCZAK, P., K. FABELSCHULTE, B. HESSDORFER, W. BRYSCH, K. H. SCHLINGENSIEPEN, A. BLESCH, and U. BOGDAHN: Transforming growth factor-beta-mediated regulation of human peripheral blood mononuclear cell proliferation as detected with phosphorothioate antisense oligodeoxynucleotides. Cellular Immunology 165(1) (1995) 125.

45. JANSEN, M., C. H. DEMOOR, J. S. SUSSENBACH, and J. L. VANDENBRANDE: Translational Control of Gene Expression [Review]. Pediatric Research 37(6) (1995) 681.

46. JULIANO, R. L. and S. AKHTAR: Liposomes as a drug delivery system for antisense oligonucleotides. Antisense Res Dev 2(2) (1992) 165.

47. KIM, S. K. and B. J. WOLD: Stable reduction of thymidine kinase activity in cells expressing high levels of anti-sense RNA. Cell 42(1) (1985) 129.

48. KITAJIMA, I., T. SHINOHARA, J. BILAKOVICS, D. A. BROWN, X. XU, and M. NERENBERG: Ablation of transplanted HTLV-I tax-transformed tumors in mice by antisense inhibition of NF-kappa B [letter]. Science 259(5101) (1993).

49. KULKA, M., C. C. SMITH, L. AURELIAN, R. FISHELEVICH, K. MEADE, P. MILLER, and P. O. TS'O: Site specificity of the inhibitory effects of oligo(nucleoside methylphosphonate)s complementary to the acceptor splice junction of herpes simplex virus type 1 immediate early mRNA 4. Proc. Natl. Acad. Sci. USA 86(18) (1989) 6868.

50. LE, D. T., L. PERROUAULT, D. PRASEUTH, N. HABHOUB, J. L. DECOUT, N. T. THUONG, J. LHOMME, et al.: Sequence-specific recognition, photocrosslinking and cleavage of the DNA double helix by an oligo-[alpha]-thymidylate cova-

lently linked to an azidoproflavine derivative. Nucl. Acids Res. 15(19) (1987) 7749.

51. MAHER, L. D. and B. J. DOLNICK: Comparative hybrid arrest by tandem antisense oligodeoxyribonucleotides or oligodeoxyribonucleoside methylphosphonates in a cell-free system. Nucl. Acids Res. 16(8) (1988) 3341.

52. MANCHE, L., S. R. GREEN, C. SCHMEDT, and M. B. MATHEWS: Interactions between double-stranded RNA regulators and the protein kinase DAI. Mol. Cell Biol. 12(11) (1992) 5238.

53. MATSUKURA, M., K. SHINOZUKA, G. ZON, H. MITSUYA, M. REITZ, J. S. COHEN, and S. BRODER: Phosphorothioate analogs of oligodeoxynucleotides: inhibitors of replication and cytopathic effects of human immunodeficiency virus. Proc. Natl. Acad. Sci. USA 84(21) (1987) 7706.

54. MERINO, E., P. BALBAS, J. L. PUENTE, and F. BOLIVAR: Antisense overlapping open reading frames in genes from bacteria to humans. Nucl. Acids Res. 22(10) (1994) 1903.

55. MILLER, P. S. , K. B. McPARLAND, K. JAYARAMAN, and P. O. Ts'o: Biochemical and biological effects of nonionic nucleic acid methylphosphonates. Biochemistry 20(7) (1981) 1874.

56. MILLIGAN, J. F., M. D. MATTEUCCI, and J. C. MARTIN: Current concepts in antisense drug design. J. Med. Chem. 36(14) (1993) 1923.

57. MOATS, S. B., B. G. RETSCH, W. A. PRICE, H. W. JARVIS, A. J. D'ERCOLE, and A. D. STILES: Insulin-like growth factor-I (IGF-I) antisense oligodeoxynucleotide mediated inhibition of DNA synthesis by WI-38 cells: evidence for autocrine actions of IGF-I. Mol. Endocrinol. 7(2) (1993) 171.

58. MORISHITA, R., G. H. GIBBONS, K. E. ELLISON, M. NAKAJIMA, D. L. H. VON, L. ZHANG, Y. KANEDA, et al.: Intimal hyperplasia after vascular injury is inhibited by antisense cdk 2 kinase oligonucleotides. J. Clin. Invest. 93(4) (1994) 1458.

59. MORISHITA, R., G. H. GIBBONS, K. E. ELLISON, M. NAKAJIMA, L. ZHANG, Y. KANEDA, T. OGIHARA, et al.: Single intraluminal delivery of antisense cdc2 kinase and proliferating-cell nuclear antigen oligonucleotides results in chronic inhibition of neointimal hyperplasia. Proc. Natl. Acad. Sci. USA 90(18) (1993) 8474.

60. MOSER, H. E. and P. B. DERVAN: Sequence-specific cleavage of double helical DNA by triple helix formation. Science 238(4827) (1987) 645.

61. NECKERS, L. M., A. ROSOLEN, and L. WHITESELL: Antisense inhibition of gene expression: a tool for studying the role of NMYC in the growth and differentiation of neuroectoderm-derived cells. [Review]. Journal of Immunotherapy 12(3) (1992) 162.

62. O'KEEFE, S. J., H. WOLFES, A. A. KIESSLING, and G. M. COOPER: Microinjection of antisense c-mos oligonucleotides prevents meiosis II in the maturing mouse egg. Proc. Natl. Acad. Sci. USA 86(18) (1989) 7038.

63. PALFNER, K., M. KNEBA, W. HIDDEMANN, and J. BERTRAM: Improvement of hammerhead ribozymes cleaving mdr-1 mrna. Biological Chemistry Hoppe Seyler 376(5) (1995) 289.

64. PERLAKY, L., Y. SAIJO, R. K. BUSCH, C. F. BENNETT, C. K. MIRABELLI, S. T. CROOKE, and H. BUSCH: Growth inhibition of human tumor cell lines by antisense oligonucleotides designed to inhibit p120 expression. Anticancer Drug Des. 8(1) (1993) 3.

65. POSTEL, E. H., S. J. FLINT, D. J. KESSLER, and M. E. HOGAN: Evidence that a trip-

lex-forming oligodeoxyribonucleotide binds to the c-myc promoter in HeLa cells, thereby reducing c-myc mRNA levels. Proc. Natl. Acad. Sci. USA 88(18) (1991) 8227.

66. RAMANATHAN, M., M. LANTZ, R. D. MACGREGOR, M. R. GAROVOY, and C. A. HUNT: Characterization of the oligodeoxynucleotide-mediated inhibition of interferon-gamma-induced major histocompatibility complex class I and intercellular adhesion molecule-1. J. Biol. Chem. 269(40) (1994) 24564.

67. RATAJCZAK, M. Z., J. A. KANT, S. M. LUGER, N. HIJIYA, J. ZHANG, G. ZON, and A. M. GEWIRTZ: In vivo treatment of human leukemia in a scid mouse model with c-myb antisense oligodeoxynucleotides. Proc. Natl. Acad. Sci. USA 89(24) (1992) 11823.

68. REED, J. C., M. CUDDY, S. HALDAR, C. CROCE, P. NOWELL, D. MAKOVER, and K. BRADLEY: BCL2-mediated tumorigenicity of a human T-lymphoid cell line: synergy with MYC and inhibition by BCL2 antisense. Proc. Natl. Acad. Sci. USA 87(10) (1990) 3660.

69. ROSSI, J. R., D. ELKINS, J. ZAIA, and S. SULLIVAN: Ribozymes as anti-HIV-1 therapeutic agents; principles, applications, and problems. Aids Research and Human Retroviruses 8(2) (1992) 183.

70. ROUGEE, M., B. FAUCON, J. L. MERGNY, F. BARCELO, C. GIOVANNANGELI, T. GARESTIER, and C. HELENE: Kinetics and thermodynamics of triple-helix formation: effects of ionic strength and mismatches. Biochemistry 31(38) (1992) 9269.

71. SBURLATI, A. R., R. E. MANROW, and S. L. BERGER: Prothymosin alpha antisense oligomers inhibit myeloma cell division. Proc. Natl. Acad. Sci. USA 88(1) (1991) 253.

72. SCHLINGENSIEPEN, K. H., R. SCHLINGENSIEPEN, M. KUNST, I. KLINGER, W. GERDES, W. SEIFERT, and W. BRYSCH: Opposite functions of jun-B and c-jun in growth regulation and neuronal differentiation. Developmental Genetics 14(4) (1993) 305.

73. SCHLINGENSIEPEN, K. H., F. WOLLNIK, M. KUNST, R. SCHLINGENSIEPEN, T. HERDEGEN, and W. BRYSCH: The role of Jun transcription factor expression and phosphorylation in neuronal differentiation, neuronal cell death, and plastic adaptations in vivo. Cellul. & Molec. Neurobiol. 14(5) (1994) 487.

74. SCHLINGENSIEPEN, R., H. TERLAU, W. BRYSCH, and K. H. SCHLINGENSIEPEN: Differential expression of c-jun, junB and junD in rat hippocampal slices. Neuroreport 6(1) (1994) 101.

75. SHAW, J. P., K. KENT, J. BIRD, J. FISHBACK, and B. FROEHLER: Modified deoxyoligonucleotides stable to exonuclease degradation in serum. Nucl. Acids Res. 19(4) (1991) 747.

76. SIMONS, M., E. R. EDELMAN, J. L. DEKEYSER, R. LANGER, and R. D. ROSENBERG: Antisense c-myb oligonucleotides inhibit intimal arterial smooth muscle cell accumulation in vivo. Nature 359(6390) (1992) 67.

77. SPAAPEN, L. J., G. T. RIJKERS, G. E. STAAL, G. RIJKSEN, M. DURAN, J. W. STOOP, and B. J. ZEGERS: The effect of deoxyguanosine on human lymphocyte function. II. Analysis of the interference with B lymphocyte differentiation in vitro. J. Immunol. 132(5) (1984) 2318.

78. SPAAPEN, L. J., G. T. RIJKERS, G. E. STAAL, G. RIJKSEN, S. K. WADMAN, J. W. STOOP, and B. J. ZEGERS: The effect of deoxyguanosine on human lymphocyte function. I. Analysis of the interference with lymphocyte proliferation in vitro. J. Immunol. 132(5) (1984) 2311.

79. STEC, W. J. and Z. J. LESNIKOWSKI: Stereospecific synthesis of P-chiral analogs of oligonucleotides. Methods in Molecular Biology 20(285) (1993) 285.

80. STEIN, C. A. and Y. C. CHENG: Antisense oligonucleotides as therapeutic agents – is the bullet really magical?. [Review] Science 261(5124) (1993) 1004.

81. STEPHENSON, M. L. and P. C. ZAMECNIK: Inhibition of Rous sarcoma viral RNA translation by a specific oligodeoxyribonucleotide. Proc. Natl. Acad. Sci. USA 75(1) (1978) 285.

82. TANG, J. Y., A. ROSKEY, Y. LI, and S. AGRAWAL: Enzymatic synthesis of stereoregular (all rp) oligonucleotide phosphorothioate and its properties. Nucleosides & Nucleotides 14(3–5) (1995) 985.

83. TIDD, D. M. and H. M. WARENIUS: Partial protection of oncogene, anti-sense oligodeoxynucleotides against serum nuclease degradation using terminal methylphosphonate groups. Br. J. Cancer 60(3) (1989) 343.

84. VALERIO, A., A. ALBERICI, C. TINTI, P. SPANO, and M. MEMO: Antisense strategy unravels a dopamine receptor distinct from the D2 subtype, uncoupled with adenylyl cyclase, inhibiting prolactin release from rat pituitary cells. Journal of Neurochemistry 62(4) (1994) 1260.

85. VOROB'EV, I.: [Nonionic analogs of oligonucleotide duplexes. Calculations using a molecular mechanics methods of the effect of configuration at the asymmetrical phosphorus atom in phosphonic and triester derivatives of d(TpCH3)6, d(TpOEt)6 on the structure and stability of their complexes with dA6]. Mol. Biol. Mosk. 24(1) (1990) 58.

86. WAHLESTEDT, C., E. M. PICH, G. F. KOOB, F. YEE, and M. HEILIG: Modulation of anxiety and neuropeptide Y-Y1 receptors by antisense oligodeoxynucleotides. Science 259(5094) (1993) 528.

87. WEISS, B., L. W. ZHOU, S. P. ZHANG, and Z. H. QIN: Antisense oligodeoxynucleotide inhibits D2 dopamine receptor-mediated behavior and D2 messenger RNA. Neuroscience 55(3) (1993) 607.

88. WHITESELL, L., D. GESELOWITZ, C. CHAVANY, B. FAHMY, S. WALBRIDGE, J. R. ALGER, and L. M. NECKERS: Stability, clearance, and disposition of intraventricularly administered oligodeoxynucleotides: implications for therapeutic application within the central nervous system. Proc. Natl. Acad. Sci. USA 90(10) (1993) 4665.

89. WOOLF, T. M., D. A. MELTON, and C. G. JENNINGS: Specificity of antisense oligonucleotides in vivo. Proc. Natl. Acad. Sci. USA 89(16) (1992) 7305.

90. YAMAMOTO, T., S. YAMAMOTO, T. KATAOKA, and T. TOKUNAGA: Ability of oligonucleotides with certain palindromes to induce interferon production and augment natural killer cell activity is associated with their base length. Antisense Res. Dev. 4(2) (1994) 119.

91. YASWEN, P., M. R. STAMPFER, K. GHOSH, and J. S. COHEN: Effects of sequence of thioated oligonucleotides on cultured human mammary epithelial cells. Antisense Res. Dev. 3(1) (1993) 67.

92. YEE, F., H. ERICSON, D. J. REIS, and C. WAHLESTEDT: Cellular uptake of intracerebroventricularly administered biotin- or digoxigenin-labelled antisense oligodeoxynucleotides in the rat. Cellul. & Molec. Neurobiol. 14(5) (1994) 475.

93. YEOMAN, L. C., Y. J. DANELS, and M. J. LYNCH: Lipofectin enhances cellular uptake of antisense DNA while inhibiting tumor cell growth. Antisense Res. Dev. 2(1) (1992) 51.

94. ZAMECNIK, P. C. and M. L. STEPHENSON: Inhibition of Rous sarcoma virus replication and cell transformation by a specific oligodeoxynucleotide. Proc. Natl. Acad. Sci. USA 75(1) (1978) 280.

95. ZELPHATI, O., G. ZON, and L. LESERMAN: Inhibition of HIV-1 replication in cultured cells with antisense oligonucleotides encapsulated in immunoliposomes. Antisense Res. Dev. 3(4) (1993) 323.

96. ZHENG, H., B. M. SAHAI, P. KILGANNON, A. FOTEDAR, and D. R. GREEN: Specific inhibition of cell-surface T-cell receptor expression by antisense oligodeoxynucleotides and its effect on the production of an antigen-specific regulatory T-cell factor. Proc. Natl. Acad. Sci. USA 86(10) (1989) 3758.

2 Large-scale Synthesis

Approaches to Large-scale Synthesis of Oligodeoxynucleotides and their Analogs

Nanda D. Sinha
Perkin-Elmer Corporation, Foster City, USA
Present address: Boston BioSystems, Bedford, USA

1 Introduction

Synthesis of oligodeoxynucleotides (oligos) and their uses in molecular biology, biochemistry and microbiology have become a routine process. Automation and efficient chemistries [1–6] developed in the last two decades have revolutionized the synthesis of oligos and its analogs. The demand of synthetic DNA for PCR primer, DNA sequencing primer and gene-walking sequencing techniques have further facilitated the synthesis, deprotection and purification of oligos in recent years [7–10]. Apart from these applications, researchers are also involved in developing newer types of application such as synthetic DNA-based diagnostic kits and therapeutic agents [11–12]. DNA diagnostic kits are being explored to detect viral and genetic disorders. Similarly the development of therapeutic agents based on synthetic natural or modified oligos are also progressing successfully. In the last ten years reseachers have focussed in the following areas for the development of DNA-based therapeutics:

(a)	Viral Infection	Human Immune Deficiency
		Hepatitis, Herpes, Influenza
		Human papiloma, CMV
(b)	Parasitic Infection	Malaria, Pneumocystis pneumoniae
(c)	Cancer	Leukemia, Lymphoma, Melanoma
		Squamous Cell Carcinoma
(d)	Genetic Diseases	Hemophilia, Cystic Fibrosis, Muscular
		Dystrophy, Sickle Cell Anemia, Diabetes

The area of interest for developing DNA-based drugs is growing rapidly. This is demonstrated by the recent increase in the number of antisense conferences held every year.

1.1 Oligonucleotides and Modified Oligonucleotides for Therapeutic Applications

There are various requirements for oligos to be effective as therapeutic agents. These requirements are 1. Nuclease resistance, 2. Cell permeability, 3. Better binding to target sequences, 4. RNase H mediated clevage of target RNA. The utilization of modified oligos have helped to achieve these requirements to some extent. Modified oligos are a class of compounds which contain (a) modified internucleotidic linkages namely phosphoromonothioate, phosphorodithioate or methylphosphonate diester linkage (b) modified sugar e.g. 2'-O-Methyl, 2'-O-Allyl or 2'-Fluro-2'-deoxyribonucleosides and (c) chimeric oligos i.e. oligomers containing natural phosphate diester and modified phosphate linkage or modified sugar or both. Of these modified oligos, phosphoromonothioated-oligonucletides are showing greater promise compared to natural and other oligomers. Over the last five years research studies using oligos have moved from the laboratory to clinical trials. The pharmaceutical industries based on antisense DNA technology like Hybridon Inc., ISIS Pharmaceuticals or Lynx Pharmaceuticals have undertaken human clinical trials using phosphorothioated oligos: Hybridon's GEM-91, ISIS's 2105 & 2922 and Lynx's OL-1 sequences.

2   General Synthesis Strategy

For the clinical trials not milligram material, but several hundred grams of highly purified oligomers are required. The present technology, which allows efficient and routine synthesis of oligos of defined sequences, is exclusively based on the concept of polymeric-support mediated synthesis strategy developed by Letsinger and Mahadevan [13] and phosphoramidite chemistry developed by Caruthers' [3] and Koester's [6] groups. Almost all oligos or modified phosphorothioated oligos syntheses are presently carried out on the solid supports, which are pre-derivatized with a desired nucleoside via a stable covalent linkage through the 3'-hydroxyl end. On completion of synthesis, the desired oligo chain is easily released from the supports without any side reaction. Generally, the synthesis is carried out from 3'- to 5'-direction by the addition of one reactive nucleoside-phosphorus derivative. In chemical synthesis terms, the addition of a reactive monomer that lengthens the chain is comprised of several steps. During the chain assembly these steps are repeated in a cyclic fashion and are defined as one coupling cycle.

2.1   General Synthesis Steps

(1) Detritylation: This step is carried out with a mild acid (2 % DCA/TCA solution [DCA = dichloro acetic acid, TCA = trichloro acetic acid]) to remove the terminal 5'-hydroxyl protecting dimethoxytrityl group prior to the addition of a reactive monomer;

(2) Washing: After the acid treatment, washing is carried out to remove excess acid and the orange-coloured dimethoxytrityl cation formed during the detritylation step. This washing should completely remove the residual acid;

(3) Coupling: This is a chain-lengthening step; a suitably protected nucleoside phosphoramidite is allowed to react with a free 5'-hydroxyl group of nucleoside/nucleotide bound to the solid supports in the presence of an activator tetrazole or its derivative;

(4) Washing: This step is needed to remove excess reagents used in the coupling step;

(5) Oxidation: This is to convert labile phosphite triester linkages (P III) between two nucleosides into stable phosphate/phosphorothioate

triester linkages (P V) using either an iodine solution or sulfurization reagent solution;

(6) Washing: To remove excess oxidising agent from the supports;

(7) Capping: During the chain-lengthening step it is possible that not every free hydroxyl group reacts with an active nucleoside phosphoramidite. However, this may react in the next coupling stage which could increase the percentage of oligomer with n-1 bases. In order to minimize this, unreacted or free hydroxyl groups are blocked with an organic acid-anhydride and base (typically with a mixture of acetic-anhydride, pyridine, N-methyl imidazole in tetrahydrofuran);

(8) Washing: To remove excess capping reagents from the supports.

These steps are repeated until the desired oligo or modified oligo sequence is assembled. Above steps are summarized in Scheme 1.

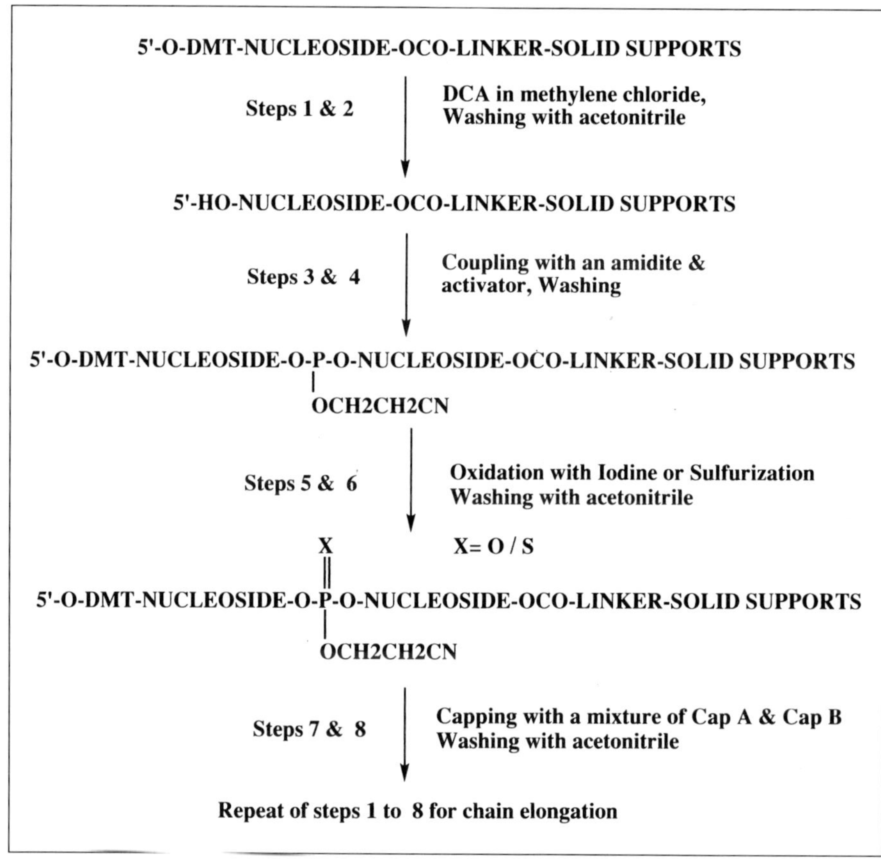

Scheme 1: Steps involved in oligodeoxynucleotide or phosphorothioated oligonucleotides synthesis on solid supports using phosphoramidite chemistry

3 Strategies and Considerations for Large-scale Synthesis

The aim of a good synthetic strategy is to obtain near-quantitative reactions in each step. The sub-micromol scale synthesis is achieved by using a small scale synthesizer at the expense of a large excess of chemicals and reagents. In large-scale oligo synthesis for therapeutic and diagnostic purposes, economic considerations inevitably become the most important factors. To make this synthesis more efficient and less expensive so that therapeutic uses of oligos become a real possibility, newer strategies require improvement in the following areas:

1. Scale-up of synthesis scales, 2. Reagent consumptions, 3. Reliable synthesis reagents, 4. Lower chemical waste generation, 5. Easy and efficient purification, and 6. Easy and reliable analysis.

Since the report on large-scale synthesis of oligo using standard CPG supports [14], several groups [15–19] either have developed or have been working to develop various methods to meet the above requirements. In the author's laboratory some success has also been achieved in scaling up and improving large-scale syntheses [20]. The methods developed for improvements are based on the following:

(a) Increased loading of nucleoside onto the existing supports or the development of better supports for higher loading in order to scale up: higher loading CPG beads and polystyrene-based Tentagel/HLP & primer supports have been developed. This has increased the scale of synthesis from 600 μmol to 4.0 mmol. (b) Reduction in the use of large excess of phosphoramidites from 4–6 equivalents to 1.5–2.0 equivalents. This has been achieved by changing the protocols of existing instruments or designing a better reactor. (c) Improvement in the quality of reactive nucleoside phosphoramidites by eliminating minor impurities to give better stability. (d) Improvement in the synthesis protocol and better reactor design have also helped to carry out synthesis using lesser amounts of other reagents and solvent, thus minimizing chemical waste generation. (e) Purification of a large amount of crude oligos are also being carried out efficiently by HPLC using a combination of reverse phase and anion exchange columns or single step purification based on the use of a polymer-based ion exchange medium. Finally, (f) easy and reliable analysis methods are being developed. These are based on anion-exchange column chromatography, capillary gel electrophoresis, ^{31}P NMR and MALDITOF / electrospray mass spectrometry techniques for the determination of purity, degree of sulfurization and molecular weight.

3.1 Automated DNA/RNA Sythesizers

Since almost all syntheses are carried out on an automated DNA/RNA synthesizer, various commercial chemical and instrument companies have developed reliable and efficient instruments to address the need of antisense pharmaceutical industries for the synthesis of large amounts of oligos or modified oligos. Presently, the following three instruments are commercially available for large-scale synthesis:

Perkin-Elmer's ABI 390 Z DNA/RNA Synthesizer
Perseptive Biosystem's 8800 Nucleic Acids Synthesizer
Pharmacia Biotech Oligo Pilot II Synthesizer

3.2   General Description of Various Instruments

Detailed descriptions of these instruments are available from the manufacturers. Some of the special features of these instruments are highlighted for quick reference purposes.

3.2.1 Perkin-Elmer's ABI 390 Z Synthesizer

The model 390 Z is driven by a flexible yet easy to use Macintosh Computer and Macintosh compatible interface software. This instrument uses a gas (argon) pressurized reagent delivery system. Generally, there are five monomer reservoirs but up to eight reservoirs can be accommodated. Separate reservoirs for oxidation with an iodine solution and sulfurization reagent solution are available to allow the synthesis of mixed phosphate and phosphoromonothioate linkages. During the synthesis, thorough mixing of the supports in the reaction vessel is performed by a vortexing mechanism.

Interface and instrument software: The 390 Z interface software contains optimized chemistry files for 25.0 μmol to 1.0 mmol synthesis scales. In these files, the first 13 procedures are pre-programmed for building a typical synthesis cycle of detritylation, coupling, capping, oxidation or sulfurization and capping. The next 13 procedures are user defined for synthesizing special analogs. It is also possible to develop customized chemistry files.

The sequence editor for this instrument is easy to use. One can enter the sequence and save it. If the oligomers to be synthesized are phosphorothioated oligos, clicking on the P = S button and then entering the sequence will allow one to do that. If the sequence has a mixed backbone, entering the sequence, highlighting the bases containing the modified backbone and clicking the P = S will also allow one to create a mixed backbone sequence.

A sequence file and chemistry file are combined in the Run Set Up file. In the Run Set Up, special cycles can be seleced for the modified bases. After selecting, the file can be saved and downloaded to the instrument. The progress of a synthesis can be monitored from the instrument's LCD or in the status window on the computer screen. The instrument also has a lock out code which can be set by the user to prevent unneccessary tampering with the synthesizer.

Documentation: The instrument contains an option for documentation of synthesis activites. One can choose to log every valve actuation, every procedure and cycle.

Conductivity cell feedback: The conductivity feedback monitor is a special feature of this instrument for monitoring the detritylation step. Complete detritylation without damaging the growing chain is essential for a satisfactory synthesis.The rate of detritylation for each base is different, therefore the time for complete detritylation is also different. The conductivity cell feedback device helps to ensure complete detritylation and eliminates the undesired exposure of nucleoside/nucleotide to dichloroacetic acid solution thus preventing depurination.

This instrument for large-scale synthesis uses HLP/Tentagel which is polystyrene beads grafted with polyethylene glycol. Some of the reagents tend to adsorb onto this support; especially dichloroacetic acid and iodine solutions. For effective removal of the excess reagents N,N- dimethylformamide is used for washings.

This instrument facilitates scale up of synthesis from the smallest research scale of 25 µmol to a production scale of 1.0 mmol, without compromising the quality of synthesis.

3.2.2 Perseptive Biosystems 8800 DNA Synthesizer

This is one of the first large-scale DNA synthesizers developed by Biosearch Inc. to address the need for large-scale synthesis of oligos. This instrument consists of different components:

(a) PC Synthesis Workstation (b) System Control Module (c) Reagents and Solvent Delivery System (d) Reaction Module

The PC workstation is an MS-DOS based computer. The menu-driven software contains chemical protocols and scaling algorithms that automate the synthesis of a wide range of synthesis scales. When a new quantity of the support is specified, the scaling algorithms are automatically performed. The program calculates the number of excess monomers, volume of solvent, duration of wash step, number of repeats, cycle time to be used in the synthesis and also implements the new parameters (user-defined parameters). The new or user-defined parameters can also be incorporated into a synthesis cycle and into synthesis files that contain sequential synthesis steps. The synthesis control file may be executed in either direct control (on-line) or download mode from the system control module.

System control module: The system control module is an internal microprocessor that controls and monitors an ongoing synthesis. In direct or on-line mode of operation, only one synthesis step at a time is sent to the system control module by the workstation. This mode of operation requires constant communication with the workstation. In down-load mode the synthesis control file is transmitted to the system control module stored in battery-protected memory and executed in a sequential manner until all synthesis steps have beeen performed. This module is the central commanding station of an 8800 DNA synthesizer. The PC workstation, the reagent and solvent delivery and reaction module are all connected to the system control module.

Reagents and solvent delivery system: This delivery module is an enclosed ventilated cabinet that contains the control valves for reagent flow and reagent reservoirs. There are four larger glass reservoirs for regular phosphoramidites and four smaller reservoirs for modified, unnatural or ribonucleoside phosphoramidites. Several larger glass reservoirs are also present for ancillary reagents such as acetonitrile, activator, deblock solution (DCA solution), capping reagents, iodine solution and sulfurization solution. The reagents and solvents in this instrument are delivered to the reaction module by pressurized argon gas.

Reaction module: The reation module contains the valve and control system for regulating reagent flow to the reaction vessel and to the waste management system. This module consists of (a) a silianized conical / cylindrical reaction vessel with a filter frit at the bottom, (b) a fail-safe sensor protection, (c) a zero dead volume control, (d) a support agitation control, and (e) a fraction collector advance circuit and waste management system.

This instrument has been upgraded in recent years to accommodate iodine and sulfurizing agent solutions. The rection vessel design has also been modified to achieve efficient mixing and washing by adopting flow through techniques. The 8800 DNA synthesizer allows one to carry out as low as 30 µmol to as high as 3.0 mmol scale synthesis using standard and high loaded CPG-supports.

3.2.3 Pharmacia Biotech's Oligo Pilot II Synthesizer

This is one of the latest large-scale DNA synthesizers available on the market. This instrument is somewhat different from the other two instruments. This oligo synthesizer utilizes a pump-driven reagent and solvent delivery system. The main features of this instrument are the following:

(a) Software for production and documentation, (b) two pump delivery system, (c) flow through reactor column, (d) on-line monitoring, and (e) waste handling and reusage of monomer solutions.

Software for production and documentation: The IBM-PC compatible workstation Oligo-Pilot software combines pre-programmed synthesis protocols for DNA/RNA and phosphorothioated DNA with full flexiblity to design and run newer methods. By adding the volume and the degree of substitution for the solid supports to be used in the synthesis, the program automatically calculates and provides the synthesis scale as well as information regarding the reagent volumes and cycles. Operating via password protection for security is an additional feature of this instrument. Oligo-Pilot II software also logs reagent volumes, lot numbers, expiration dates and compares the actual and expected runlog. It provides extended synthesis documentation and feedback control.

Two pump delivery system: The delivery module has ten standard and one optional monomer ports. The reagents and solutions are blanketed with argon gas. The reagent system is color-coded directly or via tubing into bottles. The delivery system has 6 low dead-volume motorized valves and three solenoid ports. The reagents and solvents are delivered by two pumps: the reagents with an active pumping by a piston pump (HiLoad Pump P-50) with a maximum flow rate of 50 ml/min and washing and detritylation also with an active pumping by two piston pumps (High Precision Pump P-6000) with a maximum flow rate of 100 ml/min.

Flow through reactor column: This instrument has five fixed volume (1.18 to 48.05 ml) flow through steel columns as well as an adjustable volume (from 30 to 150 ml) flow through column for carrying out 10 μmol to 4.0 mmol scale synthesis. The flow through reactor provides efficient synthesis with reduced consumption of reagents and solvent.

On-line monitor: The on-line monitoring of trityl cation due to detritylation provides a trityl value for each coupling step and the Oligo Pilot II software calculates coupling efficiencies. In the event of this value being lower than a preset value, the instrument automatically shuts down the synthesis to save chemicals and reagents.

Waste handling and reusage of monomers: This instrument has an eight port valve system which permits the separation of non-chlorinated solvents and solutions from chlorinated solutions. This facilitates the recovery of nucleoside-phosphoramidites for reuse. This may be suitable for recovering modified/unnatural nucleoside derivatives.

4   Chemicals, Reagents, Solvents and Solid Supports

In order to obtain high quality synthetic DNA, phosphorothioated DNA or modified chimeric DNA, one of the most important considerations that has to be made is the quality of chemicals, reagents and solvents. These can be divided into four different categories:

(a) Reactive nucleoside phosphoramidites, (b) an activator: tetrazole/ ethyl-S-tetrazole, (c) ancillary reagents: DCA solution, oxidising agent, capping reagents, sulfurizing agents: (3H)-1,2-benzodithiol-3-one-1,1-dioxide (Beaucage's reagent) [21], tetraethyl thiaram disulfide (TETD) [22], or bis-(O,O-diisopropoxy phosphinothionyl) disulfide (S-Tetra) [23] and solvent acetonitrile, and (d) solid supports with nucleoside: standard loading CPG supports (25–35 μmol), high loading CPG supports (80–100 μmol), tentagel (polyethyleneglycol-polystyrene based HLP supports with 100–170 μmol loading) and primer support (modified polystyrene based non-swellable supports with 20–30 and 50–70 μmol loadings).

4.1 Quality Analysis of Chemicals and their Importance

The presence of any impurity, moisture, insufficient amount of chemicals in the ancillary reagents, improperly derivatized supports or improperly stored reagents can adversely effect the quality of synthetic oligomers especially on large scales. The quality of these chemicals and reagents can be analyzed without much difficulty. ^{1}H, ^{31}P NMR and RP-HPLC are routinely used for analyzing the quality of nucleoside phosphoramidites. The presence of non phosphorus derivative can be detected by proton NMR and HPLC, while the phosphorus related impurities can be detected by phosphorus NMR. The presence of any other phosphoramidite (particularly 5'-phosphoramidite) is detrimental to oligo synthesis. Other phosphorus related impurities usually present in trace amount are derived from the hydrolysis of excess phosphitylating agent, hydrolyzed 3'-phosphoramidite or oxidized phosphoramidite. These impurities might not be detrimental to the oligo chain but would effect the stability of nucleoside-3'-O-phosphoramidite in solution, thereby reducing the effective concentration which might result in lower coupling. Non-phosphorus related impurities are solvent, moisture and nucleoside. These also affect the stability and reduce effective concentration of phosphoramidite. Ancilliary reagents may contain moisture and this may result in poor performance. The presence of moisture in capping reagents will cause their failure to adequately block the free 5'-hydroxyl groups. The presence of water can be detected by Karl-Fischer titration, moisture content higher than 50 ppm is undesirable. Appropriate amounts of iodine, pyridine and water in the oxidizing agent are also important for effective conversion of phosphite triester linkage into phosphate triester. Similarly, the stability of sulfurizing agent is also important. Incomplete conversion of P (III) into stable P (V) linkage will affect the quality of synthesis. Another source of poor quality synthesis comes from improperly derivatized solid supports. Solid supports not only contain desired nucleoside molecules but also unused amino- and silanol groups. These reactive amino-groups should be completely blocked by acetic anhydride & base. The presence of free amino group can be detected by a positive Ninhydrin test. The free amino group may also react with phosphoramidite and result in a chain with n-1 bases long oligo with 3'-phosphate. The presence of this oligomer will complicate the purification of the desired oligo chain. In addition to this, pore size of solid supports is very important. Higher loading CPG (pore size below 500 A) is suitable for oligo synthesis up to 25 to 27 bases long sequence, but is not desirable for the synthesis of longer sequences where standard loading CPG (pore size 500 A or higher) should be used. All these chemicals, reagents and supports are commercially available in very pure form, free from detectable impurities, but it is always better to check the quality of reagents prior to their use.

4.2 Commercial Suppliers of High Quality Chemicals and Reagents

There are several established suppliers such as Beckman, Cruachem, Glen Research, Perkin-Elmer's Applied Biosystems Division, Perseptive Biosystems, and Pharmacia Biotech for these chemicals, reagents and solid supports. Some of these companies have their own proprietary supports and ancillary reagents for large-scale synthesis.

High Loading CPG Supports: Glen Research, Prime Synthesis
and Perseptive Biosystems

Polystyrene Based Supports:
    HLP-Supports: Perkin-Elmer's Applied Biosystems (ABD)
    Prime Supports: Pharmacia Biotech

Sulfurizing agents:
    Beaucage's reagent: Glen Research, Pharmacia Biotech,
                                 R.I.Chemicals (Costa Mesa, CA.),
                                 A.I.C. (Newton, MA.)
    Tetraethylthiorum   Perkin-Elmer's ABD,
    Disulfide: Aldrich Chemicals

Other ancillary reagents: Although the composition of oxidizing agent, capping agents and DCA solution may vary slightly from supplier to supplier, there is no difference in their performance.

4.3 Synthesis Protocols for Large-scale Synthesis of Oligonucleotides or Phosphorothioated Oligonucleotides

The protocol provided with various DNA synthesizers are optimized for the individual instrument and these may vary from instrument to instrument because of their design. However, each protocol follows the synthetic pathway described in Scheme 1. Protocols and performance results from three different instruments are given below as a guideline. These can be modified and optimized according to individual requirements. In order to facilitate purification, it is recommended to synthesize oligomer with a terminal 5'-O-DMT. The structure of synthetic oligo and phosphorothioated oligos are given in Fig. 1.

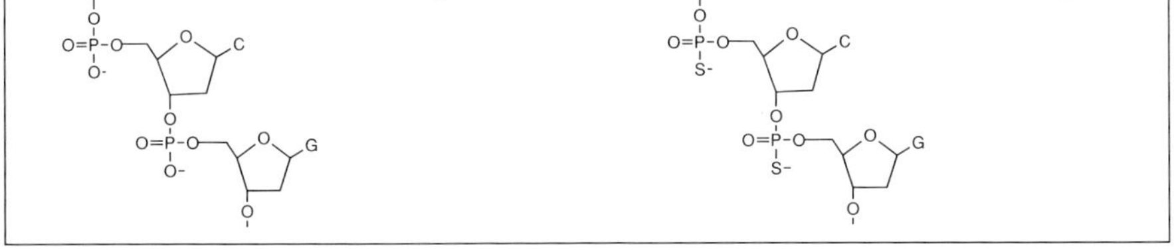

Natural Phosphate
Diester Linkages

Phosphorothioate
Diester Linkages

Figure 1: Synthetic Oligodeoxynucleotides and Phosphorothioated Oligodeoxynucleotides

4.3.1   *ABI 390 Z's Protocol for 200 to 1000 mmol Scale Synthesis using HLP Supports*

Events	Reagents/Chemicals	Amount/Duration Flow
Detrity-lation:	2 % DCA solution in Dichloromethane	Detritylation is base and sequencedependent and duration is determined by conductivity feedback.
Washing:	DMF	4 × 70 sec for 200 µmol scale 4 × 200 sec for 1.0 mmol scale
	ACN	7 × 120 sec for 200 µmol scale 7 × 300 sec for 1.0 mmol scale
Coupling:	2.0 eq. of 0.1/0.2 M amidite soln. and 0.2/0.4 M activator soln.	600 sec coupling time
Washing:	ACN	3 × 55 sec for 200 µmol 3 × 120 sec for 1.0 mmol
Capping:	Cap A and Cap B	120 sec

For oligos with regular phosphate diester linkages, capping is performed before oxidation with iodine solution.

Washing:	ACN	4 × 85 sec for 200 µmol 4 × 210 sec for 1.0 mmol
Oxidation:	0.02 M iodine soln containing pyridine, water in THF	120 sec for 200 µmol 180 sec for 1.0 mmol
Washing:	DMF	6 × 100 sec for 200 µmol 6 × 300 sec for 1.0 mmol
	ACN	5 × 130 sec for 200 µmol 5 × 210 sec for 1.0 mmol

For phosphorothioated oligo synthesis, sulfurization is carried out prior to capping in order to prevent undesirable oxidation of P-III linkage into P-V linkage (P = O) by traces of peroxide present in THF

Sulfurization:	50 mM Beaucage reagent soln or TETD soln	for 120 sec or for 900 sec
Washing:	DMF	2 × 35 sec for 200 µmol 2 × 200 sec for 1.0 mmol
	ACN	6 × 100 sec for 200 µmol 6 × 270 sec for 1.0 mmol
Capping:	Mixture of Cap A & Cap B	Duration 120 seconds
Washing:	ACN	6 × 120 sec for 200 µmol 6 × 300 sec for 1.0 mmol

Washings are performed as a combination of batch and unidirectional flow methods. Solvent is delivered into the reaction vessel from the bottom. The support and solvent is mixed by vortexing, then the solvent is emptied by gas pressure. This is followed by solvent delivered from the top, passing through the support and then out to waste. HLP support is polyethylene-glycol grafted polystyrene beads which tend to adsorb DCA soln as well as oxidizing reagents. Washing with acetonitrile alone does not remove these absorbed reagents completely and the support has to be washed with DMF. This protocol has been successfully used to synthesize up to 1.0 mmol scale [24].

Performance and Synthesis Results using HLP Supports on ABI 390 Z

Sequences	Scale (μmol)	A.C.E (%)	Yield (gm)
21-mer Phosphate diester	400	6.8	1.70
19-mer Phosphorothioate	600	97.1	1.94
18-mer Phosphate diester	800	97.6	2.81
19-mer Phosphate diester	1000	97.1	3.51

4.3.2   *Perseptive Biosystems 8800 Synthesizer Protocol up to 3.0 mmol Scale Using CPG*

The protocol developed for large-scale synthesis of oligos on this batch type reactor instrument is based on CPG supports, either standard or higher loading. Some of the steps are performed in a flow through manner to shorten the cycle time.

Events	Reagents/Chemicals	No.of Repeats/ Duration (Scale Dependent)
Detritylation	DCA in methylene chloride	4–9 rpts/1–3 min adding and mixing
Emptying	Argon	30 sec
Washing	Acetonitrile	7 rpts/1–4 min
Emptying	Argon Purge	30 sec
Washing before coupling	Dry acetonitrile	1 rpt/2–3 min
Emptying	Argon	30 sec
Coupling	100 mg/ml amidite in ACN 2.0 eq per base, 1.0 eq at a time and 0.46 M tetrazole	2 rpts /5–9 min
Washing	Acetonitrile	2 rpts/1–3 min.
Oxidation	Standard / higher iodine content solution depending on loadings and scales OR	2 rpts/3–4 min

Events	Reagents/Chemicals	No.of Repeats/ Duration (Scale Dependent)
Sulfurization	50 to 70 mM Beaucage reagent in acetonitrile	2 rpts/3–5 min
Washing	Acetonitrile	4 rpts/1–3 min
Capping	Mixture of Cap A & Cap B	1 rpt/2–4 min
Washing	Acetonitrile	4 rpts/1–3 min

Washings, capping and part of detritylation are performed in a flow through manner. During chemical reaction steps, the support is agitated with argon from time to time to ensure proper fluidization. The duration indicated does not include emptying reagents or solvent from the reactor. The amount, number of repeats or duration of reagents and solvent depends on the amount of CPG (g) which is automatically determined by the software.

Using this protocol standard, phosphorothioated and mixed oligomer up to 2–3 mmol scale have been carried out successfully [20]. Customized protocols have also been developed for this instrument to synthesize modified oligos up to 5.0 mmol scales [17–18].

Performance and Synthesis Results from 8800 Nucleic Acids Synthesizer

Sequence	Scale (mmol.)	Support CPG (gm)	Theoretical Yield	Crude Yield	% of full length	% PS
20-mer thioated	0.1	1.0	0.63 g	0.7 g	67	> 99.5
20-mer thioated	1.0	10.0	6.3 g	5.5 g	60	> 99.5
25-mer GEM-91	1.0	10.0	7.9 g	6.4 g	62	> 99.5
25-mer GEM-91	2.0	21.0	15.8 g	10.6 g	58.7	> 99.5
20-mer d(GT)10 phosphate	1.0	10.0	6.0 g	5.3 g	67	N/A
20-mer mixed 1 : 1 thioated/ phosphate	1.0	10.0	6.15 g	5.2 g	64	> 99.0

% of full length was based on HPLC analysis, % PS was based on P-NMR measurement. Average coupling efficiencies determined based on HPLC analysis vary from 96.9 to 98 %.

4.3.3   *Pharmacia Biotech's Oligo Pilot II [25]*

This instrument is different from the other two instruments' design and performance. In this instrument different columns have to be used for different scales which are available from the manufacturer, however for larger scale synthesis an adjustable column is used.

Events	Reagents/Chemicals	Amount/Duration of Flow
Detritylation	3% DCA in methylene chloride	Fast flow, duration is scale and base dependent
Washing	Acetonitrile	Fast flow, duration is scale dependent
Coupling	1.5 eq. amidite (0.2 M), 3.3 eq. excess, activator (0.45 M)	Flow & amount determined by scale and recycled
Washing	Acetonitrile	Fast flow, duration scale dependent
Oxidation	Iodine/pyridine/water in THF OR	Duration scale dependent
Sulfurization	5 % Beaucage reagent in acetonitrile; 5 fold excess	Amount & duration scaledependent
Washing	Acetonitrile	Fast flow
Capping	N-Methyl imidazole in ACN sym-collidine and acetic anhydride in acetonitrile	Duration scale dependent, determined by instrument software
Washing	Acetonitrile	Fast flow

This instrument can use CPG as well as primer support. Up to 4.0 mmol scale synthesis has been carried out successfully on this instrument [25]. Furthermore, because of the flow through mode of operation total cycle time is shorter than that of the other two instruments.

Performance and Result of Syntheses on Oligo-Pilot II

A.C.E. (average coupling efficiency) = 98.5 % to 99.5 % for scales from 10–200 µmol and > 97 % at higher scales.

Full length product at scales: 1.0 to 4.0 µmol ranges between 77 to 80 %.

Degree of sulfurization: stepwise > 99.6 % using Beaucage reagent.

Length of phosphorothioated oligos: 20 to 25 bases long.

Supports used: primer supports (polymer) and higher loading supports (CPG).

Considering the performances of all the available instruments, it can be concluded that any instrument can be successfully used to synthesize oligo sequences effectively whether with natural phosphate diester, modified phosphorothioated or mixed linkages.

5 Post Synthesis Steps

Oligos, whether normal phosphate diester, phosphormonothioated or mixed phosphate-phosphorothiated linkages, synthesized on solid supports on an automated synthesizer are obtained in a fully protected form attached to solid supports. In order to obtain synthetic oligomer in a fully deprotected form, the chain has to be released from the support and subsequently protecting groups used for preventing undesired side reaction have to be removed from the oligo chain. This process is generally termed cleavage and deprotection step.

5.1 Cleavage and Deprotection

Oligo chain release & deprotection (removal of phosphate protecting cyanoethyl, exocyclic amine protecting benzoyl & isobutyryl groups) are carried out concurrently with a concentrated NH_4OH solution either at room temperature or at 55 °C.

Protocol 1: Cleavage and Deprotection

- Transfer the support into a clean and dry beaker and cover the beaker with a clean paper towel. Place a rubber band over the paper to keep it tight.
- Place this covered beaker in a clean vacuum desiccator, carefully attach to the vacuum pump or vacuum line.
- Release the vacuum, when the support is dry (avoid contact of any acidic or basic fumes while releasing the vacuum). Weigh this material. Weight gain is due to the addition of n-1 bases with protecting groups. This gain should not be considered as the weight of synthetic oligomer.
- In order to release the chain and remove the protecting groups, place the dried material in a silianized screw-cap Schott bottle (500–1000 ml).
- Add a freshly opened bottle of concentrated NH4OH solution (30 %, 20 ml/g of supports), close the cap tightly and seal the cap with parafilm to prevent any leakage. Place the bottle into a preheated oven at 55 °C for 10 to 16 hours.
- Remove the bottle from the oven, and cool it down to –20 °C with dry ice-acetone mixture, then remove the cap carefully and decant the solution into a clean silianized flask. Wash the support with dilute ammonia solution and HPLC grade water (2–3 ×). Combine the total liquid. This step should be carried out in a well-ventilated hood.
- Freeze the total solution with dry-ice and acetone mixture and lyophilize the material under pressure. After lyophilization, the dried material can be dissolved in HPLC grade water and can be stored for analysis and purification purposes.

5.2 Purification and Analysis of the Synthetic Oligodeoxynucleotides or Modified Oligodeoxynucleotides

Although protocols for oligo syntheses are optimized to provide the best possible synthetic materials, these are not free from other unwanted materials such as amides of protecting groups (benzamide, isobutyramide) and some polymeric residue arising from cyanoethyl protecting group. In addition to these, there are mixture of shorter sequences since stepwise coupling is not 100 %. The overall yield of synthetic oligos present in deprotected materials is shown in Fig. 2. Therefore, in order to use synthetic oligos for therapeutic purposes, it is essential to purify the crude materials. Small amounts of oligos can be easily purified by preparatory gel electrophoresis or HPLC method, however the purification of large amounts of synthetic oligos require special attention. Before starting large-scale purification of oligos, it is desirable to analyze a small aliquot to determine the quality of synthesis. This can be done by using either strong anion-exchange or reverse phase HPLC methods. General parameters used for analysis by HPLC are given below.

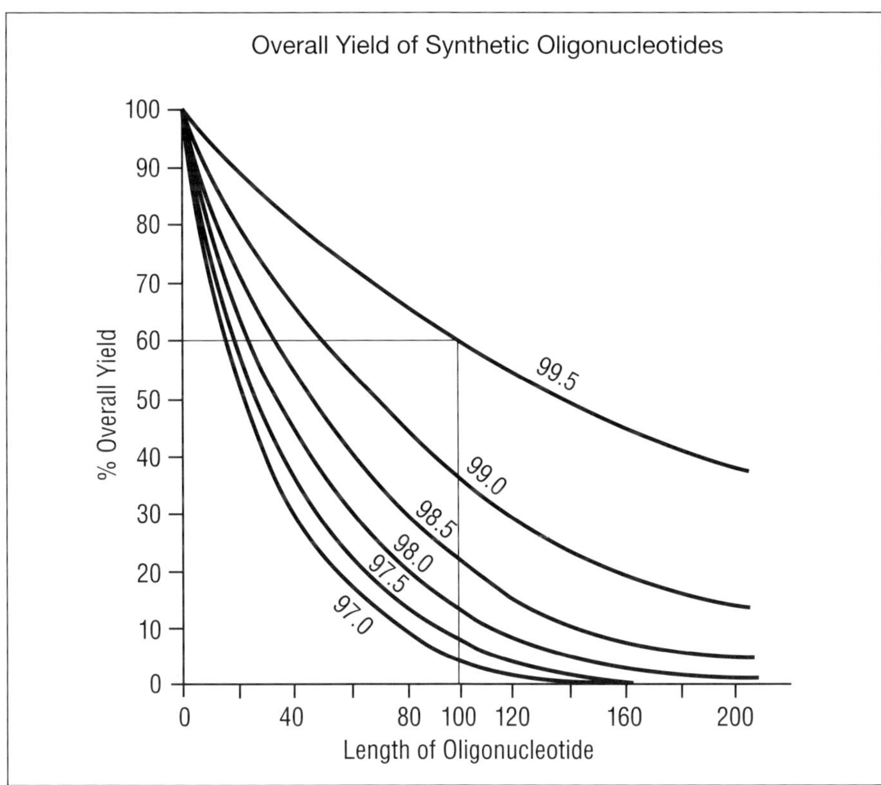

Figure 2: The reaction steps involved in the syntheses of DNA are not 100% efficient. Product yield depends upon the average coupling efficiency of the reaction steps and the overall length of the DNA synthesized.

General Parameters for HPLC Analysis

Analytical reverse phase HPLC conditions:

Buffer A	:	0.1M TEAA buffer pH = 7.0
Buffer B	:	Acetonitrile
Flow rate	:	1.0 ml / min
Gradient	:	5 % B to 50 % B over 25 min, followed by 50 % B or 5 min, then to 0 % B over 5 min.
Column	:	RP-C18 HPLC (Waters, Brownlee) or Hamilton PRP-1
Detection	:	260 nm temperature ambient

Analytical SAX HPLC conditions (SAX = Strong Anion Exchange):

Buffer A	:	20 mM ammonium phosphate, 10 % ACN
Buffer B	:	20mM ammonium phosphate, 10 % ACN and 1 M NaCl
Flow rate	:	Column dependent, Waters Gen FAX : 0.5 ml
Gradient	:	20 % B to 100% B in 60 minutes
Detection	:	260 nm , column may need to be heated at 65 °C

To determine the amount or percentage of full-length oligo or modified oligos present in the crude material, an aliquot of the sample can be analyzed by either RP-HPLC or anion-exchange column HPLC.

Protocol 2: Analytical HPCL

- In order to carry out analysis by the HPLC methods, take a sample of the crude material and clarify the sample by microfuge centrifugation using an Ultrafree-MCO 0.22 micron filter U (available from Millipore, UFC30GV00, Bedford MA.) which takes 400 ml of solution.

RP-HPLC:

- Inject about 1.5 OD units of material onto pre-equilibrated RP-HPLC set up with an RP-C18 column using the RP-HLPC parameters described above. The UV-absorbing protecting group and short or truncated sequences elute at a lower percentage of acetonitrile. The full length desired oligomer with DMT group at the 5'-end elutes with a higher percentage of acetonitrile; in the case of regular oligo only one large peak is observed however with phosphorothioated oligos usually two peaks are seen. This is due to diastereoisomers resulting from chiral phosphorothioate linkages. A typical RP-HPLC analysis chromatogram of normal and phosphorothioated are shown in Fig. 3 and 4.

Anion-Exchange:

- In order to perform the analysis by anion exchange column chromatography oligos with DMT on or DMT off can be used. Some columns require heating at 50 °C (Dionex column) or 65 °C (Waters Gen-Pac FAX)
- Inject 1.5 O.D. units of material onto an anion-exchange column which has been equilibrated with the given parameter. The desired full-length product is the last peak; a typical anion exchange HPLC analysis of the crude material is given in Fig. 5. The amount of the full-length product can be calculated from the chromatogram.

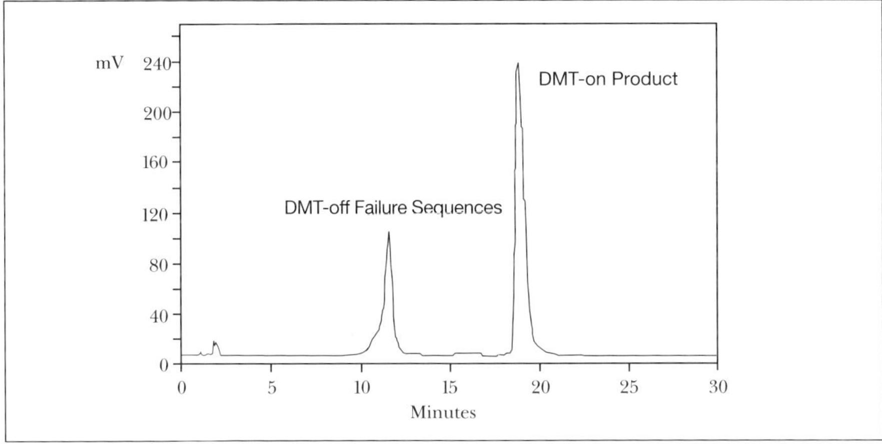

Figure 3: Reverse-phase HPCL Analysis of Crude Synthetic DNA
Column: Waters NovaPak C1 8, 3.9 × 150 mm   Flowrate: 1.0 ml/min
Eluent A. 100 mM TEAA, pH 7.0   Gradient: 5 % B to 50 % B in 25 minutes
Eluent B: acetonitrile

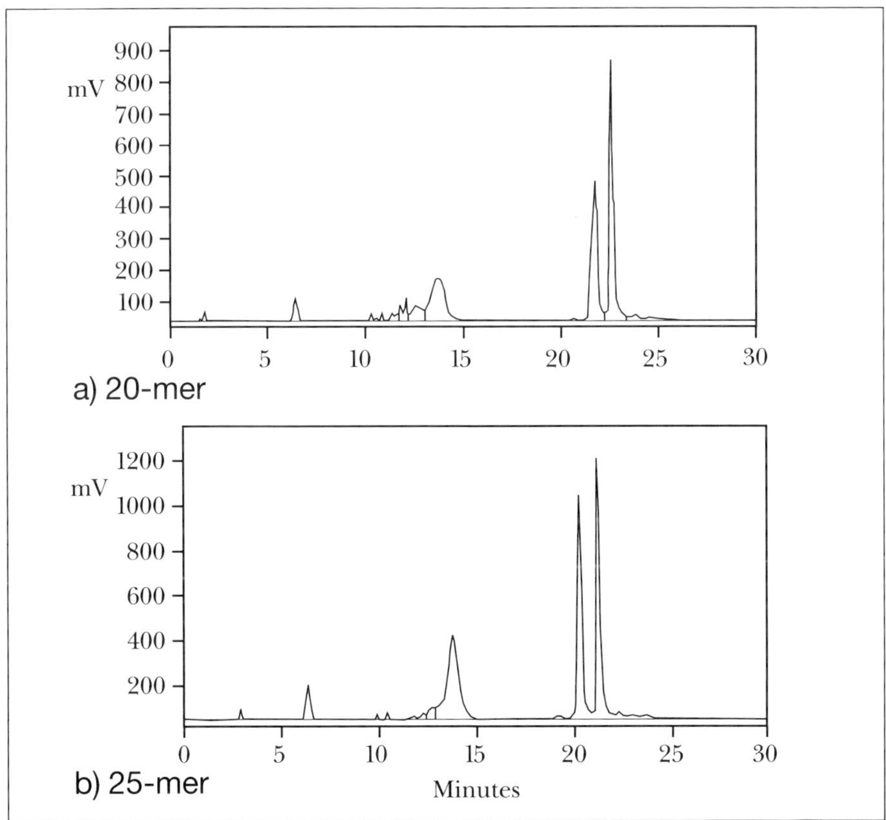

Figure 4: RP-HPLC chromatograms of crude polyphosphorothioated oligonucleotides (20- and 25-bases long) synthesized on 1.5 and 2.0 mmol scales using high loading CPG and reduced reagent consumption.

5.2.1 Large-scale Purification of Synthetic Oligonucleotides

After synthesis and deprotection, purification is one of the most important steps in the whole process. Several approaches have been developed for this purpose. Generally, a two-step purification strategy has been adopted for large scale production based on reverse phase separation followed by anion exchange chromatography and finally desalting to result in pure materials. The steps involved are outlined in the following flowchart (scheme 2) (each of the steps is further discussed below):

a   Reverse Phase HPLC Purification Conditions

This process requires a preparative HPLC system with variable higher flow rates than analytical HPLC systems. Two types of reverse phase columns; C_{18}-silica base and polystyrene base columns can be used for removing the short DMT-off failure sequences as well as protecting groups. The silica base column has a slightly higher capacity than the polystyrene base column. The latter column has other advantages such as it can be used at higher pH also. This may be neccessary when dealing with a sequence containing a higher percentage of G-residues. Therefore, selection of an RP-column should be made based on sequences. The purification protocol for these columns varies slightly from one another. These columns can be obtained either in pre-packed form or can be packed with reverse phase packing materials. Several bio-separation companies e.g. Hamilton, Phar-

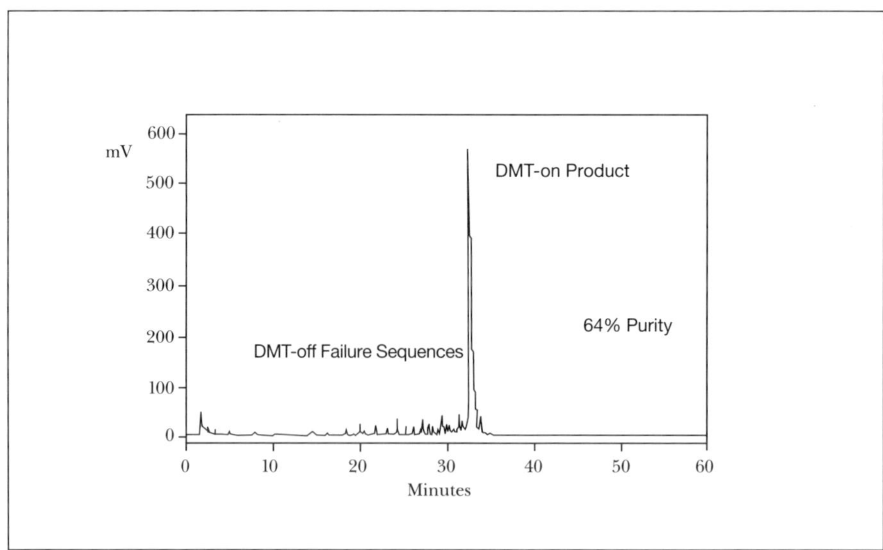

Figure 5: Anion Exchange HPCL Analysis of Crude Synthetic DNA
Column: Gen-Pak, 4.6 × 100 mm
Eluent A. 20 mM ammonium phosphate, 10 % acetonitrile
Eluent B: 20 mM ammonium phosphate, 10 % acetonitrile, 1;M NaCl
Flowrate: 0.5 ml/min

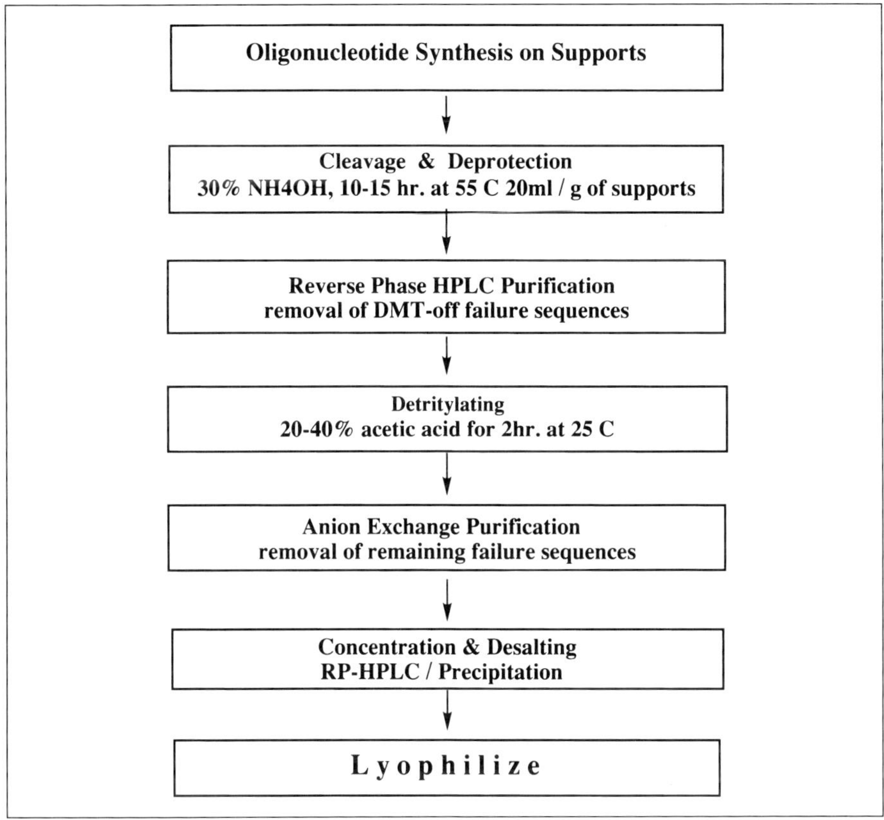

Scheme 2: Purification Flowchart for Oligodeoxynucleotides and Phosphorothioated Oligomers

macia & Waters Chromatography provide these materials for oligo purification purposes. Both columns provide satisfactory purification of synthetic oligomers. The conditions used for purification purposes (up to 1.0 mmol scales; larger batches were divided into two or more fractions) in the author's work are given below:

Protocol 3: Reserve Phase HPLC Purification

1) For C18 -silica based RP-column:

  Column : (47 × 300 mm, 37–55 micron particle size)
  Eluents : Buffer A = 0.2 M sodium acetate, pH = 7.2
     Buffer B = a mixture of buffer A and methanol (3 /7 by volume)
  Flow rate : 100 ml/min, injection volume: up to 500 ml
  Gradient : Initial settings; buffer A = 65 % & buffer B = 35 %
     0 to 40 min. same as above settings, and then from
     40 to 90 min. 35 % → 100 % buffer B

2) For PRP-1/other polymeric RP- packed column:

 Column : (50.8 × 250 mm, 12–20 micron particle size)
 Eluents : Buffer A = 3 % ammonium acetate, pH = 10.5
     Buffer B = a mixture of buffer A & acetonitrile (1/ 1 by vol.)
 Flow rate : 50 ml / min, injection volume: up to 100 ml
 Gradient : Initial settings; Buffer A = 100 % & buffer B = 0 % and
     then 0 % to 60 % buffer B in 60 min.

- In both cases, DMT-off sequences & protecting groups elute with a lower percentage of buffer B and are usually collected in one fraction (this is analyzed for the presence of any full length material prior to discarding this solution). The DMT-positive sequence which elutes with a higher percentage of buffer B is generally collected in several smaller fractions (40–50 ml).
- After analysis by either anion exchange or gel electrophoresis, the desired fractions are pooled together. This material has a purity greater than 80 %. A typical chromatogram of RP-HPLC purified product is shown in Fig. 6.
- The pooled solution is then carefully concentrated on a vacuum rotary evaporator using a previously silianized round bottom flask for further manipulation.

b Detritylation

The pooled material obtained from the above step still contains the terminal 5'-O-DMT group and small percentages of shorter sequences especially n-1, n-2, n-3 etc. The terminal 5'-O-DMT group is generally removed with an aqueous acetic acid solution (40 to 80 %) at room temperature. The choice of acetic acid concentration depends on the nature of sequence and the base at the 5'-terminal. The higher concentration of acid tends to convert thioated linkages into phosphate diester linkage and also tends to depurinate terminal adenosine nucleosides.

Protocol 4: Detritylation

- Incubate the dried pooled material with 40% acetic acid solution (100–200 ml; for 1.0 µmol) for 2 hours at room temperature. This is sufficient for detritylation and also causes minimum side reactions.
- After incubation for 2 hrs., this solution is concentrated on a rotary evaporator to remove excess acetic acid.
- The concentrated material is taken up in a pyrogen free MilliQ water (100–150 ml), and then the pH is adjusted to 8 using a 1.0 M sodium hydroxide solution. This solution can be stored at –20 °C or purified by anion exchange column chromatography to eliminate n-1, n-2, n-3 sequences.

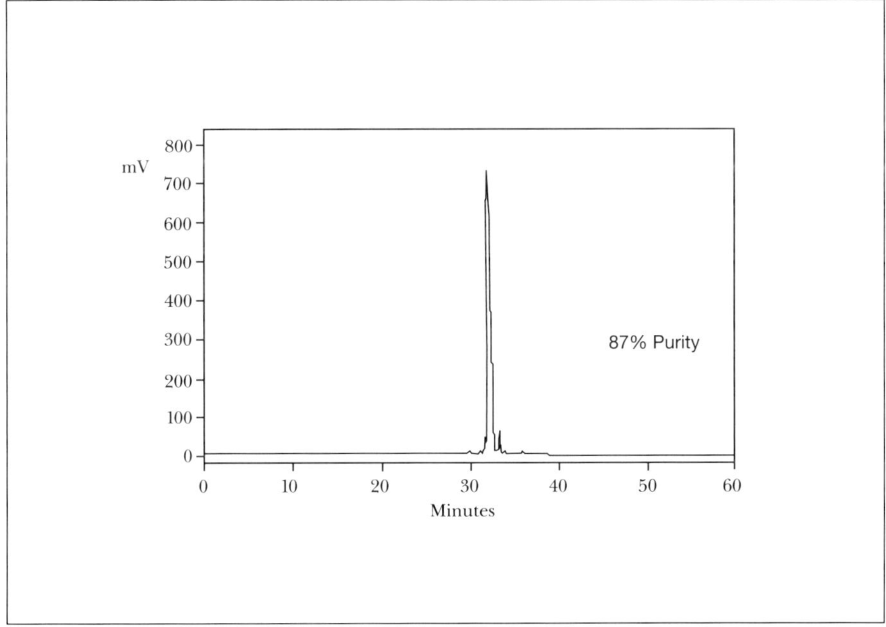

Figure 6: Anion Exchange HPCL Analysis of RP-purified Synthetic DNA
Column: Gen-PakFAX, 4.6 × 100mm
Eluent A: 20mM ammonium phosphate, 10% acetonitrile
Eluent B: 20mM ammonium phosphate, 10% acetonitrile, 1M NaCl
Flowrate: 0.5ml/min
Gradient: 20%B to 100%B in 60 minutes

c   *Anion Exchange HPLC Purification*

The anion exchange HPLC condition given below results in satisfactory purification with purity of up to 98 %. However, other anion exchange HPLC conditions described in the literatures [11, 20] can also satisfy purification needs.

Protocol 5: Anion Exchange HPLC Purification

Column: Polymeric Strong Anion Exchanger (Waters' Protein-Pac Q-15 HR™, particle size 15 micron, 50 × 100 mm)

Eluent: Solvent A = Pyrogen free MilliQ water and Solvent B = 1.5 M NaCl soln, pH adjusted to 8.0

Flow rate: 50 ml/min

Gradient: • Initial setting; 90 % A and 10 % B
   • 0 to 20 min same as initial condition
   • 20–90 min from 10 to 45 % solvent B
   • 90–100 min 45 % solvent B
   • 100–110 min from 45 to100 % B and
   • 110–120 min 100 % solvent B

- Take up reverse phase purified and detritylated material in water and adjust to pH to 8 using NaOH solution.
- Load this solution onto an anion exchange column (which has been equilibrated with the initial settings given above) at a lower flow rate.
- Collect all the materials eluted with lower salt concentrations in one flask.
- When the main peak (desired product) starts eluting at the higher salt concentration of the gradient, collect the eluent into several small fractions (50 ml fraction).
- Analyze these fractions by analytical anion exchange HPLC for purity.
- Combine the desired fractions based on the purity requirement.
- Carefully concentrate the pooled material to a small volume using a rotary evaporator under reduced pressure.

This concentrated material contains not only purified oligo or phosphorothioated oligo but also a considerable amount of free sodium chloride, which needs to be desalted.

d Desalting

The material obtained after anion exchange HPLC purification contains free NaCl, which has to be removed. Three different strategies can be used for desalting purposes: RP-HPLC, size-exclusion chromatography or cold precipitation using isopropanol.

Protocol 6: Desalting

RP-HPLC Desalting:

- Load the pooled fraction at flow rate 5–10 ml/min onto an RP-column which has been equilibrated with pyrogen free MilliQ water.
- After loading the sample, wash off free NaCl with water at a flow rate of 10 ml/min for 10 min.
- Elute the purified oligo as a single peak by increasing the percentage of methanol to 100 % in 45 min.
- Evaporate the collected material to dryness using a rotary evaporator and dissolve in pyrogen free water.
- Finally, lyophilize this solution to result in a fluffy white product.

Size-Exclusion Chromatography Desalting:

- Concentrate the pooled fraction to a small volume (30–40 ml) and load it onto a column packed with Sephadex G-15 (Sephadex G-15; 40–120 micron from Sigma, column 4.5 cm × 120 cm).
- Elute the product with sterile water and collect in void volume, then pass the solution through a sterile filter (0.2 micron pore size).
- Lyophilize this clear filtrate to obtain the purified and desalted solid material.

> **Cold Precipitation:**
>
> • Concentrate the pooled fraction (30–40 ml), add isopropanol (70–90 ml) and mix by swirling for a few minutes.
> • Keep this mixture at –20 °C for several hours and remove the precipitated material by centrifugation.
> • Wash the white solid product with cold isopropanol and dissolve in pyrogen free water and finally lyophilize to a fluffy white product.

All three desalting methods are effective in removing free NaCl present after anion exchange HPLC purification. The presence of free NaCl can be detected by conductivity measurement. RP-HPLC as well as the cold precipitation method for the removal of free salt without significant loss of valuable materials have been used in the author's laboratory.

5.2.2 *Analysis of the Purified Oligodeoxynucleotides or Modified Oligodeoxynucleotides*

Analyses of synthetic oligos whether with natural phosphate diester or modified phosphorothioate diester linkages are an important part in the development of DNA-based therapeutics. Various analytical techniques can be applied to analyze the purified oligomers. The purity of the final material is determined by analytical HPLC methods (RP or SAX) and PAGE/Capillary Zone Electrophoretic mobility. Strong-anion-exchange HPLC analysis of the purified oligo is shown in Fig. 7. In the case of phosphorothioated oligomer, the percentage of P=S linkages has to be determined. Analysis based on strong-anion-exchange HPLC can also be used to determine the percentage of P=O vs P=S linkages [27]. In addition to this, the presence or absence of P=O linkages is determined by ^{31}P NMR analysis; signal at 56–58 ppm and at 0 to -2 ppm are characteristic of phosphorothioate and phosphate linkages. Generally, greater than 99.5 % phosphorothioated linkages are present in a fully thioated oligos. An example of typical ^{31}P NMR spectra of fully thioated oligo after purification is shown in Fig. 8. Determination of base modification and sequence integrity of the purified material is also very important. In the case of regular phosphate diester, enzyme digests followed by RP-HPLC analysis are used to determine base modification. Sequence integrity determination is performed by the Maxam-Gilbert sequencing technique. However, for fully thioated oligomers analyses are performed indirectly following a method developed by Tang et. al [28]. In this method an aliquot of fully thioated oligo is converted into an oligo with regular phosphate diester linkages with sodium periodate solution treatment and then subjected to base digestion and sequencing analysis. Molecular weight determination; recently molecular weight of purified oligos can be determined by ES (electrospray) or MALDITOF (matrix-assisted-laser-dissorption time of

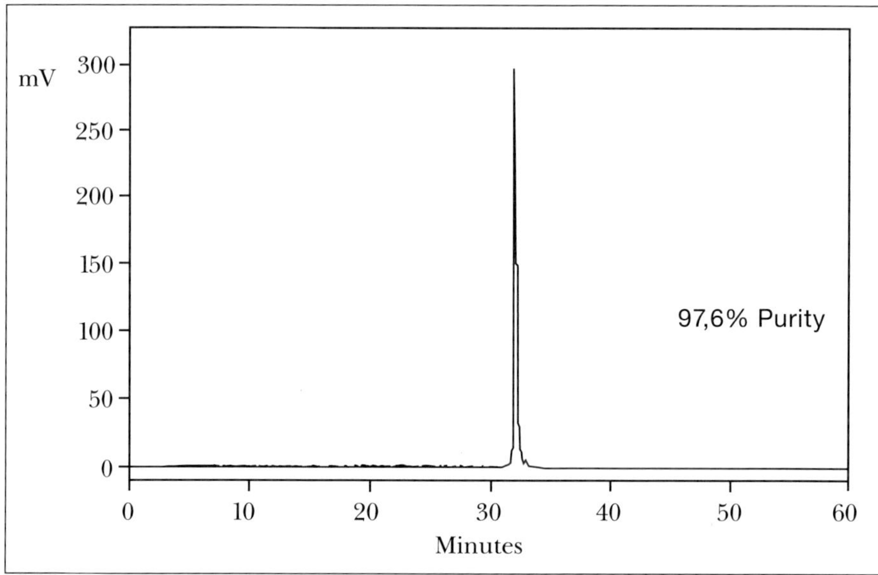

Figure 7: Anion Exchange HPLC Analysis of Purified Synthetic DNA
Column: Gen-Pak FAX, 4.6 × 100 mm
Eluent A: 20 mM ammonium phosphate, 10 % acetonitrile
Eluent B: 20 mM ammonium phosphate, 10 % acetonitrile, 1M NaCl
Flowrate: 0.5 ml/min
Gradient: 20 %B to 100 % B in 60 minutes

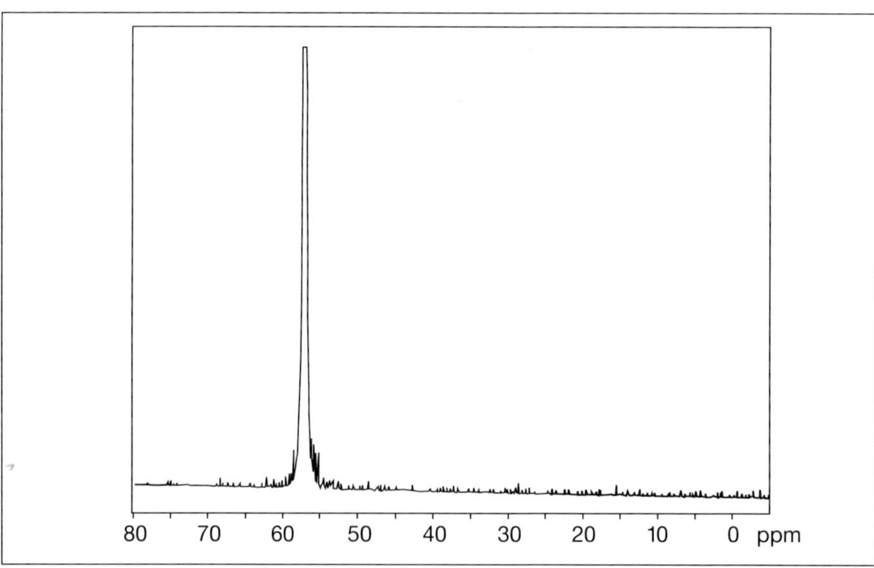

Figure 8: A Typical 31P-NMR Spectrum of a Crude Polyphosphorotioate Oligonucleotide

flight) mass spectroscopy methods (see chapter 4, Analysis of Oligonucleotides).

6   General Pre- and Post-synthesis Precautions

- Use high-quality and anhydrous solvents for all purposes. Water content should be checked: preferably water content should be lower than 50 ppm.
- If activator solution is not provided by a reputable vendor, purify commercial tetrazole by crystallizing from acetonitrile and drying over phosphorus pentaoxide under vacuum.
- Tetrahydrofuran should also be free from peroxide and the use of old oxidizing agent should be avoided.
- The reservoir for Beaucage's reagent should be silianized and this solution should be freshly made up.
- Do not use old solution of amidites and activator. Phosphoramidite solution should also be freshly made up.
- DCA solution containing DMT-cation should be collected in a well-ventilated hood.
- The supports after synthesis should be washed thoroughly with acetonitrile before drying.
- Contact with an acid / acid vapor should be avoided completely at this stage for the RP-HPLC purification strategy .
- For deprotection, always use a good-quality concentrated ammonium hydroxide solution. Use of previously opened ammonium hydroxide kept at room temperature should be avoided.
- During removal of ammonia, care should be taken to keep the solution frozen to eliminate spill during evaporation.
- Filter the solution through a 0.22 micron membrane fliter before purification.
- Do not incubate oligo with acetic acid solution for long periods.

7   Conclusion

In this chapter I have tried to provide the strategy for large-scale synthesis based on my group's work and information available in the literature. The importance of the quality of various chemicals and reagents and the nature of supports have been discussed. Information regarding commercial vendors of chemicals, reagents and instruments have been added. Brief descriptions of the various commercially available instruments with protocols and their performance have been provided for ease of reference. Methods for cleavage, deprotection, purifications have been outlined in a stepwise manner which should be easy to follow. Various analytical techniques used for analysizing the purified material have been mentioned briefly. The relevant literature references have also been included for more information.

Acknowledgement

The author would like to acknowledge the contribution of his former colleagues Steve Fry, Joan McNeal and Dr. Dennis Michaud from Perseptive Biosystems, present colleagues Dean Tsou and Dr. Krishna Upadhya at Perkin-Elmer's Applied Biosystems and information available from Pharmacia Biotech product bulletins. The author also acknowledges Dr. Brock Siegel for his interest and encouragement.

References

1. BEAUCAGE, S. L., M.H. CARUTHERS: Deoxynucleoside phosphoramidites – A new class of key intermediate for deoxypolynucleotide synthesis. Tetrahedron Letters 1859 (1981) 22.

2. McBRIDE, L. J., M. H. CARUTHERS: An investigation of several deoxynucleoside phosphoramidites useful for synthesizing deoxyoligonucleotides. Tetrahedron Letters 245 (1983) 24.

3. CARUTHERS, M. H., S. L. BEAUCAGE: Phosphoramidite nucleoside compounds. U.S. Patent No. 4,668,777.

4. SINHA, N. D., J. BIERNAT, H. KOESTER: β-Cyanoethyl N,N-dialkylamino/N-morpholino monochloro phosphoramidites, new phosphitylating agents facilitating ease of deprotection and work up of synthesized oligonucleotides. Tetrahedron Letters 5843 (1983) 24.

5. SINHA, N. D., J. BIERNAT, J. P. McMANUS, H. KOESTER: Polymer support oligonucleotide synthesis XVIII: Use of β-cyanoethyl-N,N-dialkylamino/-/N-morpholino phosphoramidite of deoxynucleosides for the synthesis of DNA fragments simplifying deprotection and isolation of the final product. Nucl. Acids Res. 4539 (1984) 12.

6. KOESTER, H., N. D. SINHA: Process for the preparation of oligonucleotides. U. S. Patent No. 4,725,677.

7. SCHULHOF, J. C., D. MOLKO, R. TEOULE: The final deprotection step in oligonucleotide synthesis is reduced to a mild and rapid ammonia treatment by using labile base-protecting groups. Nucl. Acids Res. 397 (1987) 15.

8. CHAIX, C., D. MOLKO, R. TEOULE: The use of labile base protecting groups in oligoribonucleotide synthesis. Tetrahedron Letters 71 (1989) 30.

9. McBRIDE, L. J., R. KIERZEK, S. L. BEAUCAGE, M. H. CARUTHERS: Amidine protecting groups for oligonucleotide synthesis. J. Am. Chem. Soc. 2040 (1986) 108.

10. SINHA, N.D., P. DAVIS, N. USMAN, J. PEREZ, R. HODGE, J. KREMSKY, R. CASALE: Labile exocyclic amine protection of nucleosides in DNA, RNA, and oligonucleotide analogs synthesis facilitating N-deacylation, minimizing depurination and chain degradation. Biochimie 13 (1993) 75.

11. CROOKE, S. T.: THERAPEUTIC applications of oligonucleotides. Bio/Technology 882 (1992) 10.

12. EDGINGTON, S. M.: DNA and RNA Therapeutics: Unsolved Riddles. Bio/Technology 993 (1992) 10.

13. LETSINGER, R. L., V. MAHADEVAN: Stepwise synthesis of oligodeoxyribonucleotides on an insoluble polymer supports. J. Am. Chem. Soc. 3526 (1965) 87.

14. SINHA, N. D.: Large-scale oligonucleotide synthesis using the solid supports approach. In: Agrawal, S. (ed): protocols for oligonucleotides and analogs: Methods in Molecular Biology Series, Vol 20, Humana Press, Totowa, NJ, USA (1993) 437–463.

15. GAO, H., B. L. GAFFNEY, R. A. JONES: H-Phosphonate oligonucleotide synthesis on a polyethylene glycol / polystyrene co-polymer. Tetrahedron Letters 5477 (1991) 32.

16. WRIGHT, P., D. LLOYD, W. RAPP, A. ANDRUS: Large-scale synthesis of oligonucleotides via phosphoramidite nucleosides and high loaded polystyrene supports. Tetrahedron Letters 3373 (1993) 34.

17. PADMAPRIYA, A. A., J.-Y. TANG, S. AGRAWAL: Large-scale synthesis, purification and analysis of oligodeoxynucleotide phosphorothioates. Antisense Res. and Dev. 185 (1994) 4.

18. ANDRADE, M., A. SCOZZARI, D.L.COLE, V.T. RAVIKUMAR: Efficient and economical synthesis of antisense oligodeoxyribonucleotide phosphorothioates. Bioorg. Med. Chem. Letters 2017 (1994) 4.

19. BIRCH-HIRSCHFELD, E., Z. FOELDES-PAPP, K.-H. GUEHRS, H. SELIGER: A versatile support for the synthesis of oligonucleotides of extended length and scale. Nucl. Acids Res. 1760 (1994) 22.

20. FRY, S. , N. D. SINHA: Improvements in the large scale oligonucleotide (DNA/RNA) synthesis on solid supports. In: R. Epton (ed.) Innovation and perspectives in Solid phase Synthesis Peptide, Proteins and Nucleic Acids, 3rd International Symposium, Oxford; Mayflower Worldwide Ltd., Kingwinford, West Midlands, England (1994) 135–144.

21. IYER, R. P., L. R. PHILLIPS, W. EGAN, S. L. BEAUCAGE: 3H-1,2-Benzodithiole-3-one 1,1-dioxide as an improved sulfurizing reagent in the solid-phase synthesis of oligodeoxyribonucleoside phosphorothioates. J. Am. Chem. Soc. 1253 (1990) 112.

22. VU, H., B. L. HIRSCHBEIN: Internucleotide phosphite sulfurization with tetraethyl thiaram disulfide. Phosphorothioate oligonucleotide synthesis via phosphoramidite chemistry. Tetrahedron Letters 3005 (1991) 32.

23. STEC, W. J., B. UZNANSKI, A. WILK: Bis-(O,O-Diisopropoxyphosphinothioyl)-disulfide – A highly efficient sulfurizing reagent for cost effective synthesis of oligo (nucleoside) phosphorothioates. Tetrahedron Letters 5317 (1993) 34.

24. TSOU, D., S. BEAVERS, E. HANING, A. ANDRUS: Millimole scale synthesis of oligonucleotides using phosphoramidite with HLP-supports on the Model 390Z; Poster Abstracts; presented at San Diego Conference on Nucleic Acids Beyond DNA Probes, Nov. 1993 (organized by Am. Association for Clinical Chemistry and the A.A.C.C. Molecular Pathology Division).

25. Pharmacia Biotech Oligo Pilot II™: Product Bulletin No. 18–1060–87.

26. Pharmacia Biotech Application Notes: No. 18–1102–61 and 18–1102–62.

27. Bergot, B. J., W. Egan: Separation of synthetic phosphorothioate oligonucleotides from their oxygenated (phosphodiester) defect species by strong-anion exchange high performance liquid chromatography. Journal of chrom. 35 (1992) 599.

28. TANG, J.-Y., A. M. ROSKEY, S. AGRAWAL: Method for sequencing synthetic oligodeoxynucleotide phosphorothioates. Anal. Biochem. 134 (1993) 212.

3 Labelling

Synthetic Methods for the Radioisotopic Labelling of Oligonucleotides

Sudhir Agrawal, Weitian Tan, Zhwei Jiang, Dong Yu and
Radhakrishnan P. Iyer
Hybridon Inc., Worcester, USA

1 Introduction

Oligonucleotides (oligos) have established themselves as a distinctive class of agents with a wide and ever-increasing array of applications [1]. In the broadest sense, oligos (a) are used as tools of molecular biology for probing molecular events involved in cellular function and regulation; (b) have potential as therapeutic agents against a variety of diseases; and (c) are employed as diagnostic agents for the detection of specific genes or gene products in a number of applications [1] [3]. The diagnostic applications of oligos are dependent on the ability to detect and quantify complexes (e.g., *in situ hybridization*) formed between an oligo and its target either before or after amplification e.g., by *polymerase-chain reaction.*

Overall, the ability to detect and quantitate very small levels (femtomolar levels) of a putative target, using an oligo probe, is of paramount importance to the realization of the wide array of oligo-based applications. Additionally, from a therapeutic perspective, it is critical to be able to detect and quantify products resulting from *in vivo* metabolism of oligos in order to fully delineate their metabolic pathways and pharmacokinetic profiles [4].

Important applications of labelled oligos include their use:
- as primers and probes in molecular biology applications
- in biochemical studies involving the use of oligos
- in cellular uptake studies of oligos *in vitro*
- in pharmacokinetic and metabolism studies of oligos *in vivo*

To properly define the pharmacokinetic and pharmacodynamic profiles of oligos, it is essential to determine their blood plasma levels, tissue distribution, clearance and elimination at various time-points following administration. In this regard, the sensitivity of detection of an oligo is dependent on the reporter groups or tags attached to the oligo. Oligos bearing radioisotopic labels will be most useful because of their high sensitivity of detection. The use of radioisotopically labelled oligos is especially relevant for *in vivo* evaluation of chronic or cumulative toxicity associated with the oligos or their metabolites. In this context, it may be desirable to use labelled oligos having single (e.g., ^{35}S) or multiple labels (e.g., ^{35}S and ^{14}C) of either one or multiple sites in the oligo chain. Similar considerations apply when evaluating the metabolic pathways of a given oligo. From an economic standpoint one would prefer to use ^{35}S labelled oligo as their synthesis is cheaper and relatively straightforward. However, one still needs to use ^{3}H and ^{14}C-labelled oligos to elucidate the *in vivo* metabolic fate of the nucleobases in the oligo.

Nevertheless, although tremendous advances have been made during the last few decades in the chemical synthesis of oligos, the chemical methods for incorporating radioisotopic labels into oligos are rather limited. Radioisotopic labelling of oligos, may be carried out by enzymatic means [5, 6] or chemical methods [7, 8]. In particular, it is highly desirable to have methods for the chemical synthesis of radiolabelled oligos, which would be simple and usable by chemists as well as non-chemists, in order that many potential applications of oligos are fully realized.

In this chapter, we report on the various synthetic methods which have been developed for the incorporation of radioisotopic labels into oligos, including work recently carried out, in our laboratory. Also detailed herein are experimental procedures for the preparation of reagents (where appropriate) and protocols for incorporation of ^{35}S, ^{3}H and ^{14}C labels into oligos. Additionally, with appropriate modifications, the described methods can be used for site-specific incorporation of multiple radioisotopic labels into a given oligo. Excluded from this chapter are procedures for *enzyme-mediated* incorporation of radioisotopic labels into oligos which are described in other excellent source books and articles [5,6].

2 Incorporation of the Radioisotopic [^{35}S]-label into Oligonucleotides

This labelling procedure is particularly relevant in the context of oligonucleoside phosphorothioates which are being developed as therapeutic [9–11] and diagnostic agents. The label is incorporated into the backbone of the oligo such that one of the non-bridging atoms in the internucleotidic phosphodiester linkage is replaced by ^{35}S. This can be accomplished by either site-specific incorporation of ^{35}S, as is practical using phosphoramidite chemistry [12] [13] (Schemes 1, 3, Fig. 1), or by uniform labelling as can be accomplished by using either phosphoramidite chemistry or H-phosphonate chemistry (Scheme 2). The choice of the chemistry is also dictated by the type of oligo under consideration (*vide infra*).

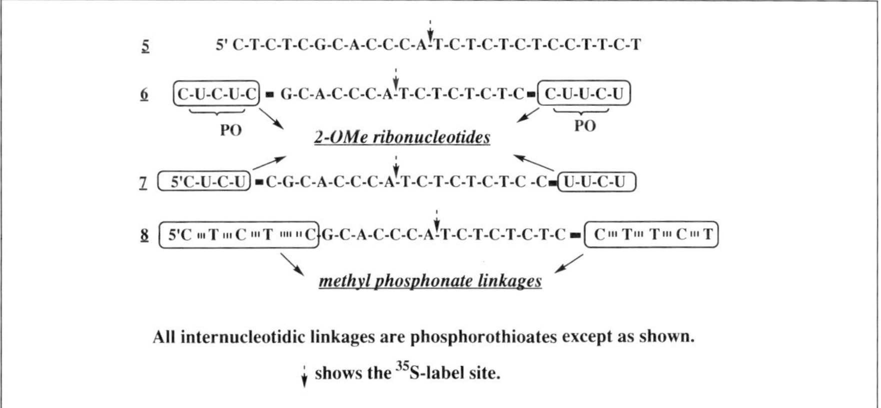

Figure 1: Site-specifically ^{35}S-label into Oligonucleotides prepared using the thiolsulfonate [^{35}S]**1**.

For the preparation of ^{35}S-labelled oligonucleoside phosphorothioates, using phosphoramidite chemistry, the labelling reagent employed is either elemental [^{35}S] or the recently introduced sulfur transfer reagent, [^{35}S]**1*** (Scheme 5). Alternatively, for the preparation of oligonucleoside phosphorothioates labelled uniformly with ^{35}S, H-phosphonate chemistry can be used in conjunction with either elemental [^{35}S] or the recently synthesized sulfur transfer reagent, [^{35}S]**2** (Scheme 4).

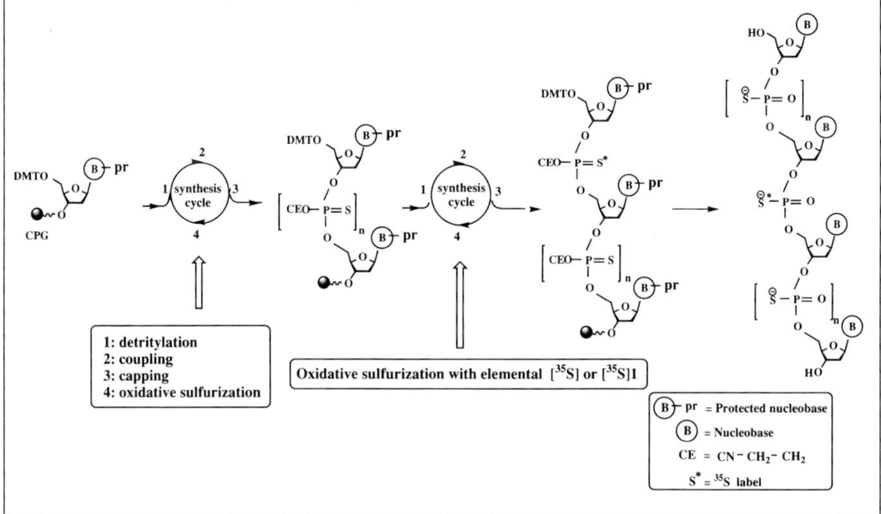

Scheme 1: Site-specific Incorporation of ^{35}S-label into Oligonucleotides using Phosphoramidite Chemistry

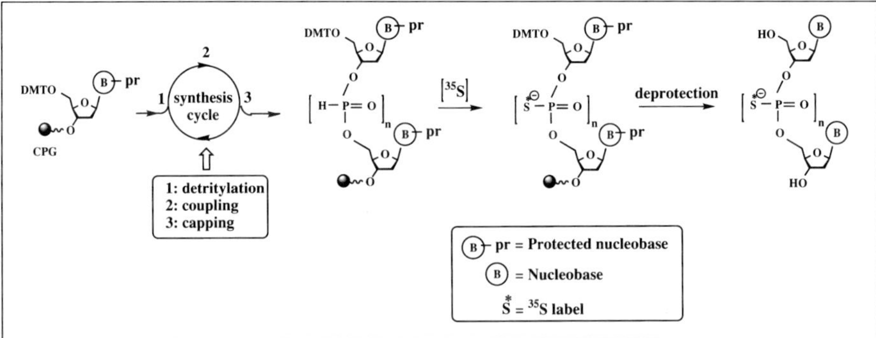

Scheme 2: Synthesis of Uniformly ^{35}S-labelled Oligonucleotides using H-phosphonate Chemistry.

---

* Bold numbers in text denote various molecules shown in Figure 1 and Schemes 3–9.

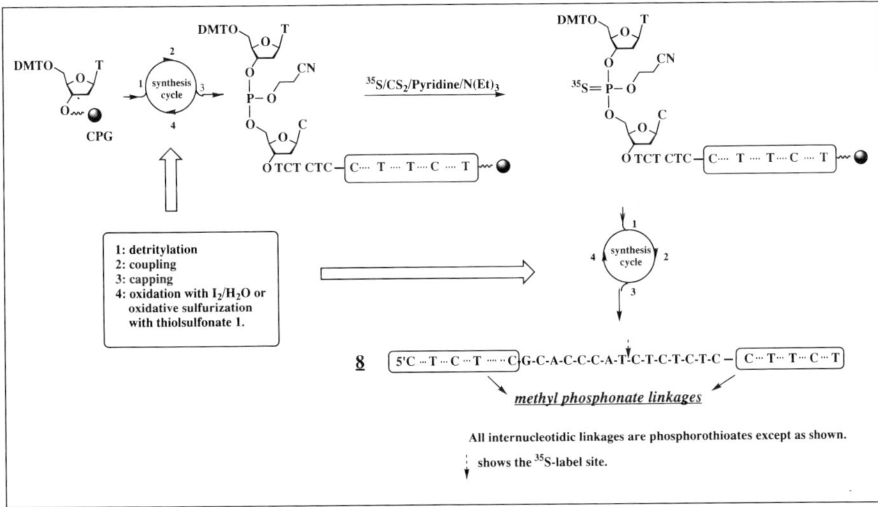

Scheme 3: Synthesis of site-specifically ³⁵S-labelled **8** using phosphoramidite chemistry and elemental [³⁵S].

Scheme 4: Preparation of ³⁵S-labelled thiobenzoic acid (**2**).

Scheme 5: Preparation of ³⁵S-labelled sulfurizing reagent (**1**).

2.1 Site-specific [^{35}S]-labelling of Oligonucleotides

As stated previously, site-specific ^{35}S-labelling can be readily accomplished using phosphoramidite chemistry.

Protocol 1: Site-specific [^{35}S]-labelling of Oligonucleotides

- A predetermined site for the incorporation the radioisotopic label is chosen.
- Solid-phase synthesis of the oligo is carried out, using phosphoramidite chemistry till the label site is reached. This can be accomplished by modifying the synthesis program in which the synthesis cycle is interrupted after formation of the internucleotidic phosphite linkage at the desired site of ^{35}S incorporation.
- Following completion of the synthesis cycle, the CPG-bound oligomer is treated with [^{35}S] in a solution cocktail consisting of toluene, and anhydrous pyridine, triethylamine and carbon disulfide. The oxidative sulfurization reaction, which converts the phosphite to a phosphorothioate linkage resulting in the incorporation of ^{35}S label, is done outside the machine to avoid inadvertent precipitation of sulfur particles in the reagent delivery lines of the synthesizer.
- To complete the oxidative sulfurization, the same cocktail of "cold" [^{35}S] or, more conveniently, with a 2 % solution of 3*H*-benzodithiole-3-one-1,1-dioxide (**1** in Scheme 5) in acetonitrile is used.
- By employing this procedure, incorporation of ^{35}S at multiple sites, in an oligo can be accomplished. The synthesis of the oligo is then completed by continuing the usual synthesis cycle.

2.1.1   Site-specific [^{35}S]-labelling of Oligonucleotides – Use of Elemental [^{35}S]

The following procedure is representative of the site-specific labelling of chimeric oligo using phosphoramidite chemistry in conjunction with elemental [^{35}S] [14]. In the example shown, ^{35}S label is introduced at the internucleotidic linkage between nucleotides 12 (C) and nucleotide 13 (T) (Scheme 3) of the target oligo.

For the synthesis of the oligo **8**, bearing the ^{35}S label site as indicated, the synthesis of the oligo is started with T-linked CPG using the appropriate phosphoramidite monomer on a 10 μmol scale.

Protocol 2: Site Specific [^{35}S]-labelling Using Elemental [^{35}S]

- The synthesis cycle is programmed on an automated synthesizer to terminate the synthesis just after the internucleotidic phosphite linkage (i.e., after the coupling reaction) between nucleotides 12 and 13 is formed. At that juncture, the column is removed from the synthesizer, and the CPG is dried in a stream of *argon*.

- The CPG is quickly placed in an eppendorf tube and treated with a freshly reconstituted mixture of [^{35}S] (4.5 mCi, 1 Ci/milligram atom, in 50 ml of toluene, Amersham) and carbon disulfide:pyridine:triethylamine (200 µl : 200 µl : 4 µl). The oxidative sulfurization is allowed to proceed at ambient temperature for one hour.
- Following this, the supernatant is carefully removed and the CPG is washed with anhydrous acetonitrile (5 × 1 ml). The CPG is next treated with a solution of "cold" **1** (1 ml, 2 % solution in acetonitrile) for 10 minutes.
- The supernatant is removed and the CPG is washed with anhydrous acetonitrile (5 × 1 µl). The CPG is next treated with the capping solution (CAP A: 300 ml, THF/2,6-lutidine/acetic anhydride, 8/1/1 and CAP B: 300 µl of 0.63 % dimethylaminopyridine in pyridine).
- Finally the CPG-bound oligo is washed with anhydrous acetonitrile (5 × 1 ml), and placed back in the synthesizer to complete the synthesis of [^{35}S]-labelled oligo.
- Following the synthesis, the crude CPG-bound oligo is treated with aqueous ammonium hydroxide (28 %, 3 ml, ambient temperature, 2 h). The supernatant solution is evaporated to dryness on a Speed-Vac concentrator to give a pale yellow residue. The residue is treated with a solution of ethylene diamine : ethanol : water (50 : 47 : 3, 5 ml) at 25 °C for 5 hours to effect complete removal of the base and phosphate protecting groups.
- Purification of the oligo is carried out using preparative PAGE (20 %, 7 M urea). The gel slices containing the labelled oligo is removed, crushed and eluted with 0.1 M ammonium acetate for 24 h. The eluent is concentrated to a small volume and loaded on a Sephadex G-25 column. The elution is monitored by scintillation counting of small aliquots of the individual fractions (0.5 ml each). Concentration of the eluent gives the [^{35}S]oligo **8** (see Scheme 3) as a white pellet.

In several experiments, performed as above, we were able to obtain 120–150 µCi of oligo, A$_{260}$ = 238 O.D., with a specific activity of 168 mCi/mmol). Analysis of the purified oligo by capillary gel electrophoresis reveals that the material is > 90 % pure (Fig. 2).

Using the above procedure, multiple incorporations of ^{35}S-label into an oligo can also be done. Similarly, various other oligos e. g., "hybrid" (**7**, Fig. 1) and "mixed" PO-PS oligos (**6**, Fig. 1) can also be labelled using the above procedure.

The label incorporation has to be performed manually in a well-ventillated hood, to minimize exposure to the toxic and poisonous carbon disulfide and to prevent inadvertent precipitation of particulate matter in the valves and delivery lines of the synthesizer. All reagents have to be dried and freshly distilled (including carbon disulfide!) before use [15].

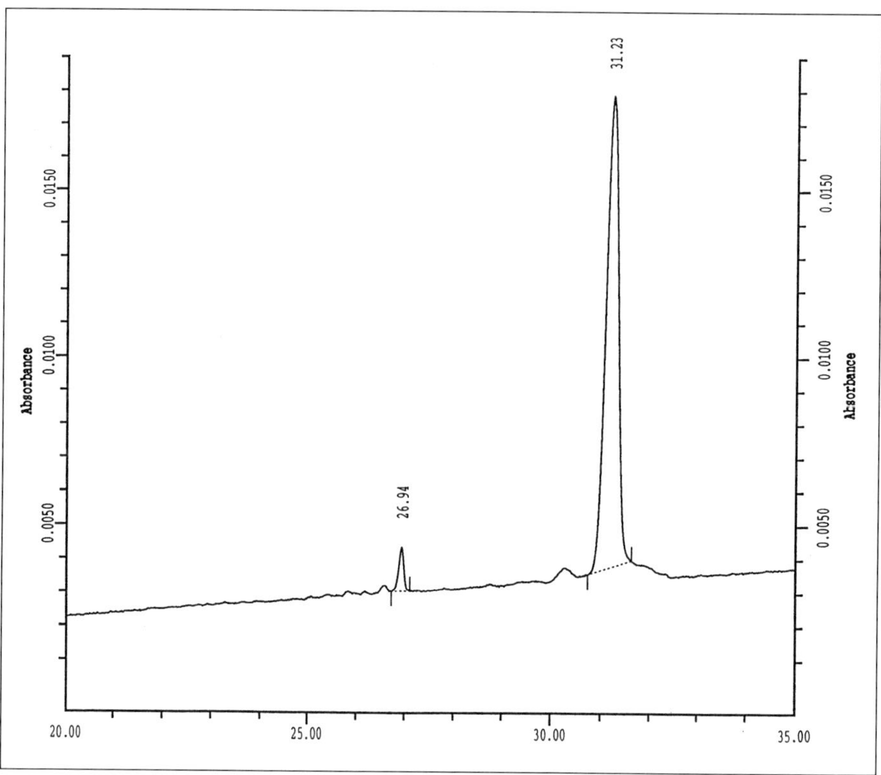

Figure 2: Capillary gel electrophoretic profile of [^{35}S]-labeled oligonucleotide **8** prepared using elemental [^{35}S]. Peak at 26.94 min is an internal standard.

2.1.2   Site-specific [^{35}S]-labelling of Oligonucleotides – Use of [^{35}S]1

Although the synthesis of ^{35}S-labelled oligos can be accomplished using [^{35}S] (as described above), the procedure is quite labour intensive and, in addition, involves the handling and use of the highly toxic carbon disulfide. These difficulties involved in using ^{35}S and the desirability to employ *similar* "chemistry" and synthesis reagents in the preparation of [^{35}S]-labelled oligos, as is used in the synthesis of oligonucleoside phosphorothioates for pharmacokinetic studies and pre-clinical evaluations, prompted us to search for a ^{35}S-labelled sulfur transfer reagent amenable to automated synthesis of ^{35}S-labelled oligos. Since phosphoramidite chemistry, in conjunction with 3*H*-1,2-benzodithiole-3-one-1,1-dioxide [16,17], is widely used in the large-scale synthesis of oligonucleoside phosphorothioates, it was desirable to use a similar protocol for the preparation of [^{35}S]-labelled oligos. This prompted us [18] to synthesize [^{35}S]3*H*-benzodithiole-3-one-1,1-dioxide reagent (**1**) as the [^{35}S] sulfur-transfer reagent (Schemes 4 and 5). Like the "cold" material, [^{35}S]**1** is a white crystalline solid and is readily soluble in acetonitrile or methylene chloride and can be used for site-specific incorporation of the ^{35}S label at any predetermined site in a given

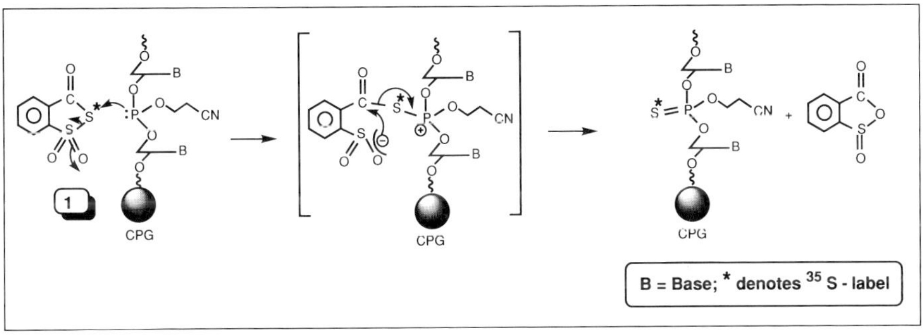

B = Base; * denotes ^{35}S - label

Scheme 6: Oxidative sulfurization of internucleotidic phosphite linkage using [^{35}S]1.

oligo. Moreover, like its "cold" counterpart, [^{35}S]1 can also be used for the site-specific incorporation of ^{35}S label into oligos which have (i) "mixed" backbones such as phosphoric diester-phosphorothioate linkages, (ii) heterogeneous backbones e.g., phosphorothioate-methyl phosphonate ("chimeric" oligos) and (iii) mixed ribonucleotide-deoxyribonucleotide population ("hybrid oligos") (Fig. 1), in conjunction with phosphoramidite chemistry.

In order to access [^{35}S]1, the requisite ^{35}S-labelled thiobenzoic acid (3) is prepared using a high temperature (125 °C) exchange reaction between thiobenzoic acid and elemental [^{35}S] (Scheme 4). [^{35}S]3 is then converted to [^{35}S]2 in presence of thiosalicylic acid (4) and sulfuric acid (Scheme 5). Upon careful oxidation with H_2O_2 in trifluoroacetic acid [^{35}S]2 is converted to [^{35}S]1. [^{35}S]1 is a white crystalline solid and readily soluble in acetonitrile and methylene chloride. For the oxidative sulfurization reaction in oligo synthesis, a solution of [^{35}S]1 (91 mCi/mmol, 1 % in anhydrous acetonitrile) is used (Scheme 6).

We have prepared a variety of oligos 5–8 (Fig. 1) (1 µmol scale) bearing a pre-determined site of incorporation of the ^{35}S-label, using [^{35}S]1 in conjunction with phosphoramidite chemistry.

Protocol 3: Site-specific [^{35}S]-labelling of Oligonucleotides – Use of [^{35}S]1

- To incorporate the ^{35}S label, the synthesis cycle is interrupted after the formation of the internucleotidic *phosphite* linkage as described before.
- The CPG is removed from the column and treated with a stock solution of [^{35}S]1 (*vide infra*) (15 µL, 30 min) followed by a solution of "cold" 1 (2 % in acetonitrile, 100 µL, 10 min). The oxidative-sulfurization is found to be quantitative as determined by "trityl assays".
- Following the synthesis, the CPG is treated with aqueous NH$_4$OH (30 %, 10 h, 55 °C) to obtain crude oligos 5–8. The crude 5–8 are then purified by preparative PAGE (20 %) and desalted on a Sephadex G-25 column.

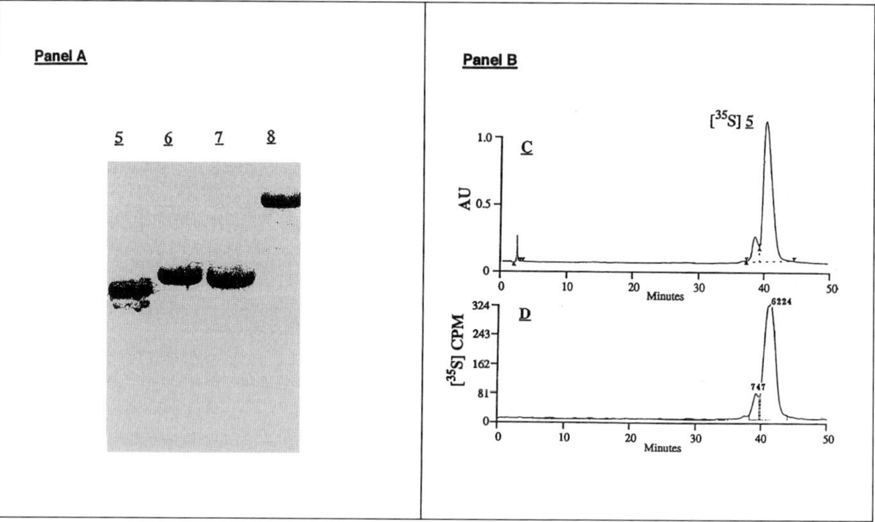

Figure 3: Panel A: Analytical polyacrylamide gel electrophoretic profiles of [³⁵S]-labeled oligonucleotides **5–8. Panel B:** Ion-exchange HPLC of [³⁵S]**5**.

Figure 3 shows the analytical PAGE profile (Fig. 3, Panel A) and ion-exchange HPLC (Fig. 3, Panel B) of the labelled oligos obtained as above. Using ion-exchange HPLC equipped with a UV detector and interfaced with flow-scintillation analysis, it can be clearly seen that the radioactive peak of the oligo is superimposable on its UV absorbing peak. The oligos **5–8** thus obtained had a specific activity of ca. 23 ~ 25 mCi/mmol. Oligos with higher specific activity can be prepared either by use of reagent [³⁵S]**1** with high specific activity or by multiple incorporations of the label into the oligo.

At the present time there is no commercial source of [³⁵S]**1** and material has to be prepared by the user according to the following typical experimental procedures.

2.1.3 *Typical Experimental Procedures for the Synthesis of [³⁵S]1*

Protocol 4A: Synthesis of [³⁵S]2

- Thiobenzoic acid (**3**) (6.5 µl, 55 µmol) and [³⁵S] (5 µCi, 1Ci/mg atom in 100 µl of toluene) are placed in an eppendorf tube (1.5 ml) and the contents heated at 97 °C for 5 h.
- The solution is evaporated to dryness under argon, and thiosalicylic acid (**4**) (5 mg) and sulfuric acid (98 %, 50 µl) are added at 0 °C. After heating at 50 °C for 3 h, the brown reaction mixture is cooled to –78 °C and quenched with water (600 µl).
- The solution is warmed, transferred to a vial and diluted with 500 µl water. The solution is extracted with methylene chloride (4 × 3 ml) and the organic layer washed with Na₂CO₃ (5 %, 2 × 2 ml).

- The organic layer is evaporated to dryness and the resulting solid is dissolved in warm hexane (2 µl) and the solution is evaporated to dryness under argon to give **2** as a yellow solid (4 mg, 2.17 µCi, 44 % yield). This material is used as such in the next step or stored at –20 °C until ready to use.

Protocol 4B: Synthesis of [^{35}S]1

- To 4 mg of **2** in an eppendorf tube, cooled to 0 °C, is added trifluroacetic acid (25 µl) and 30 % H_2O_2 (12 µl).
- The reaction mixture is kept at 42 °C for about 2 h (as monitored by TLC, silica gel, chloroform), and then quenched with ice-cold water (300 µl).
- The resulting white precipitate is washed with water (2 × 300 µl) and dried *in vacuo* to give [^{35}S]**1** (2 mg, total activity of 915 µCi, specific activity 91 µCi/mmol). A stock solution of [^{35}S]**1** in anhydrous acetonitrile (2 mg, 91 µCi/mmol in 200 µl acetonitrile) can be stored at –20 °C until ready for use in the oxidative sulfurization reaction.

For the incorporation of the ^{35}S label, using [^{35}S]**1**, the protocol employed is the same as that using elemental [^{35}S] except that the oxidative sulfurization is performed using a solution of [^{35}S]**1** in acetonitrile.

2.2 [^{35}S]-labelling of Oligonucleotides Using Elemental [^{35}S] and H-phosphonate chemistry

H-phosphonate chemistry can be used for uniform ^{35}S-labelling of oligos. Alternatively one could introduce ^{35}S-label contiguously in the phosphorothioate backbone near the 5'-end in a single site or in contiguous sites. This can be accomplished by first performing the synthesis of a certain segment of oligonucleoside phosphorothioate, starting from the 3'-end, using phosphoramidite chemistry in conjunction with 3*H*-1,2-benzodithiole-3-one-1,1-dioxide (**1**). The remaining sequence is synthesized using H-phosphonate chemistry. The oxidative sulfurization of the H-phosphonate, using elemental [^{35}S], gives the phosphorothioate segment bearing the [^{35}S] label [19].

The following labelling procedure developed for a 25-mer oligonucleoside phosphorothioate **5** is typical of this labelling procedure (19). (The label is incorporated at 5 contiguous internucleotidic linkages from the 5' end).

Protocol 5: [^{35}S]-labelling Using Elemental [^{35}S] and H-phosphonate Chemistry

- The synthesis of the oligo **5** is started from the 3'-end, on a 2 µmol scale, using phosphoramidite chemistry using the appropriate monomer precursors.

- At each step oxidative sulfurization is carried out using 3H-1,2-benzo-dithiole-3-one-1,1-dioxide.
- When the end of the segment is reached (i.e., after 19-mer has been prepared), the column is removed and the synthesis of the remaining segment (i.e., 6-mer) is carried out using H-phosphonate chemistry using the appropriate monomer precursors. All syntheses are carried out using standard synthesis programs recommended by the manufacturer.
- 10 millicuries of [^{35}S] in 80 µl of carbon disulfide/pyridine/triethylamine (10/10/1) is added to 50 mg of the CPG-bound oligo obtained as above (2 µmoles). The reaction is kept at room temperature with occasional shaking.
- 'An additional 200 µl of 5 % solution (in carbon disulfide/pyridine/triethylamine, 10/10/1) of elemental sulfur (^{32}**S!**) is added to complete the oxidative sulfurization. The reaction is kept for an additional 2 h at ambient temperature.
- The supernatant is carefully removed, and the supernatant washed with carbon disulfide/pyridine/triethylamine (10/10/1) (3 × 200 µl) followed by acetonitrile (3 × 200 µl) and dried in a speed vac.
- For removal of the base and phosphate protecting groups and cleavage from the support, the CPG-bound ^{35}S-labelled oligo is treated with 1 ml of aqueous ammonium hydroxide (28 %) at 55 °C for 12 h. The supernatant is removed. The CPG is washed with 1 ml of sterile water. The combined aqueous solution (ca. 2 millicuries) is dried in a speed vac to obtain crude labelled **5**.
- For purification, the crude **5** is dissolved in 600 µl of formamide and subjected to PAGE (20 %, 7M urea). The gel slices are removed and soaked in 0.1M ammonium acetate.
- The aqueous solution containing the oligo is filtered to remove particulate matter and loaded on a SEP-PAK cartridge which had been previously washed with acetonitrile (15 ml) sterile water (15 ml) and ammonium acetate (0.1 M, 5 ml) *in that order.* After loading on the SEP-PAK, the cartridge is eluted with sterile water (10 ml). This step removes polar contaminants from the oligo. Finally, the oligo is eluted from the SEP-PAK by washing with acetonitrile/water (2/3, 5 ml).
- The eluent is concentrated to dryness in a speed vac and redissolved in sterile water (1 ml).
- At this stage, the purified oligo is obtained in the "ammonium form". To convert it to the sodium form, the aqueous solution of the oligo, obtained as above, is loaded on to a Dowex column previously washed and already equilibrated to the sodium form. The column is washed with sterile water (2 × 5 ml).

- Each portion is separately loaded on a Sephadex G-15 column and aliquots of 0.5 ml are collected and a 1 µl portion subjected to scintillation counting. The radioactive fractions are combined, sterile filtered (if necessary), and lyophilized dry to give the ^{35}S-labelled oligo **5**.

Final radiochemical yield is ca. 11 % (based on 1.1 millicurie of activity incorporated starting from 10 millicurie of ^{35}S) and 7.5 % chemical yield (1.26 mg; expected theoretical yield: 17.0 mg). Purity of the oligo **5**, thus obtained is greater than 95 % as assessed by capillary gel electrophoresis (Fig. 4).

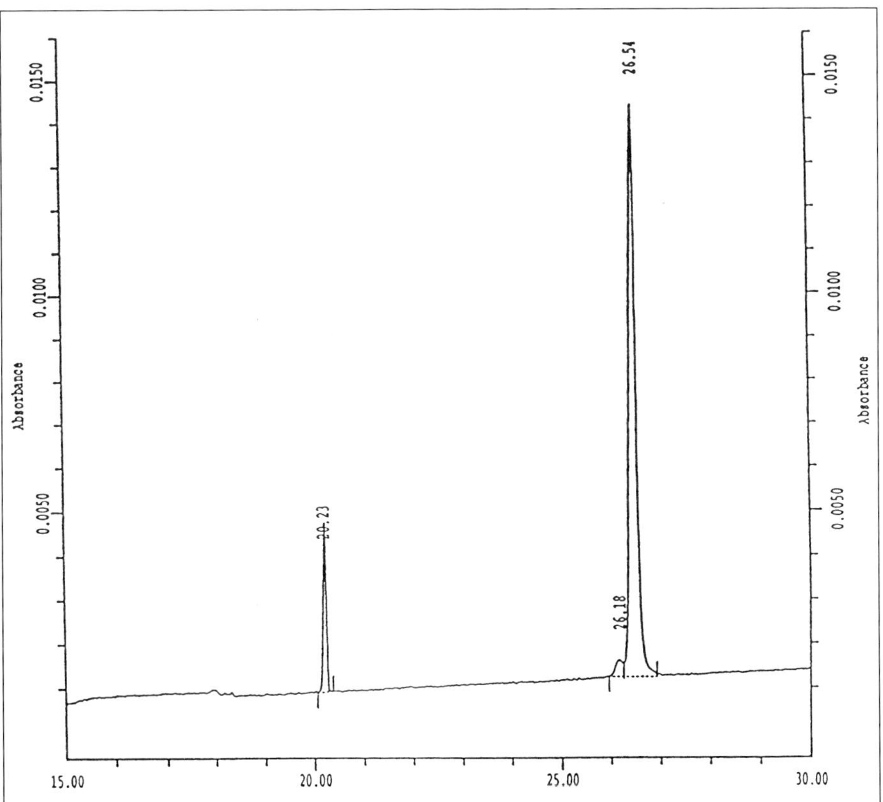

Figure 4: Capillary gel electrophoretic profile of [^{35}S]-labeled oligonucleotides **5** prepared using elemental [^{35}S] and H-phosphonate chemistry. Peak at 20.23 min is the internal standard.

3   Synthesis of [^{3}H]-labelled Oligonucleotides

The incorporation of ^{3}H into oligos can be done by exchange reaction with ^{3}H$_2$O [7] in the presence or absence of catalysts [21, 22]. [^{3}H]-labelled oligos with high specific activity can be produced this way. However, [^{3}H]-la-

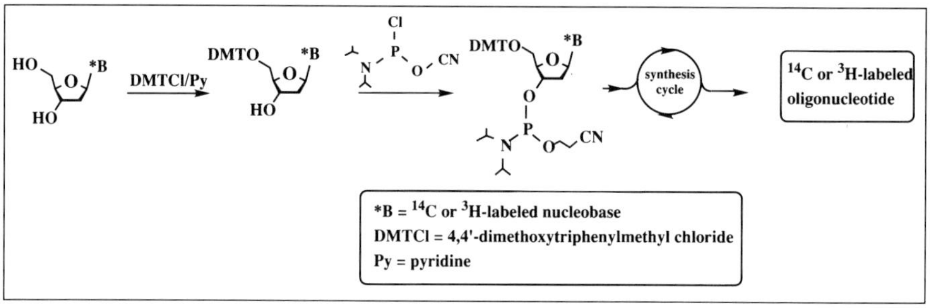

Scheme 7: Synthesis of ^{14}C and ^{3}H-labeled oligonucleotides from labeled amidite precursors.

belled oligos thus obtained are limited in their utility. For *in vivo* pharmacokinetic studies and metabolism studies, it is desirable to have labels at non-exchangeable positions and to have precise knowledge of their location within the oligo. The following section describes the methods available for preparation of [^{3}H]-labelled oligos suitable for metabolism and pharmacokinetic studies.

3.1 Synthesis of [^{3}H]-labelled Oligonucleotides Using [^{3}H]-labelled Nucleoside Phosphoramidites

In conjunction with solid-phase methodology, one could use the pre-synthesized ^{3}H-labelled nucleoside monomers for oligo synthesis. Thus, commercially available ^{3}H-labelled nucleoside can be converted in two steps to the nucleoside phosphoramidite synthons which can then be employed in oligo synthesis [23, 24] (Scheme 7). This approach is less convenient as it involves the handling, isolation and purification of sensitive and highly radioactive intermediates. Furthermore, the radioactive monomer precursors need to be diluted with the naturally abundant isotope counterpart to maintain a manageable scale of oligo synthesis. These considerations led us to develop alternate methodologies for the introduction of the ^{3}H label into oligos.

3.1.1 Synthesis of [^{3}H]-labelled Oligonucleotides Using Redox Chemistry

The essence of our new labelling approach [25] entails the chemoselective oxidation of the 5'-hydroxymethyl group of a support-bound oligo to produce the corresponding 5'-carboxyaldehyde followed by its reduction with commercially available [^{3}H]NaBH$_4$ to regenerate the 5'-hydroxymethyl group bearing the [^{3}H] label at the 5'-methylene group. Quite clearly, in this method, the oxidant and the oxidation conditions must be carefully chosen such that the desired chemoselective transformation is achieved, while leaving intact the other functional groups (including that of the solid-support), in the support-bound oligo. This can be achieved if the redox process can be conducted in neutral to slightly acidic medium. We have found that the oxidation of the CPG-bound oligo, using Moffatt-Pfitz-

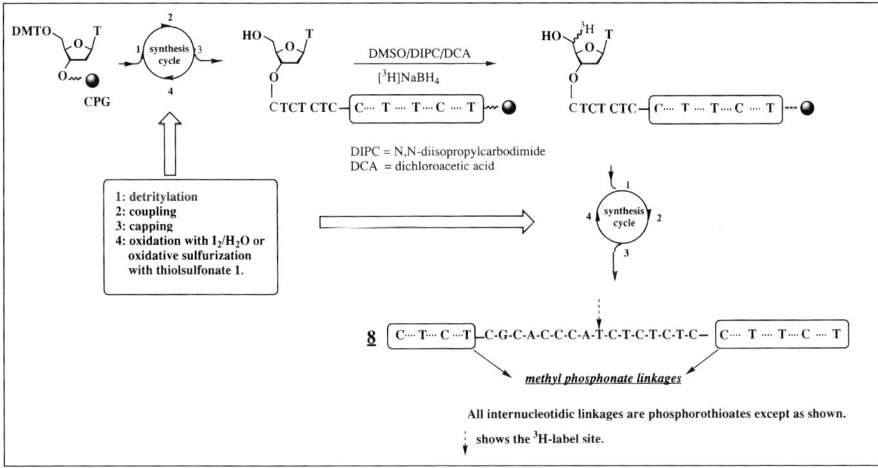

Scheme 8: Synthesis of site-specifically ³H-labeled **8** using redox chemistry.

ner reagent, and subsequent reduction with [³H]NaBH₄ in i-PrOH/1M Tris-HCl buffer, results in incorporation of the ³H label at a predetermined site without any detectable modification of the other functional groups in the oligo.

The experimental protocol is given in the following section.

3.1.2

Protocol 6: Typical Experimental Procedure for the Synthesis of [³H]-labelled Oligonucleotide 8

In the illustrative example for the synthesis of [³H]-labelled oligo **8**, ³H-label is incorporated at the 5' position of T-nucleoside corresponding to the position 13 of the 25-mer oligo (Scheme 8).

- The 13-mer is synthesized on 4 × 1 µmol scale, as before, using phosphoramidite chemistry.
- The vacuum dried CPG-bound oligo is placed in an eppendorf tube (1.5 ml) and treated with a cocktail consisting of DMSO/DIPC/DCA (250 µl/50µl/4µl) at room temperature for 2 h.
- The CPG is washed with acetonitrile (10 × 1 ml) and the support-bound "aldehyde" is transferred to a glass vial containing a solution of [³H]NaBH₄ (25 millicuries, 65 millicurie/µmol, American Radiolabelled Chemicals) in i-PrOH/1 M Tris-HCl (200 µl, 95/5) and the reaction kept at 25 °C for 2 h.
- The reaction mixture is cooled to 0 °C and a solution of NaBH₄ (0.1 M in 200 µl of 1M Tris-HCl, pH 7.0) is added. After 15 min, the supernatant is removed and the CPG is sequentially washed with 1 M Tris-HCl (pH 7.0, 0.5 ml), acetic acid (2N, 0.5 ml) and ethanol (2 × 1 ml).

- After the final wash with acetonitrile (10 × 1 ml), the CPG-bound ^{3}H-labelled 13-mer is used for continuing the oligo synthesis further. Deprotection and PAGE purification as before results in the production of [^{3}H]**8** as a white pellet (69 A$_{260}$ units, 354 microcuries, specific activity of 1.2 millicurie/micromol).

An alternative radiolabelling procedure can also be employed which uses *in situ* solid-phase phosphitylation (*vide infra*).

4 Synthesis of [^{14}C]-labelled Oligonucleotides

4.1.1 *[^{14}C]-labelled Oligonucleotides Using [^{14}C]-labelled Nucleoside Phosphoramidites*

As stated in section 3.1, one could use presynthesized [^{14}C]-labelled nucleoside monomers for the preparation of [^{14}C]-labelled oligos. Details can be found in a recent paper by Sasmor et. al. [24]. We have developed a different approach for the synthesis of [^{14}C]- and [^{3}H]-labelled oligos (*vide infra*).

4.1.2 *Synthesis of [^{14}C]- and [^{3}H]-labelled Oligonucleotides by Solid-phase Phosphitylation*

A new route to the preparation of site-specifically [^{14}C]- or [^{3}H]-labelled oligos has been recently developed in our laboratory [26]. In this approach, 5'-*O*-DMT-[^{3}H] or 5'-*O*-DMT-[^{14}C]-labelled nucleoside is employed in a coupling reaction with support-bound 5'-phosphoramidite (Scheme 9). Following the formation of a 3'-5' internucleotidic phosphite linkage, the oxidation or oxidative sulfurization reaction can be employed to convert the phosphite to phosphate or phosphorothioate. The support-bound radiolabelled oligomer can then be employed for further chain elongation in the synthesis of the target oligo.

Illustrated is the preparation of a support-bound [^{14}C]-labelled tetramer.

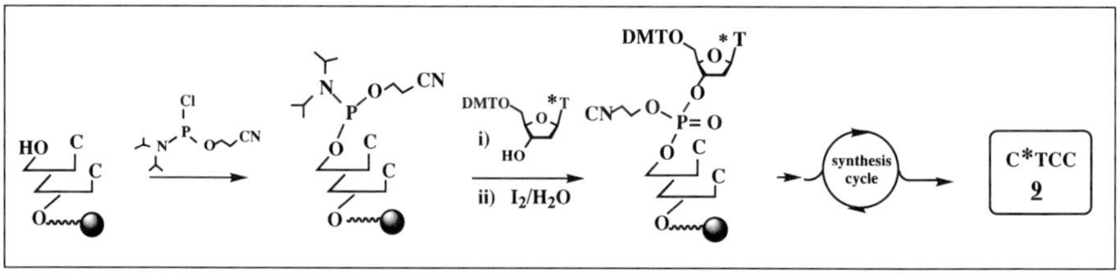

Scheme 9: Synthesis of Site-specifically ^{14}C-labelled **9** by Solid-phase Phosphitylation.

Protocol 7: Synthesis of [^{14}C]-labelled Oligonucleotides by Solid-phase Phosphitylation

- The support-bound dimer (2 micromol) is placed in an eppendorf tube and treated with a solution of 2-cyanoethyl *N,N*-diisopropyl-chlorophosphoramidite (200 µl, 0.15 M in CH_2Cl_2) and diisopropylethylamine (2.6 µl, 15 micromol) in presence of 1-methyl imidazole (2.7 mg in 8 µl pyridine) at room temperature for 1 h.
- The resultant CPG phosphoramidite is washed with methylene chloride (10 × 1 ml) and dried in a stream of argon.
- The CPG-amidite (1.5 micromol) is treated with a mixture of [2-^{14}C]-5′-*O*-dimethoxytrityl thymidine (52 mCi/mmol, 778 µCi, 75 µl, 0.2 M in anhydrous THF) and 60 µl of tetrazole solution (0.45 M in acetonitrile) for 30 min at 25 °C.
- The supernatant is removed and used for the recovery of unreacted [2-^{14}C]-5′-*O*-dimethoxytrityl thymidine. The CPG-supported phosphite is oxidized with either 0.1 M I_2 in THF/Py/H_2O to give the support-bound trimer. Extension of the synthesis cycle using this support-bound trimer gives the target tetramer.
- Cleavage from the support, deprotection and purification by PAGE, as described before, gives the [^{14}C]-labelled tetramer (**9**) (16.8 O.D. units, 22 µCi, specific activity 52 mCi/mmol).

A similar procedure can be employed for the preparation of [^{3}H]-labelled oligos.

5   Conclusion

We have outlined practical methods for the radioisotopic labelling of oligos. The methods described herein have been employed routinely in our laboratory, for the radiolabelling of oligos, for pharmacokinetic and metabolism studies. As mentioned in the introduction, with suitable adaptation of the aforedescribed methods, one could introduce multiple labels at predetermined locations within the oligo. It is also expected that these procedures will find application in the radioisotopic labelling of various nucleosides and their analogs as well as related compounds of biological interest.

References

1. AGRAWAL, S. , R. P. IYER: Modified oligonucleotides as therapeutic and diagnostic agents. Curr. Opinion in Biotech., 6 (1995), 12.
2. CROOKE, S. T., B. LEBLEU (Eds.): Antisense Research and Applications. CRC Press, Boca Raton, Florida, USA.
3. ZAMECNIK, P. C., M. L. STEPHENSON: Inhibition of Rous sarcoma virus replication and cell transformation by a specific oligodeoxynucleotide. Proc. Natl. Acad. Sci. USA 75 (1978) 280.

4. Agrawal, S. , J. Temsamani, J., W. Galbraith, J.-Y. Tang: Pharmacokinetics of antisense oligonucleotides. Clin. Pharmacokinetics 28 (1995) 7.

5. Sambrook, J., E. F. Fritsch, T. Maniatis: Molecular Cloning. A Laboratory Manual, 2nd Edition, 5.68–5.71, Cold Spring Harbor Laboratory Press, Cold Spring Harbor (1989).

6. Roychoudhury, R., Wu, R. (1980) *Methods Enzymol.* 65: 43–63.

7. Graham, M. J., S. M. Freier, R. M. Crooke, D, J. Ecker, R. N. Maslova, R. N., E. A. Lesnik: Nucl. Acids Res. 21 (1993) 3636.

8. Stein, C. A., P. L. Iverson, C. Subasinghe, J. S. Cohen, W. J. Stec, G. Zon: Preparation of ^{35}S-labelled polyphosphorothioate oligodeoxyribonucleotides by use of hydrogen phosphonate chemistry. Analytical Biochemistry 188 (1990), 11.

9. Bayever, E., P. L. Iversen, M. R. Bishop, G. J. Sharp, H. K. Tewary, M. A. Arneson, S. J. Pirruccello, R. W. Ruddon, A. Kessinger, G. Zon, J. O. Armitage: Systemic administration of a phosphorothioate oligonucleotide with a sequence complementary to p53 for acute myelogenous leukemia and myelodysplastic syndrome: Initial results of a phase I trial. Antisense Res. and Dev. 3, (1993), 383.

10. Lisziewicz, J., D. Sun, F. F. Weichold, R. Thierry, P. Lusso, J.-Y. Tang, R. C. Gallo, S. Agrawal: Antisense oligonucleotide phosphorothioate complementary to gag mRNA blocks HIV-1 replication in human peripheral blood cells. Proc. Natl. Acad. Sci. USA 91 (1994), 7942.

11. Cowsert, L. M., M. C. Fox, G. Zon, C. Mirabelli: *In vitro* evaluation of phosphorothioate oligonucleotide targeted to the E2 mRNA of papilloma virus: Potential treatment for genital warts. Antimicrob. Agents Chemotherap. 37 (1993), 171.

12. Beaucage, S. L., M. H. Caruthers: Deoxynucleoside phosphoramidites – A new class of key intermediates for deoxypolynucleotide synthesis. Tetrahedron Lett. 22 (1981), 1859.

13. Beaucage, S. L., R. P. Iyer: Advances in the synthesis of oligonucleotides by the phosphoramidite approach. Tetrahedron 48 (1992) 2223.

14. Zhang, R., R. P. Iyer, D. Yu, W. Tan, X. Zhang, Z. Lu, H. Zhao, S. Agrawal: Pharmacokinetics and Tissue Disposition of a Chimeric Oligodeoxynucleotide Phosphorotioate in Rats after Intravenous Administration. J. Pharmacol. Exptl. Ther. 278 (1996), 971.

15. Use of undistilled carbon disulfide resulted in lower yield and reduced specific activity of the final product. (Unpublished observations).

16. Iyer, R. P., W. Egan, J. B. Regan, S. L. Beaucage: 3*H*-1,2-benzodithiole-3-one-1,1-dioxide as an improved sulfurizing reagent in the solid-phase synthesis of oligodeoxyribonucleoside phosphorothioates. J. Am. Chem. Soc. 112 (1990), 1253.

17. Iyer, R. P., L. R. Phillips, W. Egan, J. B. Regan, S. L. Beaucage: The automated synthesis of sulfur-containing oligodeoxyribonucleotides usimg 3*H*-1,2-benzodithiole-3-one-1,1-dioxide. J. Org. Chem., 55 (1990), 4693.

18. Iyer, R. P., W. Tan, D. Yu, S. Agrawal: Synthesis of [^{35}S]3*H*-1,2-benzodithiole-3-one-1,1-dioxide: Application in the preparation of site-specifically ^{35}S-labelled oligonucleotides. Tetrahedron Lett., 35 (1994), 9521.

19. AGRAWAL, S. , TEMSAMANI, J., J.-Y. TANG: Pharmacokinetics, biodistribution and stability of oligodeoxyoligonucleoside phosphorothioates in mice. Proc. Natl. Acad. Sci. USA 88 (1991), 7595.

20. AGRAWAL, S. , X. ZHANG, H. ZHAO, Z. LU, J. C. YAN, R. B. DIASIO, I. HABUS, Z. ZIANG, R. P. IYER, D. YU, R. ZHANG: Absorption, tissue distribution and in vivo stability in rats of a hybrid antisense oligonucleotide following oral administration. Biochem. Pharmacol. 50 (1995), 571.

21. BERGER, M., A. SHAW, J. CADET, J. R. JONES: Deuteration and tritiation of 2'-deoxyribonucleosides in the 5' and 4' positions. Nucleosides and Nucleotides, 6 (1987), 395–396.

22. RABI, J. A., J. J. FOX: Nucleosides. LXXIX. Facile base-catalyzed hydrogen isotope labelling at position 6 of pyrimidine nucleosides. J. Am. Chem. Soc. 95 (1973), 1628.

23. SANDS, H., L. J. GOREY-FERET, A. J. COCUZZA, F. W. HOBBS, D. CHIDESTER, G. L. TRAINOR: Biodistribution and metabolism of internally ^{3}H-labelled oligonucleotides. I. Comparison of a phosphodiester and a phosphorothioate. Mol. Pharm. 45 (1994) 932.

24. SASMOR, H. M., D. J. DELLINGER, P. C. ZENK, L. P. LEE: A practical method for the synthesis and purification of ^{14}C labelled oligonucleotides. J. Labelled compounds and Radiopharmaceuticals, XXXVI (1995), 15.

25. TAN, W., R. P. IYER, D. YU, S. AGRAWAL: Site-specific synthesis of [^{3}H]oligonucleotides in high specific activity through direct solid-phase redox chemistry. Tetrahedron Lett., 36, (1995), 3631.

26. TAN, W., R. P. IYER, Z. ZIANG, D. YU, S. AGRAWAL: An efficient synthesis of radioisotopically labelled oligonucleotides through direct solid-phase 5'-phosphitylation. Tetrahedron Lett., 36, (1995), 5323.

4 Analysis

Analysis of Oligonucleotides

Markus Schweitzer and Joachim W. Engels
Universität Frankfurt, Institut für Organische Chemie, Frankfurt a.M.,
Germany

1   Introduction

Analytical methods for checking the identity and purity of oligonucleotides (oligos) have improved in parallel with methods of synthesis. A variety of sophisticated analytical methods, employing classical electrophoresis, chromatographic techniques or mass spectrometry, are now available.

During chemical synthesis the growing oligo chain is treated with several reagents. Although the synthesis conditions are optimized, each of the four reactions of the synthesis cycle may lead to formation of small amounts of side products. In addition, incomplete coupling results in contamination of the full length oligo with n-1 failure sequences. These problems can be monitored by chromatography or electrophoresis. Furthermore, the protecting groups applied in the synthesis have to be separated from the oligo product after cleavage. In these cases, HPLC is a good method to verify the purity of the product.

For antisense applications modification of the oligos is essential since unmodified DNA is readily degraded in biological systems. The introduction of modifications such as phosphorothioates, methylphosphonates, O-methyl oligos or amino groups for the attachment of reporter groups has to be monitored. Mass spectrometry is a powerful tool for this purpose and nuclear magnetic resonance (NMR) allows the determination of the average amount of sulfurization.

The stability of hybridization to the target sequence is an important criterion for antisense application. Recording of a UV-melting curve allows the determination of the Tm-value and the thermodynamic parameters of the transition. These data are particularly important if new modifications are to be evaluated. Detection of oligos in vivo is not discussed.

2   UV-Spectroscopy of Oligonucleotides

UV-spectroscopy is an easy and sensitive technique and is the first method employed in the analysis of nucleic acids. A typical UV spectrum of an oligo is shown in Fig. 1. The absorbance maximum at 260 nm and the minimum at approximately 230 nm are characteristic for DNA and RNA.

Since no separation of the molecules in the sample is accomplished by spectroscopy the measured signal is the sum of the individual absorbance spectra of all compounds present in the sample solution. Nevertheless the simple recording of a spectrum can provide useful information about the purity of the oligo (see Fig. 1). Because it is sensitive and non-destructive, UV-spectroscopy is frequently utilized as the detection method for analytical procedures like HPLC or CGE.

UV spectra of nucleic acids show pronounced hypochromicity. The absorbance of a native DNA duplex is 20–30% lower than the absorbance of single strands in random coil conformation. This phenomenon is employed to measure the melting behavior of oligos.

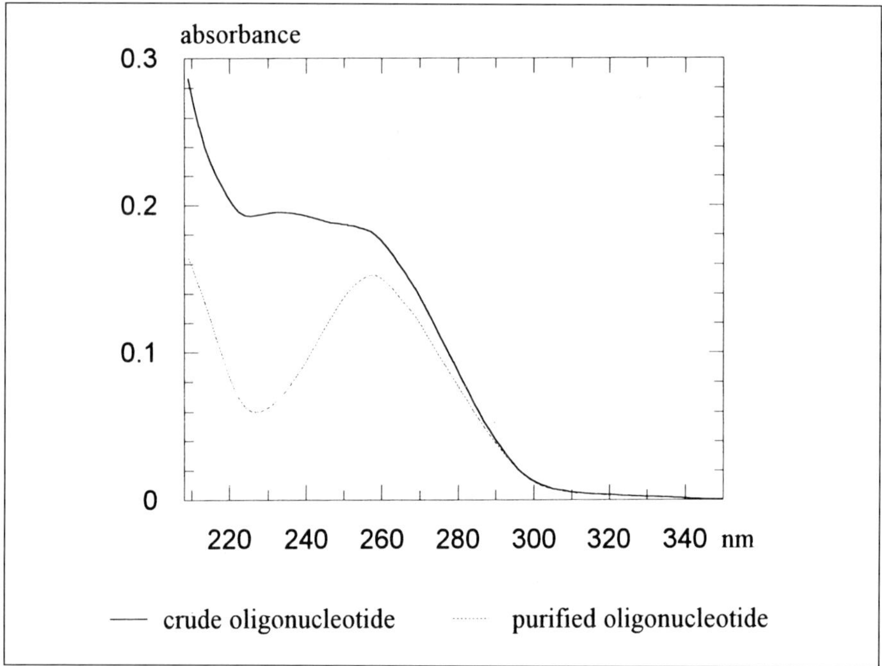

Figure 1: UV spectrum of a synthetic oligo

2.1 Quantification of Oligonucleotides

UV-spectrophotometry is the main technique for quantification of nucleic acids since the ionic character of these molecules results in an extensive hydration in the solid state. This and the possibility of salt contamination complicates the direct weighting of dry DNA samples. The determination of the absorbance or optical density (OD) is a good way to measure oligo amounts when the extinction coefficient (ε) of the molecule is known. The concentration (c) of the oligo solution is correlated with the absorbance reading (OD) according to Beer's law: $OD = \varepsilon \times c \times l$ (l = path length of the cuvette).

The concentration of an oligo solution can be determined by hydrolysis of the oligo into mononucleotides so minimizing stacking interactions between the bases. Provided that its base composition is known this is the most accurate way to determine the concentration but hydrolysis, especially of modified oligos, is not always possible. A method for the calculation of the extinction coefficient considering the nearest neighbor interactions between the bases is presented by D. M. Gray et al. [6]. An easier method that is independent of the sequence of the oligo is given by T. Brown et al. [2]. In this formula the stacking interaction is estimated and taken into account by multiplying the sum of the individual extinction coefficients by a factor of 0.9. Both methods are only valid for single stranded oligos. For duplexes or self-complementary sequences the OD measurement has to be performed above the melting point otherwise the results will be imprecise. The extinction coefficient is almost independent from the nature of the phosphate group so that these calculations are applicable for backbone modified oligos without corrections.

The measurement is routinely performed in 1 cm cuvettes with a volume of 1 ml. The absorbance at 260 nm is then referred to as OD_{260} unit.

Calculate the extinction coefficient at 260 nm as follows:

ε (260) = (a $\times$ 15.34 + g $\times$ 12.16 + t $\times$ 8.70 + c $\times$ 7.60) $\times$ 0.9 $\times$ 10^3 [M^{-1} cm^{-1}]

a, t, g, c correspond to the number of the respective bases in the oligo sequence.

Example: The 20 mer oligo with the sequence ACA CCC AAT TCT GAA AAT GG has the base composition A8, G3, T4 and C5:

ε (260) = (8 $\times$ 15.34 + 3 $\times$ 12.16 + 4 $\times$ 8.70 + 5 $\times$ 7.60) $\times$ 0.9 10^3 [M^{-1} cm^{-1}]

ε (260) = 208.8 $\times$ 10^3 [M^{-1} $\times$ cm^{-1}]

The concentration of an oligo solution is: c = $OD_{260}/\varepsilon \times 1$

A solution of the 20 mer oligo presented above with an absorbance reading of 1 OD_{260} in a 1 cm cuvette is 4.8 µM. 1 ml of this solution contains 4.8 nmol oligo.

2.2 Melting Curves of DNA

The melting temperature (Tm) of a duplex formed by an antisense oligo with a target sequence (DNA or RNA) is a good measure for the stability of the complex. Particularly, the introduction of modifications like phosphorothioates or methylphosphonates into such oligos affects the hybridization with the target molecule. In order to evaluate these effects, and to determine the hybridization characteristics of an oligo, a melting curve has to be recorded.

UV-spectroscopy is most frequently used due to the sensitivity of the method, the small amount of sample needed (typically 5 nmol for a single curve) and the ease of sample preparation. It should be mentioned here that the melting theory is not restricted to a single measuring technique so every measurable signal that changes during the melting process (e.g. circular dichroism or chemical shift in NMR) can be used to monitor the transition.

In UV-melting experiments the change of absorbance with increasing temperature is measured [16]. The cause of this hypochromicity is the stacking interaction between the chromophores of the nucleobases (sugar and phosphate backbone do not contribute to the UV-absorbance) resulting in a lower absorbance than expected for the sum of the extinction coefficients of the free nucleotides. The UV-absorbance increases as a duplex (with stacked bases) melts into two single strands with random coil conformation. The extent of hypochromicity is wavelength dependent. For AT basepairs maximum hypochromicity is observed at 260 nm while for CG basepairs the maximum is at 280 nm. The optimum wavelength is somewhere between these two values and depends on the base composition of the investigated oligo. Some commercial UV-spectrometers offer the possibility to record melting curves at several wavelengths simultaneously.

Once a curve is recorded it has to be analyzed to extract the Tm value and, if desired, the thermodynamic parameters of the transition. Several procedures for data analysis have been reported [1] [10] [16], but they are not discussed here. If simply the melting point is to be determined an easy method of analysis is to calculate the numerical differentiation of the data (forming d(absorbance)/d(T)) and to take the maximum of the resulting curve as Tm-Value. This approximation is quite good for symmetric transitions but gives poor results if applied to asymmetric melting curves. A typical melting curve and its derivative is presented in Fig. 2. Since the duplex formation is a bimolecular process it is dependent on the concentration of the oligo strands. A good way to obtain thermodynamic parameters of the transition is to plot $1/\mathrm{Tm}$ versus $\ln(c)$. The resulting plot should be linear, and the slope should correspond to $R/\Delta H^\circ$ and the intercept to $\Delta S^\circ/\Delta H^\circ$.

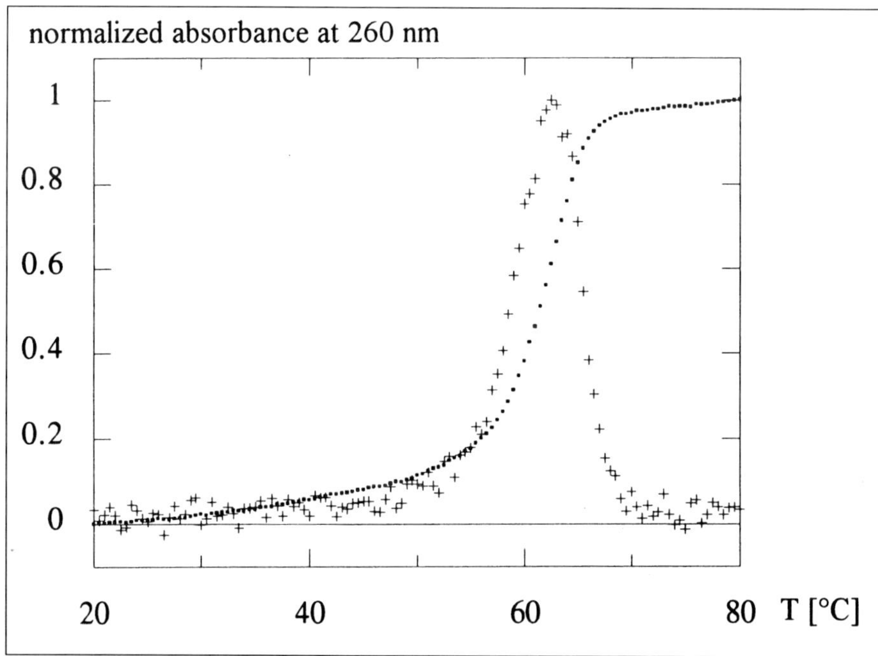

Figure 2: Normalized UV-melting curve [■] and its derivative d(absorbance)/d(T) [+] of a 20 mer oligo. The transition was monitored at 260 nm and the Tm-Value was determined to 62 °C.

Furthermore, duplex to random coil transition is dependent on the ionic strength and the pH of the buffer. For the antisense oligos under investigation salt concentration and pH that correspond to the physiologic conditions are convenient. EDTA is present in most buffers for melting experiments to chelate divalent cations thus preventing the sample from nuclease degradation.

2.2.1

Protocol 1: Recording a UV-Melting Curve

Materials required:

Chemicals: NaCl, $NaH_2PO_4 \cdot H_2O$, $Na_2HPO_4 \cdot 2H_2O$, disodium EDTA dihydrate, water.

Apparatus: UV-spectrophotometer with a thermostat, cuvettes with 1–10 mm path length.

- Prepare an aqueous buffer with the desired ionic strength and pH value. For 100 ml of a 140 mM NaCl, 10 mM phosphate and 1 mM EDTA buffer dissolve 818 mg NaCl, 85 mg $NaH_2PO_4 \cdot H_2O$, 69 mg $Na_2HPO_4 \cdot 2H_2O$ and 37 mg disodium EDTA dihydrate and make up to 100 ml with double distilled (or equivalent pure) water. The pH should be 7.0, otherwise adjust it with 0.1 M aqueous NaOH or HCl. This buffer is susceptible to bacterial growth. It should be stored in a cool dark place and freshly prepared every few weeks. Autoclaving is possible but may change the pH value.
- Dissolve equal amounts (by mol) of the single strands in 1 ml buffer. Degas the solution by bubbeling nitrogen or argon through it for several minutes. Dissolved oxygen may form bubbles at elevated temperatures leading to spurious results.
- Transfer the sample solution to a cuvette with an appropriate path length. The final absorbance reading of this solution relative to air should not exceed 2. The choice of different cell width allows the recording of melting curves at several concentrations. This is particularly important if the thermodynamic parameters of the transition shall be determined.
- Heat the solution to 90 °C for 5 minutes and let equilibrate at the starting temperature for at least 15 min. Alternatively, this procedure can be performed in the spectrometer.
- Transfer the cuvette to the spectrometer and insert the temperature probe through a Teflon stopper with a hole. Carefully seal the probe with paraffin film. Any evaporation of solvent must be avoided. Place a sealed cuvette with buffer as a reference cell in the spectrometer.
- Record the melting curve at 260 nm (or 274 nm if the sequence contains several CG basepairs) by heating the sample at 0.75 °C/min and acquir-

ing absorbance data every 0.5 °C. Most microcomputer-controlled UV-spectrometers equipped with a thermostated cell holder (e.g. a Cary 1 instrument, Varian) offer software which performs the acquisition and storage of a melting curve automatically. Since a melting curve should be an equilibrium measurement it is advisable to check this assumption by recording the curve at a lower heating rate or by measuring a curve on cooling the sample.

- Plot the absorbance versus temperature and analyze the data according to literature procedures. To obtain the melting point of a symmetrical transition form d(absorbance)/d(T) and determine the maximum of this curve.

3 Analysis of Oligonucleotides by High Performance Liquid Chromatography (HPLC)

High performance liquid chromatography (HPLC) is the most widespread tool in the analysis of oligos [13]. Chromatography is the only separation technique which can be applied to modified antisense oligos with neutral backbones such as methylphosphonates or O-methylated oligos.

For all applications a commercial HPLC system composed of a gradient pump, an UV absorbance detector and a data collection/chromatography control unit is appropriate. The HPLC separation is fast and the chromatogram can be monitored on-line, quantified and stored. The retention times are reproducible and a scale up for the preparative purification of synthetic oligos is possible. This is a great advantage of HPLC compared to capillary or gel electrophoresis where large-scale purification is not possible or at least laborious.

Among HPLC techniques two main forms are distinguished according to the separation principle applied: reversed phase (RP) and anion exchange chromatography. In RP-HPLC the oligo is bound by hydrophobic interaction to a nonpolar matrix and eluted with a gradient of increasing organic solvent content. For anion exchange HPLC a positively charged ion exchange material is used as the stationary phase. The oligo binds to this material with the negatively charged phosphorodiester backbone. The elution is accomplished by a gradient of increasing ionic strength.

3.1 Reversed Phase HPLC

Reversed phase HPLC is frequently used for the analysis of crude oligo synthesis mixtures. Because the separation is based on the hydrophobic interactions with the column material it is advisable to leave the terminal dimethoxytrityl (DMTr) protection group still attached to the oligo. This nonpolar group retards the migration and allows a good separation of the full length product from truncated sequences which do not carry a DMTr group. With preparative HPLC separations the DMTr group is cleaved from the oligo after purification by acidic treatment and removed by extraction with diethyl ether. In every case care should be taken to prevent undesired detritylation of the oligo during chromatography. Analysis of trityl bearing oligos occasionally reveals splitting into two product peaks. Short oligos ($<$ 10 nts) can also be analyzed with the DMTr group off but the resolution quickly decreases with increasing length of the molecule.

An important application of RP-HPLC is the analysis of backbone modified oligos which cannot be accomplished by other techniques. Neutral oligo analogs do not bind to ion exchange resins and do not migrate in an electric field so that RP-HPLC is the only method for their separation. The modification of the backbone can introduce chirality at the phosphorous atom. Methylphosphonates or phosphorothioates consist of 2^n diastereomers (n = number of modifications in the oligo) if no attempts for a diastereoselective synthesis were undertaken. This large amount of diastereoisomers can lead to a broadening of the peaks.

A wide variety of RP-columns for oligo analysis, differing in the type of packing material and particle diameter, is commercially available. For each of these columns a slightly different gradient is recommended. Furthermore, the choice of gradient depends on the nature and length of the oligo to be analyzed. Triethylammonium acetate is most commonly used as elution buffer. This volatile compound can be removed in vacuo from the purified oligo upon preparative separations. A typical gradient is given in protocol 2. For an analytical HPLC chromatogram approximated 0.1–0.2 OD_{260} units corresponding to 0.5–1.0 nmol of a 20 mer oligo are required. The sample solution must not contain any particles which may cause a blockage of the column. To ensure this every sample should be filtered prior to injection. Fig. 3 shows a RP chromatogram of a crude 20 mer phosphorodiester oligo with DMTr group on.

3.1.1

Protocol 2: RP-HPLC Analysis of Oligonucleotides

Materials required:
Chemicals: Triethylamine, acetic acid, water, acetonitrile (chromatography grade).
Apparatus: HPLC-System, RP C18 column.

- Prepare a 1 M stock solution of triethylammonium acetate, by carefully mixing on ice, 140 ml triethylamine and 58 ml acetic acid. Make up to 1 l with distilled water and adjust the pH to 7.0 with triethylamine or acetic acid. Dilute this solution with a ninefold volume of water and check the pH value to obtain solvent A. This solution should be filtered through an 0.45 µm membrane filter prior to use.
- Use chromatography grade acetonitrile as eluent B.

- Use a RP C18 (5 µm 250 × 4 mm) or equivalent column with a flow rate of 1.0 ml/min for analytical separations. The analysis should be performed at ambient temperature with detection of the oligo at 254 nm.
- Dissolve 0.2 OD_{260} units DMTr-ON oligo (corresponding to approximately 1 nmol of a 20 mer) in 100 µl water and filter this solution through an 0.45 µm membrane filter if necessary.
- Inject the sample at 5 % B and run a linear gradient to 45 % B for 40 min. Increase to 100 % B within 10 min, hold for 10 min and return to start conditions.

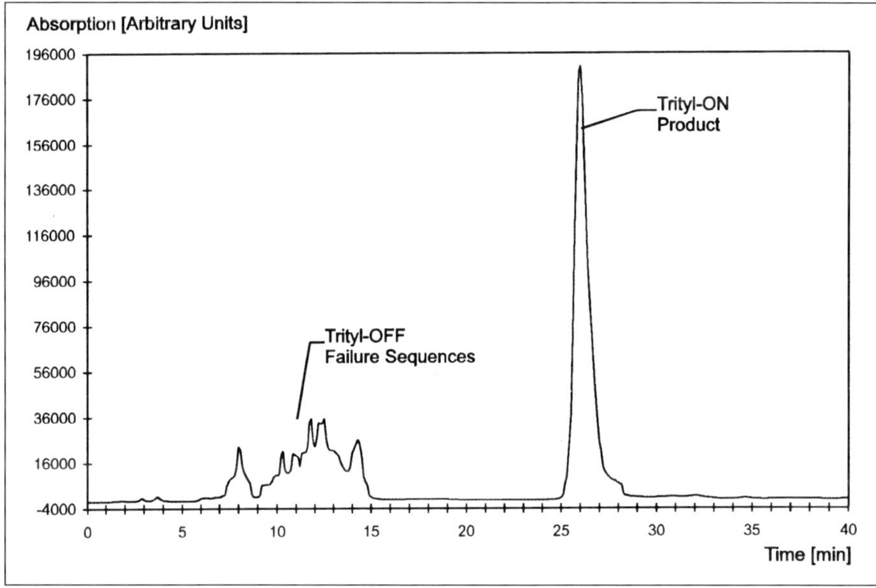

Figure 3: RP-HPLC chromatogram of a crude 20 mer phosphorodiester oligo with DMTr group on. The trityl group retards the elution of the full length product. The chromatogram was recorded on a LiChro CART RP 18 column (Merck) with the gradient given in protocol 2.

3.2 Anion Exchange HPLC

Anion exchange HPLC is a very effective method of analysis for nucleic acids [4, 11]. The resolution of n from n-1 chain length can be achieved for up to 30 mer oligos. Attempts to increase the resolution for up to 50 mer samples have been successfully undertaken. The individual resolution depends significantly on the column type chosen, gradient and pH. Analytical and preparative anion exchange columns are available from several suppliers. They are filled with silica particles or organic polymers functionalized with positively charged groups. Strong anion exchange material remain ionized at pH values up to 12 whereas weak exchangers possess lower pK values and are operated at neutral pH values. The elution of the sample is accomplished with a gradient of increasing salt content in the solvent. Some columns allow the addition of organic solvent to the mobile phase which gives a better separation of certain analytes by minimizing hydrophobic interactions between the sample and the column packing material. Metal ions may contaminate the exchange resin of certain columns so that these should be used with metal free HPLC systems.

For anion exchange HPLC the oligos are applied with the DMTr group off. The separation relies on the interaction of the negatively charged phosphorodiester groups with the positive charges of the exchange material. Short failure sequences with less negative charges elute prior to the full length product. The retention time is almost independent of the base composition of the oligo.

Phosphorothioates bind strongly to anion exchange columns. An elution can be obtained by increasing the ionic strength (e.g. 2.0 M NaCl) and by prolonged chromatography time. For these compounds weak anion exchange columns have been employed successfully. The broadening of the peaks due to the diastereomeric nature of the thioates is not so pronounced as that observed by RP-HPLC.

Preparative separations can be accomplished by anion exchange HPLC but the sample is recovered along with large amounts of salt from the elution buffer. This has to be removed by alcohol precipitation, gel filtration, RP desalting or dialysis before the oligo is ready for use.

The sample preparation for anion exchange HPLC is similar to RP-HPLC. The oligo is dissolved in distilled water and filtered. Depending on the nature and purity of the sample $0.1–0.2$ OD$_{260}$ units should be used for each injection.

Protocol 3 gives a typical procedure for an anion exchange HPLC separation and Figure 4 shows the chromatogram of a 16 mer phosphorodiester and a 15 mer phosphorothioate oligo.

3.2.1

Protocol 3: Analytical Anion Exchange HPLC of Oligonucleotides

Materials required:
Chemicals: Phosphate buffer, water, NaCl, acetonitrile (chromatography grade).
Apparatus: HPLC-System, anion exchange column (e.g. Waters Gen-Pak FAX).
- Prepare the following elution buffers:
- Buffer A: 10 mM aqueous phosphate pH 6.8 with 20 % (v/v) acetonitrile.
- Buffer B: buffer A containing 1.5 M NaCl
- Dissolve 0.1 OD$_{260}$ units of oligo (corresponding to approximately 0.5 nmol of a 20 mer) in 100 µl water and filter this solution.
- Use a anion exchange column (e.g. Waters Gen-Pak FAX) with a flow

rate of 0.75 ml/min for analytical separations. Detect the oligo at 254 nm.
- Inject the sample at 5 % B and increase to 40 % B within 40 minutes. Go to 100 % B, hold for 10 minutes and return to start conditions. Phosphorothioate oligos require higher ionic strength for elution. A typical gradient is 30 % B to 100 % B in 40 minutes. Adjustments in the gradient are necessary depending on column type and nature of the oligo.

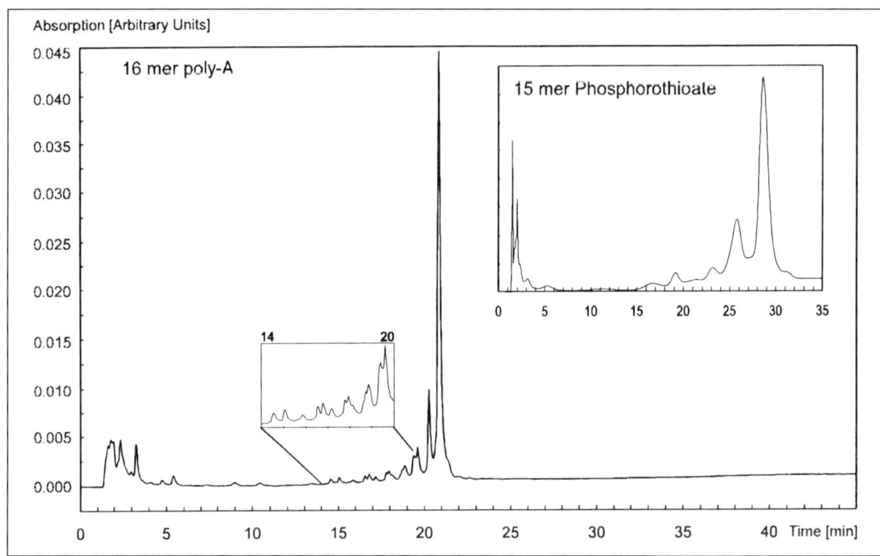

Figure 4: Anion exchange HPLC chromatogram of crude d(A)$_{16}$ recorded on a Waters Gen-Pak FAX column with the gradient of Protocol 3. The first eluting peaks represent the failure sequences and depurinated molecules. The insert shows an anion exchange chromatogram of a crude 15 mer all phosphorothioate oligo on the same column performed with a gradient of 0.45 M NaCl to 1.5 M NaCl in 10 mM phosphate buffer pH 6.8 containing 20 % acetonitrile in 40 minutes (30 % to 100 % buffer B protocol 3).

4   Electrophoretic Techniques

The electrophoretic analysis of oligos [17] is based on the migration of charged molecules under the influence of an electric field. During electrophoresis the molecules are separated on their ratio of mass-to-charge due to interaction with a supporting medium. For DNA fragments of the size of antisense oligos, cross-linked polyacrylamide is commonly used as supporting gel; for larger molecules agarose gels are suitable. Only molecules that carry a charge under electrophoresis condition can be analyzed by this technique. Therefore, electrophoresis is not applicable for modified antisense oligos with neutral backbones like all-methylphosphonates or phosphate methylated DNA. There have been few changes in the basic protocols for gel electrophoresis since the introduction of the method in the middle of this century. Recently, capillary gel electrophoresis (CGE) has arisen as a new and potent alternative to the traditional electrophoretic techniques. Here a capillary with a diameter of 25 to 100 µm and a length of 20 to 50 cm is filled with the sieving medium and electrophoresis is carried out at a high voltage.

4.1 Polyacrylamide Gel Electrophoresis (PAGE) on Slab Gels

Electrophoresis of nucleic acids on slab gels still remains a powerful tool for the analysis of oligos. The main advantage of this method is a good resolution of the full length oligo from truncated sequences. Furthermore, it combines the possibility of running several samples on one gel with a high sensitivity, especially in combination with radioactively labeled oligos. The required instrumentation is rather simple and inexpensive in comparison to HPLC or CGE but the protocols are quite laborious and automatization is not possible.

A polyacrylamide (PAA) gel is obtained by the polymerization of monomeric acrylamide in the presence of a bifunctional cross-linking agent, usually N,N'-methylene bisacrylamide ("bisacrylamide"). The polymerization is a radical mechanism and is initiated by the addition of ammonium persulphate (APS) and N,N,N',N'-tetramethylethyenediamine (TEMED) as accelerator. The mixture is poured between glass pates of varying size forming a slab. After polymerization the PAA forms a three dimensional mesh whose effective pore size is influenced by the total concentration of acrylamide in the gel mixture. During electrophoresis the electric field drives the negatively charged oligos towards the positive anode. The interaction with the stationary PAA matrix retards this migration depending on size and shape of the moving species. For short oligos high concentrations of PAA (20 %) are necessary while lower concentrations allow the separation of oligos up to 1000 base pairs (3%). Electrophoresis is usually carried out in the presence of urea as denaturing agent. Under these conditions the oligos are separated by their length and sequence specific effects are minimized. Analysis at low temperature in absence of denaturing agents may provide further information like secondary sturcture or duplex formation. The sample is applied to wells (formed by the insertion of a comb prior to polymerization of the mixture) on the upper side of the gel and migrates towards the lower side during electrophoresis. Depending on the type and amount of oligo used, different techniques for the visualization of the molecules may be performed.

The simplest method for the visualization of oligos is UV-shadowing. Following electrophoresis, the gel is removed from the glass plates and placed on a surface coated with a fluorescence marker and irradiated with UV light (254 nm). A sheet of silcia gel used for thin layer chromatography (e.g. Silica gel F_{254}, Merck) serves well for this purpose. The oligo absorbs the light and appears as dark spot on the fluorescent background.

A further method used for the detection of oligos is staining the gel. Several different staining techniques are commonly used. A protocol for silver staining will be presented later on. This method is based on the exchange of counterions of the oligos with silver ions. These are then reduced to metallic silver and the deposited silver grains appear as dark bands while the polyacrylamide remains colorless. Fig. 5 shows a silver stained analytical gel of a crude and a purified 20 mer phosphorothioate oligo. "Stains all" is another dye suitable for the visualization if oligo samples in PAA gels.

The most sensitive method of detection is by autoradiography of radioactively labeled oligos [18]. The γ irradiation of ^{32}P is detected with a phosphorus imager or by covering the gel with an X-ray film. The densitometric analysis of the darkening of the film allows for quantification of the oligo.

For the analysis of oligos up to 40 mers a gel with 10×10 cm size and 0.6–1 mm thickness has proven to be most useful in our laboratory. Apparatus for running such gels may be obtained from several laboratory equipment suppliers (e.g. Biometra and Hoefer Scientific).

The gel is prepared by initiating the polymerization of a buffered, urea containing solution of monomeric acrylamide and pouring this solution between the plates of a preassembled gel unit. The concentration of PAA is chosen according to the length of the oligo to be analyzed. In general, 10–15 mers may be analyzed on 20 % PAA gels, 15–25 mers on 16 % and 25–40 mers may be run on 12 % PAA gels. A buffer (TBE) composed of tris-hydroxymethylaminomethane (Tris), boric acid and ethylendiaminetetra-acetate (EDTA) at pH 8.3 is most commonly used for the gel itself and the

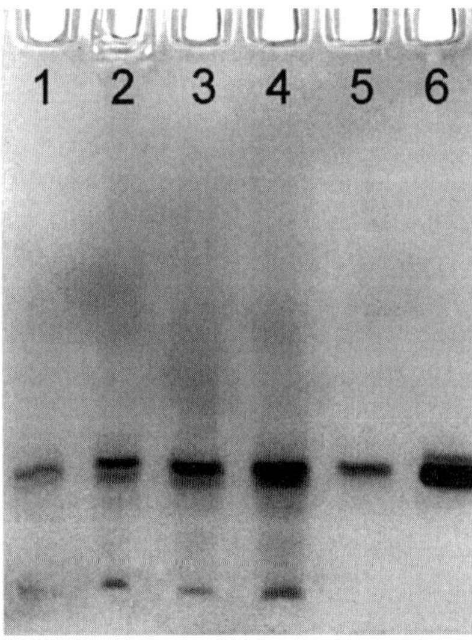

Figure 5: Silver stained denaturing polyacrylamide gel. Lanes 1–4: four different concentrations of a crude 20 mer phosphorothioate oligo (0.125, 0.25, 0.5 and 1 nmol sample dissolved in 10 µl formamide). Lane 5: purified 20 mer phosphorothioate oligo. Lane 6: 20 mer phosphorodiester oligo with the same sequence. Note that the diester oligo stains stronger than the phosphorothioate compounds.

electrophoresis buffer. Several other buffers have been described; for further information see ref. [17].

4.1.1

Protocol 4: Preparing an Analytical Denaturing Polyacrylamide Gel

Caution:

Acrylamide is a strong neurotoxin. Be careful when handling this compound and follow the safety advice of the supplier. Polyacrylamide gel is not toxic unless it contains free monomeric acrylamide.

The high voltages employed during electrophoresis may be hazardous. Be sure that all apparatuses are correctly assembled and not damaged.

Materials required:

Chemicals: Acrylamide, N,N,N′,N′-methylene bisacrylamide, water, tris base, boric acid, disodium EDTA dihydrate, ammonium persulphate, N,N,N′,N′-tetramethylethyenediamine (TEMED), bromophenol blue, formamide, urea.

Apparatus: Electrophoresis apparatus (e.g. Biometra or Hoefer Scientific), power supply unit.

- Prepare a 40 % (w/v) stock solution of acrylamide containing 2 % (w/v) bisacrylamide by dissolving 400 g acrylamide and 20 g N,N,N′,N′-methylene bisacrylamide in distilled water. Add distilled water to a total volume of 1 l.
- 10 × TBE: 109 g (0.89 mol) Tris base, 55 g (0.89 mol) boric acid and 7.5 g (0.025 mol) disodium EDTA dihydrate are dissolved and made up to 1 l with distilled water.
- Prepare a small amount of 10 % (w/v) ammonium persulphate (APS) solution (e.g. 1 g APS added up to 10 ml with distilled water). This solution should be freshly prepared at least every week.
- Dissolve 5 mg bromophenol blue in 10 ml formamide as tracking dye.
- Prepare a urea (final concentration 7 M) containing solution of desired acrylamide concentration by mixing urea, 10 × TBE and acrylamide stock solution according to the following table. Add water to a final volume of 50 ml and degas because oxygen is an inhibitor of polymerization.

final PAA concentration	20 %	16 %	12 %
urea (final concentration 7 M)	21 g	21 g	21 g
acrylamide stock solution	25 ml	20 ml	15 ml
10 × TBE buffer	5 ml	5 ml	5 ml
distilled water (approx. vol.)	1 ml	6 ml	11 ml

- Clean the glass plates with ethanol and wipe them to dryness with a clean tissue. Assemble and seal the gel unit. Possible leakages can be sealed with a melted 1 % agarose solution.
- Add 300 µl 10 % APS solution and 30 µl TEMED, mix well and pour the solution between the glass plates being carefully not to introduce air bubbles.
- Immediately insert the sample well former (comb) and allow to polymerize for at least 60 minutes. The rate of polymerization depends on temperature and amount of initiator and accelerator added. If the gel polymerizes too fast reduce these amounts or chill the solutions before use. A slow polymerization sometimes can be accelerated by heating the gel with a warm stream of air from a hair dryer.
- Transfer the gel to the electrophoresis apparatus and fill the reservoir with $1 \times$ TBE buffer (obtained from $10 \times$ TBE by dilution with a ninefold volume of distilled water). Remove the comb and carefully clean the sample wells of PAA particles. Before loading the samples it is advisable to run the gel for 15–30 min at 200 V.
- Sample preparation: 0.5 nmol desalted oligo (corresponding to approximated 0.1 OD of a 20 mer) are lyophilized and dissolved in 5–10 µl formamide or bromophenol blue containing formamide. Self-complementary sequences may require short heating to 80 °C followed by rapid cooling to 0 °C. Load the samples on the gel including at least two lanes with tracking dye.
- Electrophorese at 250 V (or a constant current of 20 mA) until the bromophenol blue marker reaches the bottom of the gel (usually 1.5 h depending on gel concentration). Bromophenol blue usually co-migrates with a 6 mer oligo on a 20 % PAA gel, with a 12 mer on 10 % PAA and with a 35 mer on 5 % PAA.
- Stop electrophoresis and visualize the oligo by UV-shadowing, staining or autoradiography.

4.1.2

Protocol 5: Silver Staining of a Polyacrylamide Gel

Materials required:
Chemicals: trichloroacetic acid, double distilled water, $AgNO_3$, Na_2CO_3, formaldehyde (35 %), acetic acid, methanol, glycerol.
Apparatus: glass vessels, platform shaker.
Note: Double distilled (or equivalent pure) water should be used for all steps of the staining procedure. The best results are obtained when the staining is carried out in glass vessels under constant agitation with a platform shaker.

- Remove the glass plates and transfer the gel into a suitable dish.
- Soak the gel for 10 minutes in a solution of 10 g trichloroacetic acid (TCA) in 40 ml water followed by two 5 minute washes with water.
- Incubate the gel for 20 minutes with a solution of 125 mg $AgNO_3$ in 125 ml water (0.1 % (w/v) solution).
- Rinse the gel thoroughly (but briefly) with water.
- Develop the gel to the desired darkness of the bands with a freshly prepared solution of 3.75 g Na_2CO_3 and 64 µl formaldehyde (35 %) in 125 ml water.
- Stop the development by immersion of the gel in 1 % (v/v) acetic acid.
- Now the gel can be photographed (Fig. 5) and dried. Before drying, the gel should be soaked for several hours or overnight in a mixture of 40 % methanol, 3 % glycerol and 57 % water.

4.2 Capillary Gel Electrophoresis (CGE)

Capillary electrophoresis (CE) is a special form of electrophoresis where the charged molecules are separated in a capillary glass with a lumen diameter of 25 to 100 µm [3] [5] [7]. For the separation of oligos capillaries filled with a sieving gel (CGE) are almost exclusively used, although there have been some publications describing the separation of DNA in open capillaries. The development of sieving media is still in progress. Several types of polymer have been described such as the traditional cross-linked PAA, linear PAA without cross-linking agent, cellulose, agarose, hydroxyethylcellulose, and others. The relatively short lifetime and sensitivity to handling and temperature changes are the major drawbacks of cross-linked PAA. Linear PAA and other polymers are more stable and allow up to 100 runs per capillary. For the separation of oligos several suppliers offer ready-made capillaries, but they can also be self prepared although these procedures are quite complicated and the results may not be reproducible. Commercial instruments for CE/CGE have been available since 1988. Basically, they consist of a high voltage power supply, a thermostated capillary holder with exchangeable buffer and sample reservoirs, a detector and a data acquisition system. The sample is injected electrokinetically by short immersion of one end of the capillary into the oligo solution and application of an electric field. The molecules migrate into the capillary, which is then shifted to a buffer reservoir, and electrophoresis is typically performed at 300 V/cm corresponding to 12 kV on a commercial 40 cm PAA capillary. The sample molecules are resolved according to their ratio of mass to charge by the same principles valid for PAGE on slab gels. At the end of the capillary a window in the coating material allows direct detection of the oligos. Several methods have been described for this but UV- or laser induced fluorescence (LIF) are the most common ones. Coupling with an ESI mass spectrometer is possible and can provide valuable information about the chemical composition of the detected peaks.

CGE is a good method for the analysis of all phosphorothioate oligos since these molecules bind strongly to ion exchange HPLC columns and thus, would require high salt buffers for elution. The main advantage of this method is high resolution (single base resolution up to 300 nucleotides), and chemically modified oligos like phosphorothioates, 5'-aminohexyl or biotinylated oligos are well resolved from unmodified molecules. The analysis is rapid and automation is possible. The method is limited by salt or buffer ions present in the sample that may interfere with the migration of the oligos. Although attempts to perform preparative separations with this technique have been made [3], CGE is still only an analytical method. At present, the purification of large amounts of oligos is not possible with CGE.

Due to the great diversity of capillaries and apparatus we recommend following the manufactors description for the analysis of antisense oligos on CGE. Figure 6 shows an electropherogram of a crude 20 mer phosphorothioate oligo.

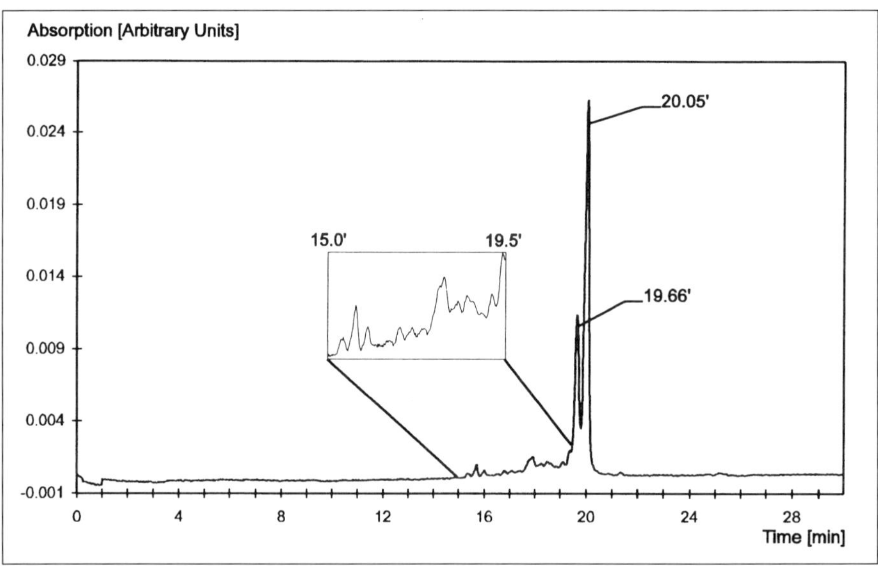

Figure 6: Capillary gel electrophoresis of a crude 20 mer all phosphorothioate oligo. Failure sequences appear between 15.5 and 19.7 minutes. The electropherogram was recorded with 300 V/cm on a Beckman P/ACE 2100 system with a Beckman eCAP capillary (effective length 30 cm, cross-linked PAA).

5   Mass Spectrometry of Oligonucleotides

Mass spectrometry (MS) is a valuable analytical technique that can provide additional information to chromatographic and electrophoretic data. The determination of molecular mass allows the verification of length and base composition of the molecule. Fragmentation of the sample and analysis of the resulting pattern permits direct sequencing of short (< 20 nts) oligos. Furthermore, the introduction of modifications, particularly those introduced directly during the synthesis process, can be monitored by MS.

Today the analysis of oligos is mainly performed by electrospray ionization (ESI) and matrix-assisted laser desorption ionization (MALDI) in combination with various mass separation and detection techniques. The application of these methods to the analysis of oligos has been developed during the last ten years and led to a rapid replacement of other ionization techniques.

5.1 Electrospray Ionization Mass Spectrometry (ESI-MS)

Electrospray [9, 15] is a special form of atmospheric pressure ionization. The sample is delivered in solution through an HPLC flow system to a stainless steel capillary tube set at a high negative or positive voltage. The strong electric field at the tip of this capillary induces an aerosol composed of small solvent droplets carrying an excess of negative or positive charge on their surface. The formation of this spray is assisted by a stream of gas (nitrogen) that passes the capillary concentrically. This special form of electrospray is sometimes referred to as ion spray. In the ion source a stream of warm drying gas (nitrogen) withdraws solvent molecules from the droplets resulting in a consecutive concentration of charges on the surface and subsequent fragmentation of the droplets. The charged particles are introduced through a small hole into the high vacuum area of the source where remaining solvent molecules are completely stripped off the ions. The exact mechanism of this fragmentation and ionization process is still under investigation. At the end of this process single solvent-free ions carrying several charges remain. They are accelerated and focused in the source, conducted to the spectrometer, and analyzed according to their mass to charge ratio.

The formation of multiple charged ions is a characteristic of ESI. It allows the analysis of high molecular mass compounds on quadrupole mass filters whose upper mass limit is typically some thousands m/z units (Da). The charge distribution pattern depends on the characteristics of the sample molecule. A typical ESI spectrum (Fig. 7) is composed of a series of peaks representing differently charged ions. The molecular mass of the oligo can be determined from each of these peaks provided that the charge state of the series is known. Commercial instruments offer analysis software that allow the determination of the molecular mass from at least two ion series. The spectrum can then be transformed into a virtual spectrum of single charged ions by a mathematical algorithm. ESI is a gentle ionization technique that induces little fragmentation of the sample molecules. This is the main reason for the impact of this method on the analysis of high molecular mass compounds like biopolymers.

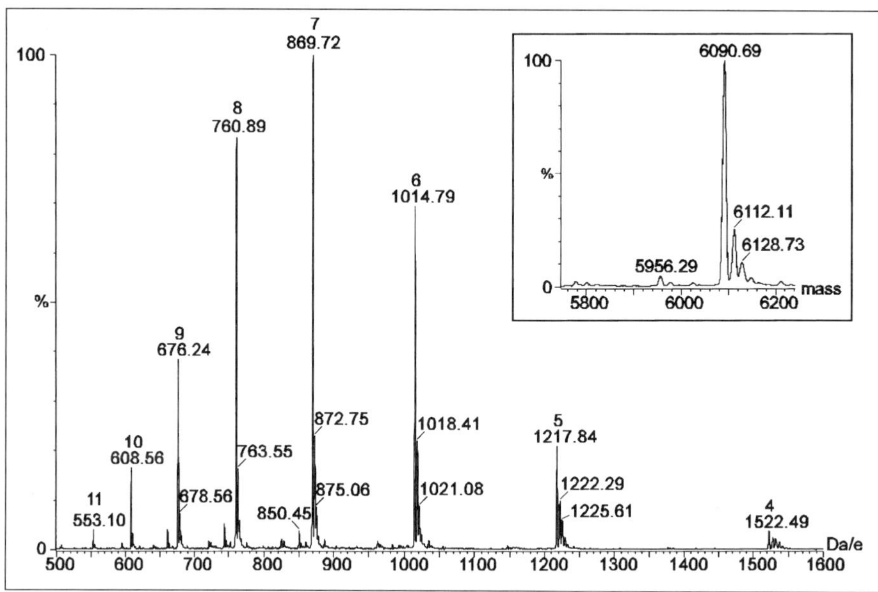

Figure 7: ESI mass spectrum of 20 mer phosphorodiester oligo with the base composition $A_8T_4G_3C_5$ ([M-H]⁻ calculated: 6093.0). The 11⁻ to 4⁻ charged ion series are shown. In the insert the transformed spectrum is displayed. The peak at 6112.11 Da is the first sodium adduct and the peak at 5956.29 is depurinated (loss of adenine) oligo. The spectrum was recorded on a Fisons Platform II instrument according to protocol 6.

For oligos, which carry negative charges on the phosphate backbone, negative ionization mode is chosen. Problems arise when some of the negative charges are neutralized due to adduct formation with alkalimetal cations like sodium [21]. This will result in the detection of a series of peaks representing the molecular ion, the molecular ion plus one sodium atom, the molecular ion plus two sodium atoms and so on ([M-H]⁻; [M-2H+Na]⁻; [M-3H+2Na]⁻; ...). Although these adducts allow the determination of the charge state of the series (which corresponds to 22 (= M[Na⁺]-M[H⁺]) divided through the measured distance of two peaks of one series) sodium is not acceptable in oligo samples because of the loss of sensitivity when the signal spreads over several peaks. With the analysis of longer oligos the maximum of the distribution shifts to higher adducts so that the [M-H]⁻peak may be absent in the spectrum. These problems may be reduced when the cations in the oligo sample are exchanged to ammonium ions which dissociate into ammonia and H⁺ in the vacuum of the spectrometer. The ammonia form of oligos may be obtained by ammonium acetate precipitation, treatment with ammonium loaded ion exchange resin, dialysis with 50 mM ammonium acetate solution or gel permeation chromatography. Sometimes the direct addition of aqueous ammonia to the sample solution may also help reducing the extent of sodium adduct formation.

5.1.1

Protocol 6: Recording ESI-MS of Oligonucleotide Samples

Materials required:

Chemicals: Ammonium acetate, ethanol, isopropanol, methanol, double
distilled water, aqueous ammonia (25 %).

Apparatus: Centrifuge, mass spectrometer equipped with ESI source.

- Ammonium acetate precipitation: Dissolve between 0.5 and 2.5 OD_{260} units of oligo in 100 µl distilled water and add 34 µl 10 M ammonium acetate. Add a 2.5 fold volume ethanol/isopropanol 50/50 (v/v), mix thoroughly, cool to –20 °C and pellet by centrifugation at 12000 rpm. Discard the supernatant and wash the pellet carefully with ice cold 70 % ethanol. Dry in vacuo and dissolve the oligo in distilled water.
- Set the solvent of the electrospray flow system to methanol/water 50/50 (v/v).
- Calibrate the spectrometer in the negative mode (Mass range: ~ 400–1500) according to the instrument manual. Several mass standards like NaI or oligosaccharide mixtures serve well for this purpose.
- Dissolve 2 nmol oligo (corresponding to approximated 0.4 OD of a 20 mer) in 100 µl methanol/water 50/50 (concentration = 20 pmol/µl). If you are not sure whether this solution contains any particles, filter it through a small 0.45 µm pore size membrane filter to avoid blockages of the flow system.
- Inject an aliquot of this solution (at a flow rate of 10 µl/min) and tune the spectrometer according to the manufactors instructions. On a Fisons Instruments VG Platform II with a standard electrospray probe, the following settings yielded good results: capillary voltage: 2.5 kV; HV lens 0.2 kV; cone voltage: 40 V; skimmer offset: 7 V; lens 1: 70 V; lens 2: 20 V; lens 3: 30 V; lens 4: 150 V; source temperature: 45 °C.
- Accumulate several scans until the signal to noise ratio does not improve any more or until the whole sample has been injected. Combine, smooth and peak detect the spectrum as usual and transform it to obtain the exact molecular mass ($[M-H]^-$).
- If the spectrum shows pronounced sodium adduct formation add up to 10 % (v/v) of concentrated aqueous ammonia (corrosive!) to the sample solution (if necessary to the solvent too) and record a new spectrum. Be careful on addition of ammonia to base labile oligos as e.g. methylphosphonates.

5.1.2

Protocol 7: Calculating the Molecular Mass of an Oligonucleotide

Calculate the molecular mass [M] of an oligo as follows:

$[M] = a \times 249.230 + t \times 240.215 + g \times 265.229 + c \times 225.204 + 2.016 + (n{-}1) \times 62.972 + (n{-}1) \times 1.008$

where:

- a, t, g, c correspond to the number of the respective bases in the oligo sequence and n is total number of nucleotides.
- The summand 2.016 represents the two hydrogen atoms at the strand ends.
- 62.972 is the mass of the phosphorodiester group. Replace this factor with 79.039 for each phosphorothioate linkage, with 95.106 for each phosphorodithioate linkage and with 62.008 for each methylphosphonate linkage.
- 1.008 is the mass of the proton as counterion to the negatively charged phosphate backbone. Calculate with 22.990 to obtain the mass of the sodium salt and 18.039 for the NH_4^+ salt of the oligo. Note that methylphosphonates are neutral so that no counterion should be considered for these compounds.
- Example: The molecular mass of the free acid of a 20 mer phosphorothioate oligo with the base composition A8, G3, T4 and C5 calculates to:
- $[M] = 8 \times 249.230 + 4 \times 240.215 + 3 \times 265.229 + 5 \times 225.204 + 2.016 + 19 \times 79.039 + 19 \times 1.008$
  $[M] = 6399.3$

5.2 Matrix-assisted Laser Desorption Ionization (MALDI)

A new mass spectrometric technique which has had a rapid impact on the analysis of oligos is MALDI. This ionization method is frequently used in combination with time-of-flight (TOF) mass detection [8] [14] [23]. For a MALDI-TOF spectrum (Fig. 8) the sample solution is mixed with a concentrated solution of a matrix compound. This mixture is spotted on a target, usually a stainless steel disk. Upon evaporation of the solvent the matrix crystallizes on the surface thereby including sample molecules. The target is then inserted into the high vacuum of a mass spectrometer. A laser pulse is fired on the matrix crystals leading to evaporation and ionization of matrix and sample molecules. The ions are accelerated in an electric field and move towards a detector placed at the end of a flight tube. The time from the laser pulse until the impact of the ions at the detector is used to determine their mass to charge ratio. Low molecular mass ions are accelerated to higher speed and reach the detector prior to the heavy ions. The raw spectrum is a plot of the measured ion intensity versus time after the laser pulse. The measurement of compounds with known molecular mass allows

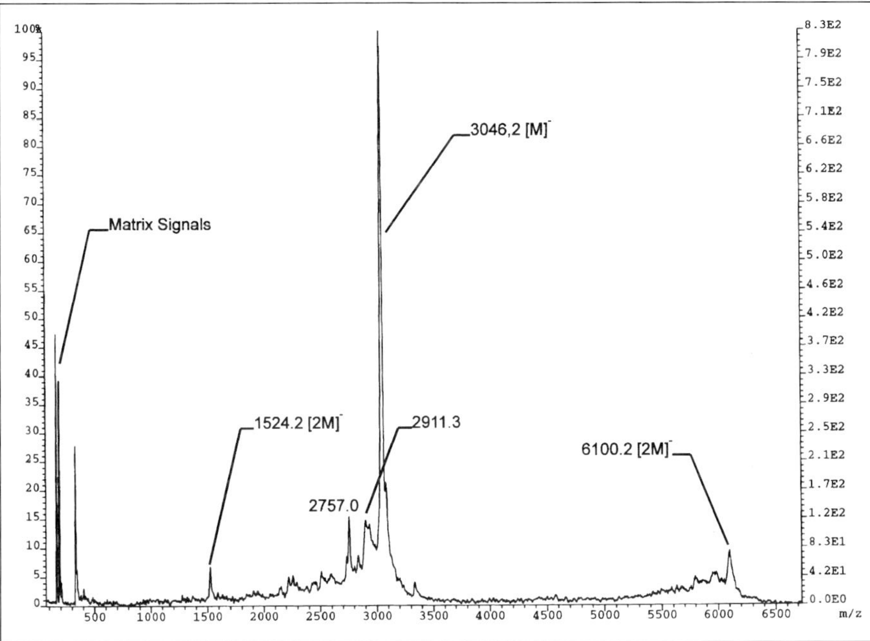

Figure 8: MALDI mass spectrum of a 10 mer phosphorodiester oligo with the base composition $A_4G_3C_3$ ([M-H]⁻ calculated: 3045.0). The compound is contaminated with n-1 (2757.0 Da; $C_2G_3A_4$), and depurinated material (2911.3 Da, loss of adenine). The spectrum was recorded on a VG TofSpec in the linear mode with 2,4,6-trihydroxyacetophenone as matrix according to protocol 8.

the transformation into an intensity versus mass spectrum. The choice of the matrix substance depends on the wavelength of the applied laser. A great diversity of compounds has been described. For UV laser instruments, (337 nm N_2 Laser) 2,4,6-trihydroxyacetophenone [14] or 3-hydroxy-picolinic acid are good matrices.

A variation of this TOF technique is the electrostatic reflection of the ions at the end of the flight tube. The registration is then performed at a second detector near the front of the tube. The advantage of the reflectron mode is an enhanced resolution and a suppression of undesired matrix peaks in the spectrum. Furthermore, this technique offers the possibility to investigate fragments of the primary ions which have formed during the first flight period (post source decay). The direct sequencing of oligos up to 21 mers has been accomplished by this method.

MALDI-TOF is a valuable technique due to the small amount of sample required, the ease of sample preparation and the high mass range covered by the TOF mass detector. The analysis of enzymatically synthesized RNA up to 150 kDa (461 nts) has been published [8]. As with ESI, the sodium adduct formation is critical for MALDI, but considerable advantages have been achieved by the use of ammonium citrate or ammonium loaded anion exchange resin to convert the oligo into the ammonium form. Further improvement of mass resolution and sensitivity is expected from new forms of MALDI-TOF such as delayed ion extraction MALDI [22]. The sequence analysis of an 50 mer DNA was successfully accomplished by the combination of this technique with the Sanger dideoxy method.

5.2.1

Protocol 8: Recording a MALDI-TOF Mass Spectrum of Oligonucleotides

Chemicals: 2,4,6-trihydroxyacetophenone, dibasic ammonium citrate, ethanol, double distilled water.
Apparatus: MALDI-TOF mass spectrometer.

- Prepare a 0.5 M solution of 2,4,6-trihydroxyacetophenone by dissolving 93 mg in 1 ml ethanol.
- Dissolve 22 mg dibasic ammonium citrate in 1 ml double distilled (or equivalent pure) water forming a 0.1 M solution.
- Dissolve 0.8 nmol oligo (corresponding to approximated 0.2 OD of a 20 mer) in 10 µl sterile double distilled water.
- Combine 10 µl of the matrix solution with 5 µl of the citrate solution, mix thoroughly, add 1 µl of the oligo solution and mix again.
- Spot 1 µl of this solution onto the target and allow the solvent to evaporate, leaving the white fluffy needles of the crystallized matrix. Insert the target into the spectrometer and measure spectra in the negative mode. Oligothymidylic acid (d(pT)$_{10}$, d(pT)$_{15}$ or d(pT)$_{20}$) can be used for external or internal calibration. The samples on the target should be measured as quickly as possible since oligo targets degrade rapidly.

6   Nuclear Magnetic Resonance (NMR) of Oligonucleotides

Nuclear magnetic resonance is not routinely employed in the purity analysis of oligos. The large amount of sample required, the laborious sample preparation (exchange of H_2O to D_2O), the long experiment duration and the difficulties on assigning the resonances of large molecules restrain the application of this technique to the structural and dynamic investigation of certain oligos. A great diversity of one-, two- and multidimensional experiments has been developed for this purpose. They shall not be discussed here.

Nevertheless, ^{31}P-NMR spectra can provide valuable information on the extent of modification of antisense oligos [20]. In particular, the introduction of phosphorothioate moieties during the synthesis cycle is prone to incomplete sulfurization. The resulting oligo is a mixture of phosphorothioate and phosphorodiester. Chromatographic and electrophoretic techniques may fail in resolving these species. Mass spectrometry should

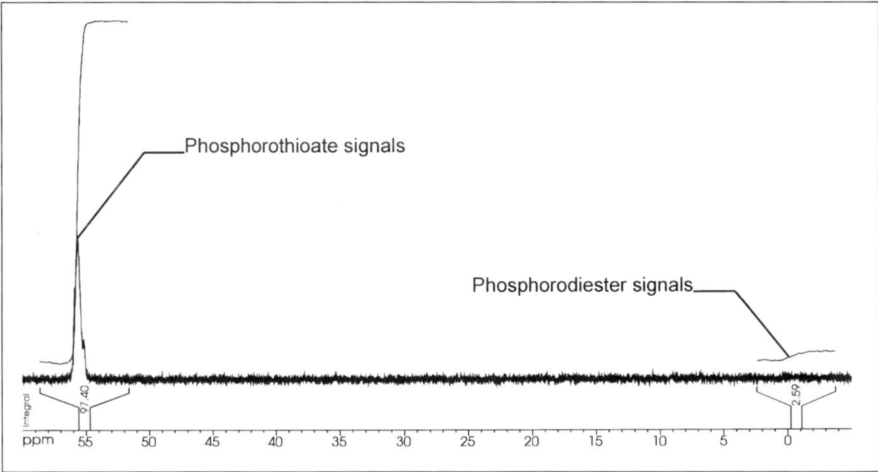

Figure 9: ^{31}P NMR spectrum of a 20 mer all phosphorothioate oligo in D$_2$O. The integral of the signals at 55 ppm and 0 ppm shows that the average rate of sulfurization is 97 %.

allow this separation but the resulting isotope pattern gets quite complicated, particularly at increased phosphorodiester content. The determination of the extent of sulfurization is then difficult due to distribution of the signal over several peaks. The measurement of the extent of sulfurization is important when new sulfur transferring reagents and new synthesis scales or cycles are evaluated.

^{31}P-NMR spectroscopy allows the differentiation of the two species by observation of the chemical shift. The shift of each phosphate fragment is approximately independent from the other groups present in the molecule. Two sets of signal appear in a typical ^{31}P-NMR of a phosphorothioate oligo (Fig. 9) representing the phosphorodiesters at ~ 0 ppm and the phosphorothioates at ~ 55 ppm (~ 110 ppm for phosphorodithioates). The integration of these peaks gives the average rate of sulfurization of the oligo. It should be mentioned here that the chirality of the phosphorothioate moieties leads to the formation of 2^n (n = number of thioates in the molecule) diastereomers that give signals at different chemical shift. The signals of this large amount of diastereomers (e.g. 524288 diastereomers for a 20 mer thioate oligo) are generally not resolved in the spectrum but result in a broadening of the peaks.

Acknowledgement

We thank Kerstin Jahn-Hofmann, Beate Conrady, Renate Konrad, Ilona Prieß, and Hannelore Brill for the synthesis of oligonucleotide samples and for assistance on recording chromatograms, electrophreograms, and mass spectra. Dr. E. Uhlmann, Hoechst AG is gratefully acknowledged for providing anion exchange chromatograms.

References

1. ALBERGO, D. D.; MARKY, L. A.; BRESLAUER, K. J.; TURNER, D. H.: Thermodynamics of (dG-dC)$_3$ double-helix formation in water and deuterium oxide. Biochemistry *20*, 1409–1413 (1981).

2. BROWN, T.; BROWN, D. J. S. ; IN ECKSTEIN, F. (ed.): Oligonucleotides and analogues. A practical approach. IRL Press, Oxford, (1991).

3. COHEN, A. S. ; NAJARIAN, D. R.; PAULUS, A.; GUTTMAN, A.; SMITH, J. A.; KARGER, B. L.: Rapid separation and purification of oligonucleotides by high-performance capillary gel electrophoresis. Proc. Natl. Acad. Sci. USA *85*, 9660–9663 (1988).

4. DRAGER, R. R.; REGNIER, F. E.: High-performance anion-exchange chromatography of oligonucleotides. Anal. Biochem. *145*, 47–56 (1985).

5. ENGELHARDT, H.; BECK, W.; KOHR, J.; SCHMITT, T.: Capillary electrophoresis: methods and possibilities. Angew. Chem. Int. Ed. Engl. *32*, 629 (1993).

6. GRAY, D. M.; HUNG, S. -H.; JOHNSON, K. H.: Absorption and circular dichroism spectroscopy of nucleic acid duplexes and triplexes. Methods in Enzymology *246*, 19–34 (1989).

7. KHUR, W. G.; MONING, C. A.: Capillary electrophoresis. Anal. Chem. *64*, 389R-407R (1992).

8. KIRPEKAR, F.; NORDHOFF, E.; KRISTIANSEN, K.; ROEPSTORFF, P.; LEZIUS, A.; HAHNER, S.; KARAS, M.; HILLENKAMP, F.: Matrix assisted laser desorption/ionization mass spectrometry of enzymatically synthesized RNA up to 150 kDa. Nucl. Acids Res. *22*, 3866–3870 (1994).

9. LITTLE, D. P.; THANNHAUSER, T. W.; MCLAFFERTY, F. W.: Verification of 50- to 100-mer DNA and RNA sequences with high-resolution mass spectrometry. Proc. Natl. Acad. Sci. USA *92*, 2318–2322 (1995).

10. MARKY, L. A.; BRESLAUER, K. J.: Calculating thermodynamic data for transitions of any molecularity from equilibrium melting curves. Biopolymers *26*, 1601–1620 (1987).

11. METELEV, V.; AGRAWAL, S : Ion-exchange High performance liquid chromatography analysis of oligodeoxyribonucleotide phosphorothioates. Anal. Biochem. *200*, 342–346 (1992).

12. NORDHOFF, E.; KARAS, M.; CRAMER, R.; HAHNER, S. ; HILLENKAMP, F.; KIRPEKAR, F.; LEZIUS, A.; MUTH, J.; MEIER, C.; ENGELS, J. W.: Direct mass spectrometric sequencing of low-picomole amounts of oligodeoxynucleotides with up to 21 bases by matrix-assisted laser desorption/ionization mass spectrometry. J. Mass Spectrom. *30*, 99–112 (1995).

13. OLIVER, R. W. A. (ed.): HPLC of macromolecules. A practical approach. IRL Press, Oxford, (1989).

14. PIELES, U.; ZÜRCHER, W.; SCHÄR, M.; MOSER, H. E.: Matrix-assisted laser desorption ionization time-of-flight mass spectrometry: a powerful tool for the mass and sequence analysis of natural and modified oligonucleotides. Nucl. Acids Res. *21* , 3191–3196 (1993).

15. POTIER, N.; DROSSELAER, A. V.; CORDIER, Y.; ROCH, O.; BISCHOFF, R.: Negative electrospray ionization mass spectrometry of synthetic and chemically modified oligonucleotides. Nucl. Acids Res. *22*, 3895–3903 (1994).

16. PUGLISI, J. D.; TINOCO JR., I.: Absorbance melting curves of RNA. Methods in Enzymology *180*, 304–325 (1989).

17. RICKWOOD, D.; HAMES, B. D. (eds.): Gel electrophoresis of nucleic acids. IRL Press, Oxford, 2nd ed. (1990).

18. SAMBROOK, J.; FRITSCH, E. F.; Maniatis, T.: Molecular cloning, a laboratory manual. Cold Spring Harbor Laboratory Press, 2nd ed. (1989).

19. SMITH, R. D.; LOO, J. A.; EDMONDS, C. G.; BARINAGA, C. J.; UDSETH, H. R.: New developments in biochemical mass spectrometry: electrospray ionization. Anal. Chem. 62, 882–899 (1990).

20. STEIN, C. A.; SUBASINGHE, C.; SHINOZUKA, K.; COHEN, J. S.: Physicochemical properties of phosphorothioate oligodeoxynucleotides. Nucl. Acids. Res. *16*, 3209–3221 (1995).

21. STULTS, J. T.; MARSTERS, J. C.: Improved electrospray ionization of synthetic oligodeoxynucleotides. Rapid Commun. Mass Spectrom. *5*, 359–363 (1991).

22. VESTAL, M. L.; JUHASZ, P.; MARTIN, S. A.: Delayed extraction matrix-assisted laser desorption time-of-flight mass spectrometry. Rapid Commun. Mass Spectrom. *9*, 1044–1050 (1995).

23. WU, K. J.; SHALER, T. A.; BECKER, C. H.: Time-of-flight mass spectrometry of underivatized single-stranded DNA oligomers by matrix-assisted laser desorption. Anal. Chem. *66*, 1637–1645 (1994).

5 Assay Systems and Controls

Evaluating the Efficacy and Specificity of Antisense Oligonucleotides

Wolfgang Brysch
Biognostik GmbH, Göttingen, Germany

1   Introduction

Meaningful assays and proper controls are an indispensable part of any antisense experiment.

Like the other sections of this book, the following chapter will focus on the use of synthetic oligonucleotides (oligos) targeted against cellular mRNAs or viral sequences.

Synthetic oligonucleotides – chemically modified or unmodified – can cause a number of different and often unexpected biological effects in cultured cells and in-vivo. For the researcher working with antisense oligonucleotides in such biological systems, this poses the difficult task to differentiate between those effects which are due to the desired sequence-specific inhibition of the targeted mRNA and the undesired sequence-related and non-sequence-related effects that antisense oligonucleotides can potentially cause.

In order to choose the appropriate assays and controls for a particular antisense experiment, it is important to be aware of the specific and non-specific side-effects that could play a role in the system under investigation.

1.1 Sequence-specific Effects

Sequence-specific effects are due to the hybridization of the antisense oligo with the targeted cellular mRNA (or viral DNA/RNA) in a sequence dependent complementary fashion. Binding is exclusively via hydrogen bonds where A-T (A-U) base-pairs form 2 hydrogen bonds and C-G base-pairs form 3 hydrogen bonds (see chapter 1). Only effects which are the result of such sequence specific binding are regarded as true "antisense" effects.

1.2 Sequence-related Side-effects

Sequence-related side-effects result from the particular sequence of an oligo but are not caused by hybridization to the desired target mRNA. Sequence-related side-effect may have different causes including:

1. Sequence homologies: The most common cause for sequence-related side-effects is the partial or total homology of the mRNA target sequence with other mRNA sequences in related or even completely unrelated genes. Such homologies may result in the partial suppression of other genes and lead to a misinterpretation of the observed biological effects [18]. Especially sequences with a high G-C content are prone to form stable hybrids with unrelated genes since a perfect match of only 8 or 10 consecutive G-C bases is often sufficient to cause an inhibitory effect. Undesired sequence-related effects can occur even with seemingly "nonsense" sequences like the homopolymers poly-dA or poly-dC. Tracts of 10 or more identical bases are not uncommon in the genome,

especially in non-coding but transcribed sequences. Such tracts may therefore form potential targets for undesired but nevertheless highly sequence-related side-effects of homopolymer controls.

2. Sequence motifs: Certain short sequence motifs seem to cause unwanted side-effects in most cell types. One of the best known motif is the G-quartet, i.e. 4 G-bases in a row [19] [7]. Other motifs only seem to play a role in certain cell types under certain conditions [8]. If non-specific effects due to certain motifs are suspected, control oligos which bear this motif in different sequence contexts can be used to clarify this point (see section 3.3, below)

3. Higher-order structures: Depending on their sequence, oligonucleotides – especially longer ones – can form secondary structures which may bind to various proteins. This phenomenon is deliberately utilized in the so-called aptamer approach [17] for the design of novel agonist/antagonist compounds. With antisense oligonucleotides, the formation of secondary- or higher-order-structures always bears the risk of unwanted protein binding [16].

Stem-loop structures and dimerization of oligonucleotides may form double stranded pieces of DNA, which can serve as potential binding sites for transcription factors. Again, this phenomenon has been exploited to design novel competitive blocking agents for transcription factors [14] but in oligonucleotides which are designed as antisense reagents, such double-stranded binding sites should be avoided.

The effects described above are often classified as non-sequence-specific. In our view, this is a misnomer. Even though these effects are not caused by Watson-Crick type base pairing with a target mRNA, the effects are the result of the specific sequence of oligo. Every oligo does have a sequence and the fact that this sequence may not make ‚sense' to the researcher does not mean that effects caused by this oligo are non-sequence specific. Thus, we usually refer to such effects as ‚sequence-related'.

1.3 Substance-related Effects

Antisense oligonucleotides are macromolecules which are mostly between 14 and 25 bases in length. Depending on their size, sequence, chemical modification and the presence of ancillary chemical groups coupled to the oligo, these molecules may have a number of desired and undesired biological effects which result from the gross chemical properties of the compound, irrespective of the individual base sequence. These effects, which are due to the chemical substance ‚oligonucleotide', should be differentiated from effects which are related to the sequence of the individual oligo. Negative controls as described below are usually employed to demonstrate that a biological effect is not caused by a non-specific substance-related side-effect.

2   Assay Systems

The antisense-mediated gene inhibitory effect of every oligo should be confirmed by appropriate biochemical and biological assays. As it is not possible to directly visualize the sequence-specific hybridization of an antisense oligo to its target, circumstantial evidence from one or more indirect assays has to serve as ‚proof' that an antisense mechanism is the reason for the observed biological effects.

One or a combination of the following procedures can be employed to produce such circumstantial evidence:

2.1 Measurement of Target mRNA Levels – Northern Blots

Unmodified DNA-oligos and some types of chemically modified oligonucleotides such as phosphorothioates and phosphoramidates are potential substrates for RNase H, an enzyme that cleaves the RNA strand of a DNA/RNA hybrid [5]. In biological systems, where RNase H plays a role, for example in xenopus oocytes [6], and when using compatible substrate oligonucleotides, the demonstration that the targeted mRNA species is selectively diminished following antisense treatment is a strong indicator for a sequence-specific antisense effect. In such a case it is important to show that the levels of other mRNA species are not or at least significantly less affected. Such reference mRNAs should fulfill the following two criteria:

1. No portion of the reference mRNA should have any (complementary) homology to the antisense sequence in order to exclude direct cross-hybridization of the antisense oligo to the reference mRNA.
2. The reference mRNA should exhibit relatively stable expression levels under different conditions. This will minimize the chance that a specific antisense inhibition, for example of an oncogene with subsequent cellular growth arrest, causes a physiological down-regulation of the reference mRNA which is then misinterpreted as a non-specific side-effect.

The specific inhibition of target mRNA levels is strong evidence for a true sequence-specific antisense effect. Conversely however, unchanged target mRNA levels do not necessarily mean that an antisense oligo did not work [2]. In the latter case other assays must be used to confirm an antisense effect.

2.1.1   Summary: Northern Blots

Advantages:
- standard method, highly reproducible
- hybridization probes (oligonucleotides) readily available

Disadvantages:
- only applicable for RNase H mediated antisense effects

2.2 Measurement of Protein Levels

It is the aim of any antisense approach, irrespective of the type of antisense molecule used, to specifically inhibit the expression of a protein. Thus, measuring the level of this protein at different time points following antisense treatment provides the most straight forward evidence for the efficacy of an antisense compound. A direct proof that the protein product of the targeted gene is down-regulated or completely suppressed should be given wherever possible. Suitable assays for changes in specific protein levels are western blots, ELISAs and immunoprecipitation.

Western blots are the preferred method, since they provide information on the approximate molecular weight of the detected protein and allow differentiation between signals produced by antibody-binding to the specific protein and signals caused by cross-reactivity of the antibody with other proteins. The drawback of western blots is that a relatively large number of cells is needed for each assay and that the method is laborious, making repeated measurements at different time points or experimental conditions difficult.

ELISA: If an established ELISA is available for the protein under investigation, this is a good alternative for testing many different samples. Only small amounts of protein are needed for each assay, the method is rapid and in many cases even quantitative. A disadvantage is that specific signals cannot be differentiated from signals that are caused by cross-reactivity of the primary antibody.

If many time points or experimental conditions are to be tested, a combination of both methods is recommended: First run a western blot and an ELISA for 2 or 3 experimental points in parallel to demonstrate the specificity of the antisense oligo and to show that there is a good correlation between the results obtained by either method. Further experiments can be done using ELISAs alone.

A very important factor for protein assays in antisense experiments is the amount of time allowed for between the beginning of antisense treatment and the lysis of cells for protein extraction. From our own experience, it takes most cells about 6–8 hours to incorporate sufficient amounts of (phosphorothioate) oligos for an effective inhibition. After that period, antisense inhibition should have an effect only on the amount of newly synthesized protein. Thus, it is very helpful to know the approximate half-life and/or turnover rate of the protein that is to be inhibited. A continuous inhibition for one or two half-life periods is often necessary to get a clearly detectable decrease in protein levels. On the other hand, waiting too long can mean that the antisense-oligo is not active any longer and temporarily decreased protein levels have returned to normal.

Immunoprecipitation is a good alternative if the half-life and turnover of the target protein is very uncertain. Here only proteins that are newly synthesized after the addition of radio-labelled methionine are detected. Thus, any time-point after adding the labelled methionine will reflect the degree of specific protein inhibition at that time. The fact that cells have to be grown in the presence of radio-labelled methionine limits the use of immunoprecipitation to cell culture facilities which are equipped with safety measures to handle such materials.

2.2.1 Summary: Western Blot

Advantages:
- direct evidence of protein inhibition
- very specific
- shows approximate MW of detected protein
- semi quantitative

Disadvantages:
- a suitable antibody must be available
- large number of cells/assay needed
- laborious and time consuming

2.2.2 Summary: ELISA

Advantages:
- direct evidence of protein inhibition
- small number of cells per assay needed
- many samples can be processed simultaneously
- quantitative

Disadvantages:
- a suitable antibody and protein standard must be available
- no information on size of detected protein
- specificity depends on quality of ELISA (kit)

2.2.3 Summary: Immunoprecipitation

Advantages:
- direct evidence of protein inhibition
- detects only newly synthesized proteins
- small number of cells/assay needed
- semi-quantitative

Disadvantages:
- a suitable antibody must be available
- use of potentially hazardous materials (35S-methionine) in cell cultures

2.3 Biochemical Assays

The level of many proteins with a catalytic function can be assayed by monitoring their biochemical activity with suitable substrates. Examples for such proteins are enzymes or receptors that have a protein kinase domain [2]. Biochemical assays are usually very sensitive, so cells grown in a single microtiter well are often sufficient for one measurement and even small deviations in activity can be detected. Furthermore most assays can be easily performed in larger numbers. These factors make biochemical assays a good choice in cases where many different conditions are to be tested, for example to establish dose-response curves for an antisense drug.

However, biochemical assays bear the risk of misinterpretation. They should be evaluated with caution and a clear understanding of the enzyme/substrate system is required. Many substrates can be processed by different (iso-)enzymes or catalytic protein domains. Thus, most biochemical assays are not absolutely target specific. Furthermore the activity of enzymes and other catalytic proteins is often up- or down-regulated transiently via reactions like phosphorylation and dephosphorylation. In that case changes in biochemical activity do not reflect changes in protein level.

The validity of every biochemical assay should be demonstrated in conjunction with at least one direct method like northern- or western blots or by a positive antisense-control. In cases where there is good correlation between the results obtained by each method under some exemplary experimental conditions, biochemical assays can be regarded as valid for the measurement of antisense effects.

2.3.1   Summary: Biochemical Assays

Advantages:
- sensitive
- many samples can be processed easily
- small number of cells/assay needed
- semi-quantitative

Disadvantages:
- only suitable for proteins with a catalytic or enzymatic domain
- changes in activity can be misinterpreted as changes in protein level
- limited specificity (isoenzymes)

2.4 Binding Assays

Receptor-binding assays can be viewed as special forms of biochemical assays. Binding assays monitor the degree of binding of a ligand to a protein-binding-site (receptor) [20]. The specificity of the method depends largely on the specificity of the ligand for its receptor. The affinity between receptor and ligand can vary significantly under different conditions [3], thus care must be taken to ascertain that a change in the number of binding sites actually reflects a change in receptor-protein levels. Often radio-labelled ligands are used which can be either detected quantitatively by scintillation counting or visualized *in situ* by autoradiography.

2.4.1   *Summary: Binding Assays*

Advantages:
- specific (depending on ligand)
- rapid
- allows quantitation and visualization

Disadvantages:
- only useful for receptor proteins
- (radio-labelled) ligand needed
- changes in affinity can mimic a decrease in receptor density

2.5 Cell Proliferation

A major application of antisense techniques is the control and inhibition of pathologic cell growth. Examples are the development of novel anti-cancer drugs or the suppression of smooth-muscle cell proliferation which often causes a restenosis of blood vessels after catheter dilatation [12]. Thus, proliferation assays have been widely used to demonstrate the efficacy of an antisense oligo. Proliferation assays are easily performed either by manual counting on small samples or they can be automated using a cell counter. Proliferation assays are not limited by the need for special reagents like specific antibodies or receptor ligands. However, changes in cell proliferation rates are a very general phenomenon which can be induced by a wide variety of events, ranging from specific antisense-mediated gene inhibition of sequence-related and unrelated side-effects, to the simple fact that an oligo preparation may be toxic due to inadequate purification. Oligonucleotides or their break-down products may interfere with some indirect proliferation assay such as ^{3}H-thymidine and BrdU incorporation [9] (see also chapter 6, Cell Culture Protocols). The most reliable assay seems to be the direct counting of cell numbers. Despite the lack of specificity, cell proliferation assays are very valid to prove the efficacy of an antisense compound if they are combined with other methods like protein assays or a combination of negative and positive controls.

2.5.1   *Summary: Cell Proliferation*

Advantages:
- easy to perform
- can be automated
- no need for specific antibodies or receptor ligands

Disadvantages:
- not very specific

2.6   Morphological Analysis

In some instances, morphological changes of cultured cells can be used to monitor an antisense effect. This applies mainly to the inhibition of structural proteins [10] or factors which induce gross morphological changes like cell-differentiation [11]. In cell-cultures, morphological changes can be monitored over an extended period without disturbing the culture itself thus allowing repeated measurements in the same culture (see also chapter 8, Neurobiology and chapter 12, Oncology). If the specificity of an antisense oligo for the inhibition of its target mRNA has been demonstrated by other more direct methods, morphological analysis can be an excellent way to monitor antisense effects over extended periods under many different conditions.

2.6.1   Summary: Morphological Analysis

Advantages:
- easy to perform
- repeated measurements in same cell-culture possible

Disadvantages:
- not very specific
- sequence-related and toxic effects can also cause morphological changes

2.7   Functional Assays

The inhibition of many proteins can be monitored by functional assays. Especially in neurobiology, where the shape, magnitude and duration of many electrophysiological signals is characteristic for certain ion-channels or receptors, the changes of such signals may be closely correlated with changes in respective protein levels [4] [1].

Antisense inhibition of a receptor can be functionally monitored by comparing cells that have been treated with a specific antisense, a control-oligo and no oligo. If these cells are then exposed to the respective receptor agonist (i.e. a growth-factor, cytokine, neurotransmitter, hormone, etc.) a strongly diminished or absent cellular response to this stimulus, if present only in the antisense-treated cells, is highly indicative for a specific inhibition of the respective receptor protein [15]. Functional assays are a key procedure in the development of antisense drugs since it is at this point that an oligo must prove to be superior to other drugs either in efficacy or specificity.

2.7.1   Summary: Functional Assays

Advantages:
- specific
- highly relevant parameter in drug development

Disadvantages:
- gene function must be known beforehand

3   Controls

Appropriate controls are an integral part of every antisense experiment and a prerequisite for the publication of antisense work in scientific journals [13]. Controls are necessary to evaluate which of the observed biochemical or biological effects are caused by a sequence specific antisense mechanism and which effects are due to sequence-related and substance-related side-effects.

There are two types of control experiments: negative and positive controls.

Negative controls serve to rule out that the (chemically modified) antisense-oligo causes pronounced side-effects in the treated cells or organism. Negative controls are normally designed in such a way that they are unable to specifically hybridize to a genetic target in the cell. Thus, all biological effects observed in such control experiments have to be attributed to non-specific side-effects like cross-hybridization, protein binding or toxicity.

Common design criteria for negative control oligonucleotides are:
1. equal length and hybridization characteristics as the corresponding antisense oligo
2. identical chemical modifications in antisense and control oligos
3. reduced or no ability to hybridize to the antisense target
4. no (complementary) homology to other cellular mRNAs

Positive controls are oligonucleotides which provide additional evidence that a true antisense effect is the cause of the observed biological effects.

Many different types of controls have been proposed and used in the past, each with its own advantages and disadvantages. As there is no single ideal control, a combination of two or more different controls is recommended.

The following paragraphs describe the most widely used and accepted types of controls.

3.1 Sense Controls

These are oligonucleotides which have a sequence complementary to that of the antisense oligo (Fig. 1). Sense controls are identical to the mRNA target which is also ‚sense' and therefore unable to hybridize to this target.

<table>
<tr><td>antisense sequence</td><td>5' TCAGGAACCCCTTGCAGA 3'</td></tr>
<tr><td>sense control</td><td>5' TCTGCAAGGGGTTCCTGA 3'</td></tr>
</table>

Figure 1: Antisense sequence complementary to human PKC-gamma (top) and its respective sense control (bottom). Note that the base composition changes and a GGGG motif appears in the control sequence. Note that both oligos are presented in a 5'-3' orientation.

This type of control has been widely used, especially in early antisense work. Sense oligos are readily designed and usually have hybridization characteristics which are very similar to their antisense counterparts due to the equal content of G-C and A-T bases. For these reasons, they have long been regarded as the ideal and logical type of control. However, in many cases other researchers and ourselves found that sense controls produced biological effects of their own at a frequency that could not be attributed to substance-related side-effects alone. One possible explanation was that sense oligos may interact with the antisense-strand of the DNA. But given the highly complex structure of nuclear DNA it seems very unlikely that a short oligo can separate the double helix, and bind to the one strand tightly enough to stop the transcription of a gene. We have therefore analyzed the sense counterparts of many of our antisense sequences which had been shown to be specific with other types of controls. We found that many sense sequences had an unexpectedly high (complementary) homology to completely unrelated genes. This is due to the non-random nature of sense sequences and may explain the biological activity of many sense controls. It has been shown that certain sequence motifs like GGGG can cause sequence-related side-effects. Even if such motifs are avoided in the design of the antisense oligo, they may appear in the respective sense control. Before using a sense control, a homology search in a sequence data base like EMBL or GeneBank should be conducted in order to assess the likelihood of unintended cross-hybridization with other mRNA species. If no such cross-homologies are found and the sequence does not contain problematic motifs, sense oligos can be a valid and recommended type of control.

3.1.2 *Summary: Sense Controls → Negative Control*

Advantages:
- readily designed
- hybridization properties nearly identical to antisense counterpart
- widely accepted

Disadvantages:
- high incidence of non-specific cross hybridization
- questionable interaction with DNA antisense strand

3.2 Random Controls

Randomized controls, which are also termed ‚nonsense' or ‚scrambled', can be obtained by mixing up the bases of an antisense oligo in a randomized fashion (Fig. 2). The rationale behind this approach is, that a random sequence would lack any specific information and that the biological effects of such an oligo would reflect the non-sequence specific or substance-related side-effects of an oligo. In this respect it is important to realize that even a random sequence may have unexpected homologies and that there is probably no such thing as a completely ‚nonsense' sequence.

Random sequences may also have different hybridization properties compared to the antisense oligo. A balanced distribution of A/T and G/C bases along the length of an antisense oligo may become very lop-sided in a random sequence – for statistical reasons, this risk is more common in short oligos.

Some authors have proposed a complete random mixture of oligonucleotides of length n as a control, where n is the length of the corresponding antisense oligo. These oligos are synthesized with all four bases (A, C, G and T) present in each coupling step. This yields a mixture of all possible sequences of length n, i.e. 4^n different oligos. In theory, oligos with problematic sequences would comprise only a minute fraction and the gross biological effect of the mixture would just reflect the substance-related side-effects. We have tested such mixtures of 14- and 18-mer phosphorothioate oligonucleotides in a number of different mammalian cell lines at concentrations between 1 and 4 µM. This treatment showed massive signs of toxicity in all cells (unpublished observations). Our hypothesis to explain this effect is that a complete random mix contains oligos which hybridize to virtually every mRNA present in the cell. Even if only a fraction of each mRNA is blocked, this is obviously sufficient to cause massive toxicity. There have been reports that have not detected any toxicity with random mixtures of unmodified oligos. This may have been due to insufficient stability of these oligos, which were probably degraded before they could have such detrimental effects on the cells.

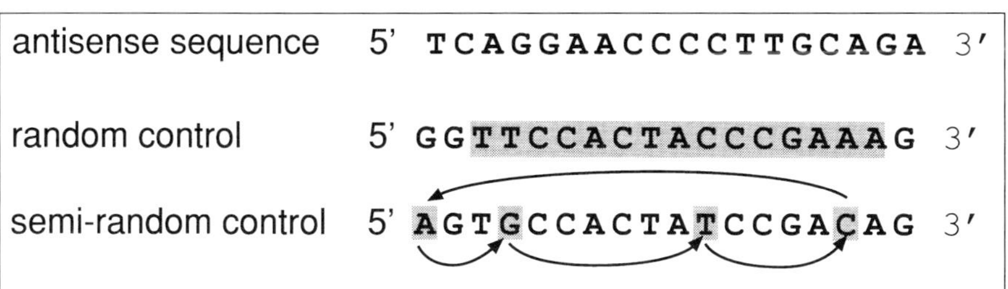

Figure 2: Antisense sequence (top) and randomized control (middle). Highlighted bases show a 100% complementary homology to the c-myc gene. Semi-random control (bottom) The c-myc homology is disrupted by circulating the position of 4 bases within the sequence (arrows).

In our own laboratory we had good success with controls that are semi-random. These controls are obtained by generating 40–50 randomized sequences for the respective antisense oligo. For each of these sequences a homology search is run against all eukaryote sequences in the EMBL database. The random sequences with the lowest degree of cross-homology to sequences in the database are then compared to the antisense oligo and the sequence which most closely resembles the antisense oligo with respect to the distribution of A/T and G/C bases, sequence motifs, etc. is chosen. Often the base-sequence is adjusted manually to closely resemble the characteristics of the antisense oligo. A good method to further randomize a sequence manually, is to pick 3 or 4 bases and move their position in a circular fashion (Fig. 2 , semi-random control). A final homology search against the EMBL database should be done to ensure that manual changes do not bring about new cross-homologies. Such (semi-)random controls can often be used for different antisense oligos of equal length and G/C content. They are an excellent first control to check for substance-related side-effects of antisense oligos in a cell line or *in vivo*.

3.2.1 Summary (Semi-)Random Controls → Negative Control

Advantages:
- good indicators of substance-related side-effects
- may be used for multiple antisense sequences

Disadvantages:
- laborious design

3.3   Reverse Controls

Reverse controls are obtained by reversing the antisense sequence with respect to its 5' – 3' orientation (Fig. 3). This sequence is unable to hybridize to the antisense mRNA target.

In comparison to sense controls, reverse oligos have several advantages:
1. The base composition remains exactly the same.
2. The hybridization characteristics and free energy parameters remain the same.
3. Most sequence motifs, which may cause non-specific effects, like a G-quartet or palindromes are conserved. Thus, reverse controls can be

| antisense sequence | 5' TCAGGAACCCCTTGCAGA 3' |
| reverse control | 5' AGACGTTCCCCAAGGACT 3' |

Figure 3: Antisense sequence (top) and reverse control (bottom) which is obtained by reversing the 5' → 3' orientation of the antisense.

used to determine whether these motifs do indeed produce non-specific effects in the biological system under study.

4. Reverse sequences seem to have a lower tendency towards "accidental" cross-homologies as compared to sense controls. This became apparent when we analyzed over 200 of our antisense-sequences which targeted mammalian mRNAs. The sense and reverse counterparts of these antisense sequences were checked for cross-homologies with related and unrelated sequences and it was surprising to see that the incidence of cross-homologies was significantly lower with the reverse sequences (unpublished results). However, the absence of unwanted cross-homologies must be checked for each individual reverse-control by a data base search as described for random controls.

The fact that reverse controls usually have exactly the same binding- and hybridization characteristics as the respective antisense sequence, makes them an excellent and highly valid type of control, and are often used in our laboratory as second line control after initial experiments with a random control.

3.3.1   Summary: Reverse Controls → Negative Control

Advantages:
- readily designed
- same hybridization characteristics as antisense
- problematic motifs are often retained

Disadvantages:
- must be individually prepared for each antisense sequence
- cannot be used if cross-homologies occur

3.4 Mismatch Controls

Mismatch controls are generated by introducing one or more deliberate mismatches into an antisense sequence (Fig. 4). Theoretically the more mismatches with respect to the target mRNA an antisense sequence has, the less stably it should bind and hence inhibit the expression of the target gene. Generally this assumption holds true, but it is of importance where the mismatches are located, what type of bases are mismatched and in which sequence context the mismatch occurs. In the following we will discuss each of these factors separately, but it should be kept in mind that each

antisense sequence	5' **TCAGGAACCCCTTGCAGA** 3'
mismatch control	5' **TCCGGAACCACTTCCAGA** 3'

Figure 4: Antisense sequence (top) and mismatch control (bottom). The highlighted bases have been changed with respect to the antisense sequence.

mismatch will cause a combination of these effects and the results cannot be predicted in detail.

1. Location of the mismatch: Mismatches around the center of oligonucleotides have more effect than peripheral mismatches. A mismatch at the 5' or 3' end will only shorten the oligo by one base causing little effect especially in oligos of 18 bases or longer. On the other hand, a mismatch in the center of a 15mer can completely abolish the efficacy of an antisense oligo. If more than one mismatch is introduced, spacing these out along the oligo will have more effect than a cluster of mismatches.

2. Type of base: It is of importance which base is changed and for what base it is exchanged for. As a rule of thumb the following base changes will have decreasing effects. G to A/T > C to A > G/C to C/G > A/T to G > A to C > A/T to T/A > C

Mismatch controls can also be used to estimate the specificity of an antisense oligo. If one mismatch in an antisense sequence completely abolishes the biological activity, partial hybridization of the antisense oligo to unrelated mRNAs is unlikely to cause any side-effects. If, however, pronounced effects are still observed with an antisense sequence bearing 3 or more mismatches, the specificity of this antisense is very questionable.

3.4.1 Summary: Mismatch Controls → Negative Control

Advantages:
- readily designed
- can be used to estimate specificity of the corresponding antisense

Disadvantages:
- must be individually designed for each antisense sequence

3.5   Non-expressing Control Cells

Sequence-related or non-specific effects of an antisense oligo can be evaluated by treating cells which do not express the mRNA against which the oligo is targeted. In such cells the antisense oligo should have no biological effect.

The test system to be used should be carefully chosen in order to obtain meaningful results. Ideally the control cells should be identical to the original cells except that they do not express the target gene. As this will hardly occur in natural systems, less related cell types usually serve as the control system. One alternative is to use cells from a different tissue which do not express the target gene, the other is to use the same cell-type from a different species. The latter alternative however is only suitable if the target mRNAs in both species are sufficiently divergent to exclude the possibility of an effective cross-reaction of the antisense oligo. Non-expressing control cells are useful to estimate non-specific side-effects that an antisense compound may have and are thus especially suited for the early phases of antisense drug development.

3.5.1 Summary: Non-expressing Control Cells → Negative Control

Advantages:
- the antisense sequence serves as its own control
- no separate oligo necessary

Disadvantages:
- suitable control cells may be difficult to find
- control cells may have different biological properties

3.6   Second Antisense

A second oligo, directed to a different part of the target mRNA can serve as a positive control. Especially in those cases where there is no suitable antibody available to show reduction of the respective protein product, and if there is no reduction in target mRNA levels, a second oligo can provide good circumstantial evidence for a true antisense mechanism. In order to draw such conclusions a number of factors should be carefully considered:
1. Both antisense oligonucleotides should be effective in inhibiting target mRNA expression. If a biological effect cannot be reproduced with a second oligo, the reason can be that the original oligo did not act via a true antisense mechanism or that the second oligo is poorly designed and therefore ineffective. Thus, a number of secondary oligos may need to be tested.
2. Both oligos should target different parts of the mRNA and share no sequence homology.
3. Often two oligos are not equally effective. In this case the magnitude of the biological response may be different but the type of response, such as reduced proliferation or cell-differentiation, should be identical.
4. Each oligo should be individually tested against an appropriate negative control.

3.6.1 Summary: Second Antisense → Positive Control

Advantages:
- good alternative if gene product cannot be assayed directly

Disadvantages:
- often need to design several oligos to find a pair of active ones
- each oligo must be tested against a negative control

3.7   Antisense Induction of Opposite Biological Effects

Many oligonucleotides are designed to target genes that enhance cell proliferation, the aim being to inhibit pathologic cell growth as in cancer or restenosis of dilated blood vessels. As detailed above, cell proliferation rate is a relatively non-specific parameter by which to judge the specificity of an antisense effect. Reduced cell growth can be the result of a specific gene inhibition but just as well be a non-specific toxic effect. One way to demonstrate that the cells under investigation do indeed respond specifically to the inhibition of a target gene is to inhibit a gene which has the opposite function in a parallel experiment. We have used the tumor-suppressor gene p53 as such a control gene. The inhibition of a (tumor-)suppressor should theoretically result in an increased cell proliferation rate. Such an effect could be demonstrated by comparing 3T3 fibroblasts that were treated with either an anti-c-myc oligo or an anti-p53 oligo. Inhibition of the proto-oncogene c-*myc* reduced proliferation rate by almost 80 %, while inhibition of p53 resulted in a doubling cell proliferation as compared to untreated cells (Fig. 5). Whenever relatively non-specific biological responses like proliferation or differentiation are assayed, the use of an opposite-effect antisense can greatly substantiate the observed results [11].

3.7.1 Summary: Antisense Induction of Opposite Biological Effects → Positive Control

Advantages:
- good circumstantial evidence for sequence-specific antisense mechanism
- one control target (i.e. a tumor-suppressor) can be used for different primary targets (i.e. oncogenes)

Disadvantages:
- a gene-target with an antagonistic function must exist
- the control gene-target must be active in the cells under investigation
- efficacy and specificity must be established for the control

4   Conclusions

The choice of appropriate assay systems and controls is the basis for a meaningful interpretation of antisense experiments. The assays chosen should enable the investigator to evaluate the efficacy of an antisense oligo with respect to the aim of the study. Valid controls are needed to confirm the specificity of the observed effects.

The requirements and stringency of control experiments may vary, depending on the aim of the research. In basic research, when antisense inhibition is used to determine the basic function of a gene and its protein product, very stringent controls are needed in order to make sure that conclusions drawn from the biological effects of an antisense oligo are indeed due to the specific reduction of target gene expression. In this case the absence of sequence-related or substance-related side-effects must only be demonstrated for the cell-type or biological system used in the study. Valid

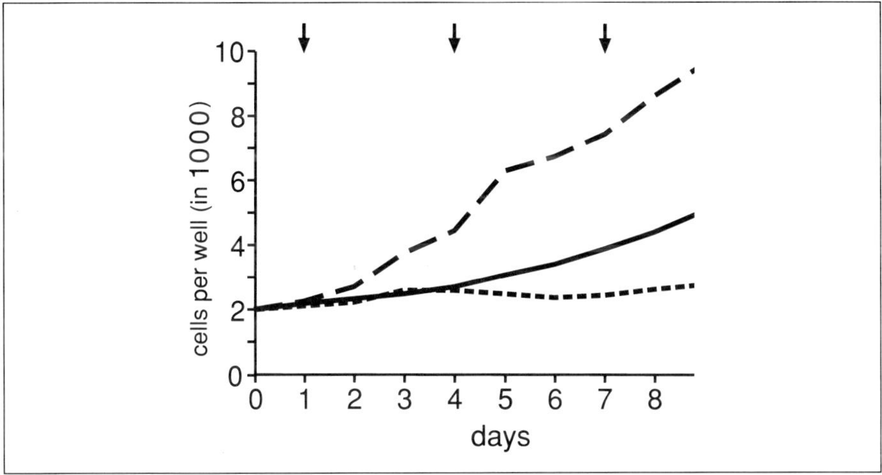

Figure 5: Proliferation rate of 3T3 cells treated with an antisense c-myc oligo (short dashes) or an antisense p53 oligo (long dashes). Solid line = untreated cells. Proliferation rate was assayed by counting cell numbers. 2µM of the respective antisense oligo was added to the cell culture medium at the times indicated by arrows on top of graph.

conclusions about the physiological or pathophysiological role of the target gene can be drawn from such data.

However, if the aim of the research is the development of a new drug, efficacy in the target cells or organ along with the absence of side-effects in other tissues become the most important parameters. As long as an antisense drug is active and its side-effects are tolerable with respect to the target disease, the proof that the mode of action is via a true antisense mechanism may become secondary. The fact that the antisense technique is one of the most rational approaches to drug development today should not lead us to the conclusion that an antisense drug may only be used if there is unequivocal and complete proof that its mode of action is exclusively via a true antisense mechanism. Just as for many other drugs, the successful and beneficial application of antisense therapeutics will probably precede the discovery of all mechanisms of action involved.

Thus, the optimal combination of assay systems and controls has to be individually adapted, depending on the aim of each antisense project.

References

1. BRUSSAARD, A. B. and R. E. BAKER: Antisense oligonucleotide-induced block of individual GABAA receptor alpha subunits in cultured visual cortex slices reduces amplitude of evoked inhibitory postsynaptic currents. Neurosci. Lett. 191(1–2) (1995) 111.
2. BRYSCH, W., E. MAGAL, J. C. LOUIS, M. KUNST, I. KLINGER, R. SCHLINGENSIEPEN, and K. H. SCHLINGENSIEPEN: Inhibition of p185c-erbB-2 proto-oncogene expression by antisense oligodeoxynucleotides down-regulates p185-associated tyrosine-ki-

nase activity and strongly inhibits mammary tumor-cell proliferation. Cancer Gene Therapy 1(2) (1994) 99.

3. BYLUND, D. B. and M. L. TOEWS: Radioligand binding methods: practical guide and tips. Am. J. Physiol. (1993).

4. CAMPBELL, V., N. BERROW, and A. C. DOLPHIN: GABAB receptor modulation of Ca^{2+} currents in rat sensory neurones by the G protein G(0): antisense oligonucleotide studies. J. Physiol. Lond. 470(1) (1993) 1.

5. CROUCH, R. J. and M. L. DIRKSEN: Ribonucleases H. In: R.J. Roberts, ed. Nucleases. Cold Spring Habor Laboratory, Cold Spring Habor (1982) 211.

6. DAGLE, J. M., D. L. WEEKS, and J. A. WALDER: Pathways of degradation and mechanism of action of antisense oligonucleotides in Xenopus laevis embryos. Antisense Res. Dev. 1(1) (1991) 11.

7. KADONGA, J. T.: Purification of sequence-specific binding proteins by DNA affinity chromatography. Methods Enzymol. 208 (1991) 10.

8. KRIEG, A. M., A. K. YI, S. MATSON, T. WALDSCHMIDT, G. A. BISHOP, R. TEASDALE, G. G. KORTZKY, et al.: CpG motifs in bacterial DNA triggers direct B-cell activation Nature 374 (1995) 546.

9. MATSON, S. and A. M. KRIEG: Nonspecific suppression of [3H]thymidine incorporation by "control" oligonucleotides. Antisense Res. Dev. 2(4) (1992) 325.

10. OSEN-SAND, A., M. CATSICAS, J. K. STAPLE, K. A. JONES, G. AYALA, J. KNOWLES, G. GRENNINGLOH, et al.: Inhibition of axonal growth by SNAP-25 antisense oligonucleotides in vitro and in vivo. Nature 364 (1993) 445.

11. SCHLINGENSIEPEN, K. H., R. SCHLINGENSIEPEN, M. KUNST, I. KLINGER, W. GERDES, W. SEIFERT, and W. BRYSCH: Opposite functions of jun-B and c-jun in growth regulation and neuronal differentiation. Developmental Genetics 14(4) (1993) 305.

12. SHI, Y., A. FARD, A. GALEO, H. G. HUTCHINSON, P. VERMANI, G. R. DODGE, D. J. HALL, et al.: Transcatheter delivery of c-myc antisense oligomers reduces neointimal formation in a porcine model of coronary artery balloon injury. Circulation 90(2) (1994) 944.

13. STEIN, C. A. and A. M. KRIEG: Problems in interpretation of data derived from in vitro and in vivo use of antisense oligodeoxynucleotides. Antisense Res. Dev. 4 (1994) 67.

14. TANAKA, H., P. VICKART, J. R. BERTRAND, B. RAYNER, F. MORVAN, J. L. IMBACH, D. PAULIN, et al.: Sequence-specific interaction of alpha-beta-anomeric double-stranded DNA with the p50 subunit of NF kappa B: application to the decoy approach. Nucl. Acids Res. 22(15) (1994) 3069.

15. TSENG, L. F., K. A. COLLINS, and J. P. KAMPINE: Antisense oligodeoxynucleotide to a delta-opioid receptor selectively blocks the spinal antinociception induced by delta-, but not mu- or kappa-opioid receptor agonists in the mouse. Eur. J. Pharmacol. 258(1–2) (1994).

16. VENCZEL, E. A. and D. SEN: Parallel and antiparallel G-DNA structures from a complex telomeric sequence. Biochemistry 32(24) (1993) 6220.

17. WANG, K. Y., S. MCCURDY, R. G. SHEA, S. SWAMINATHAN, and P. H. BOLTON: A DNA aptamer which binds to and inhibits thrombin exhibits a new structural motif for DNA. Biochemistry 32(8) (1993) 1899.

18. WOOLF, T. M., D. A. MELTON, and C. G. JENNINGS: Specificity of antisense oligonucleotides in vivo. Proc. Natl. Acad. Sci. USA 89(16) (1992) 7305.

19. YASWEN, P., M. R. STAMPFER, K. GHOSH, and J. S. COHEN: Effects of sequence of thioated oligonucleotides on cultured human mammary epithelial cells. Antisense Res. Dev. 3(1) (1993) 67.

20. ZANG, Z., W. FLORIJN, and I. CREESE: Reduction in muscarinic receptors by antisense oligodeoxynucleotide. Biochemical Pharmacology 48(2) (1994) 225.

II. Gene Function Analysis and Development of Therapeutic Agents

6 Cell Culture Protocols

Antisense Oligonucleotides in Cell Culture Experiments

Reimar Schlingensiepen and Ingrid Klinger
Max-Planck-Institut für Biophysikalische Chemie, Göttingen, Germany

1   Introduction

Cell culture may be regarded as the most convenient system to investigate antisense oligodeoxynucleotides (oligos) in living organisms. It combines advantages of an isolated *in vitro* system with at least some important aspects of live tissues. For example, cultured cells may interact through intercellular junctions or via paracrine loops. Primary neurones may form synapses [15] while fibroblasts migrate and multiply to close artificial "wounds" [33]. With regard to antisense experiments this means that both changes within the cell and within a cell population may be studied at the same time. For most gene function analyses with antisense oligos it is advantageous that cultured cells are isolated from other cell types, the interstitium or the immune system, since complex interactions may complicate the evaluation of functional and morphological changes. The interaction of different cell types can be studied in co-cultures. For example, the interaction between tumor cells and immune cells may be assessed in a cross-talk system (see chapter 7, Immunology and chapter 12, Oncology).

Functional assays and data on the specific inhibition of a targeted protein within a cell culture system provides the most important indicator for the utility of antisense compounds as *in vitro* agents. Data from cell culture experiments may often serve as a step towards animal models, drug development and clinical investigation. Still, cell culture experiments must be interpreted with caution as a model for *in vivo* applications because of a lack of natural tissue formation and the lack of complex local and systemic interactions. Nevertheless, it has been shown repeatedly that cell culture experiments may serve as excellent models for *in vivo* experiments (see chapters 8, 9, 12, and 13).

This chapter provides an introduction to the most widely applied antisense experiments in cell culture. Protocols are written in a more general style to be adaptable for most experimental settings.

1.1   General Considerations for Antisense Experiments in Cell Culture

Choice of the cell line: Since the functional and structural changes evoked by antisense inhibition may be complex it is advisable to choose a cell line (or primary cells) routinely cultured in your lab with well-known characteristics under different experimental conditions. Ideally, doubling time, response to different concentrations of fetal calf serum (FCS), duration for cell adhesion to occur, contact inhibition or different morphologies at different growth states, should be well characterized. New strategies should be tested in such a reference cell line before applying them to other cell lines.

State of the cells: For continuous cell lincs it is preferable that they have been passaged only a few times to avoid overgrowth by mutated clones with altered characteristics. Otherwise, comparison of different experiments

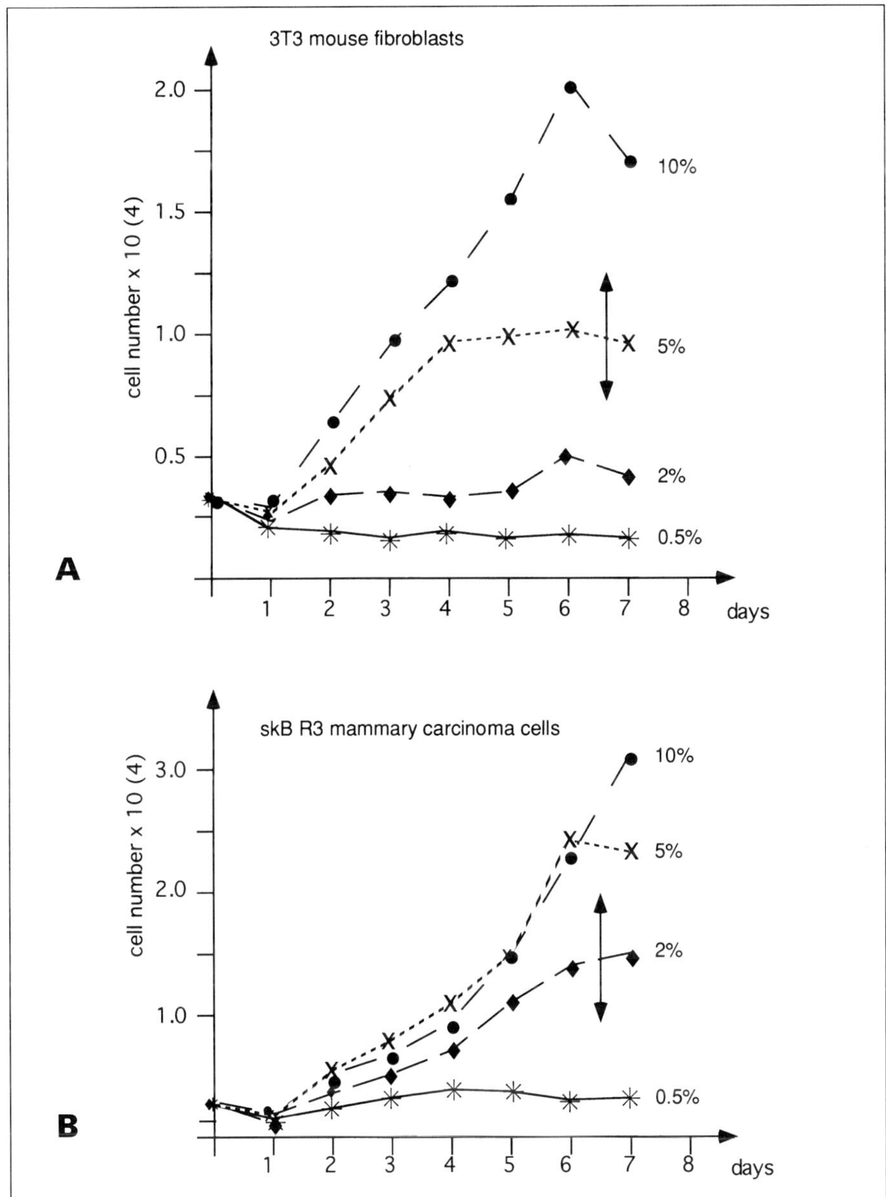

Figure 1: Cell growth of two different cell lines treated with different amounts of fetal calf serum (FCS). **A:** NIH 3T3 mouse fibroblasts are maximally growth stimulated with 10 % FCS and almost growth arrested with 0.5 % FCS. Addition of 5 % FCS leads to sub-maximal growth stimulation allowing for the assessment of both inhibitory and stimulatory effects of antisense oligos (arrow). **B:** The growth response of SK-Br 3 mammary carcinoma cells to different concentrations of FCS is shown to demonstrate that many malignantly transformed cells require an adaptation to low concentrations of FCS to achieve a sub-maximal proliferation rate. At 2 % FCS cell growth may be further stimulated or reduced by antisense oligos (arrow).

may be difficult. Accordingly, cells should be kept in the logarithmic growth phase and regularly transferred at subconfluency. Cells must be regularly checked for infection with mycoplasma since these may interfere with antisense compounds (for discussion see section 6).

Choice of cell type: If a certain gene is to be blocked it is best to choose a cell type with a known level of expression of the gene. Even an amplification of a gene does not necessarily lead to an expression at a sufficient level to be detected. If the expression level is unknown or low it is advisable to perform experiments in parallel with a reference cell line with a known expression.

Time frame: Antisense oligos are designed to inhibit protein synthesis but not protein function. Thus, the inhibitory effect may not become apparent before preexisting protein has decayed. For proteins with a low turnover (e.g. many cell receptors or ion channels) this means that a significant reduction of the protein level may not be detectable until several days after the first application of the oligos. Conversely, other classes of proteins like immediate-early genes have half-lives of minutes to a few hours. Consequently, the time frame for experiments has to be adapted to the half-life of the protein. Secondly, the duration of the cellular process that is to be evoked has to be taken into consideration. For example, it may take only a short time to modulate cell proliferation but it may take days or weeks for a cell to differentiate (see also chapter 8, Neurology).

Growth factor concentration: Generally, cells are supplemented with an amount of growth factors allowing (sub-) maximal proliferation. For cell growth assays growth stimulation should be reduced to be able to assess both inhibitory and desinhibitory effects of the antisense compound (see section 3). For many cells a reduction to 50 % of normal FCS concentration is suitable but in some malignant cells serum concentration has to be decreased further to slow down proliferation. Ideally, a dose-response diagram should be worked out if cell growth has not yet been well characterized (see Figure 1).

All the following experiments are described for a "standard" cell line: a medium sized continuous cell line with a doubling time of around 20–30 h, routinely grown in 10 % FCS (but 5 % FCS in growth assays) in standardized medium, supplemented with antibiotics (penicillin and streptomycin), kept at 37 °C, 5 % CO_2. These conditions should be adaptable for most other adherent or non-adherent cell lines.

2   *In vitro* Pharmacology

2.1   Toxicity Assays

For any antisense compound it is crucial to determine the optimal effective range where maximal selective inhibitory activity meets minimal non-specific toxicity. Polyanions like dextransulfate or oligodeoxynucleotides are known to cause growth inhibition in cell culture at high concentrations [32]. For oligos it has been shown that toxicity increases also with length [3]. In our laboratory, comparison of different concentrations of a highly purified 14 mer anti-p53 and control phosphorothioate oligomers (S-ODN) revealed that the inhibitory efficiency increases at concentrations between 0.5 and 2 µM without apparent toxicity. Between 2–5 µM no significant increase in specific activity was observed, but at concentrations from 20 µM to 50 µM an increasing non-specific toxicity emerged. Similar results were obtained by others [9]. In our hands a 2µM concentration has proven to be adequate to start with for most cell types to achieve good inhibitory effects and minimal or no toxic side-effects [2] [35] [29] [36] [37]. Depending on the cell type, on the target and on the purity of the compound it is often possible to scale down the oligo concentration to 0.1–0.5 µM.

2.1.1 Trypan Blue Exclusion Cytotoxity Assay

Cytotoxicity is conveniently measured by trypan blue exclusion. Viable cells transport the dye back into the extracellular space and remain unstained. Dead cells take up the dye passively and may thus be detected. Trypan blue is an acidic dye which binds easily to proteins. Thus, no serum should be present during testing. Its ability to enter a cell depends also on the pH. Uptake is maximal at pH 7.5. Trypan blue itself is toxic and may cause cell death within 5 min. Thus, cells should always be counted immediately after addition of the dye. Weakly stained cells should be considered as non-viable.

Protocol 1: Viability and Toxicity Assay – Trypan Blue Exclusion

Special material and equipment:
- 0.5 % w/v Trypan blue/0.9 % NaCl, sterile filtered (0.22 micron filter). *Trypan blue is harmful by skin contact. It causes cancer in animals. Take appropriate precautions!*
- PBS, pH 7.5 without calcium/magnesium
- Inverted microscope or Neubauer counting chamber (hemocytometer)

- Plate cells in 24-well flat bottom microtiter plates at 50–70 % confluency in supplemented medium. Prepare identical plates for each oligo or time point. Allow to adhere for some hours or overnight.
- Add specific antisense and control oligonucleotides in increasing amounts to four replicate wells. Suggestions for concentrations are: S-ODN: 0.5, 2, 5, 10, 20, 50 µM. Me-ODN (methylphosphonates): 10, 20, 50, 100, 200, 500 µM.
- Unmodified ODN should be tested preferentially in a serum free environment since nucleases cause degradation within minutes. If addi-

tional carriers are used (e.g. cholesterol, poly-L-lysine, liposomes) determine toxicity of carrier alone at increasing concentrations in parallel.
- Incubate cells for 24 h. In a second set of experiments, different time points should be chosen (12 h, 24 h, 48 h).
- Discard supernatant medium, wash cells once in PBS to stop incubation and to remove excess protein. Some adherent cell lines do not attach well enough to be washed with PBS. In this case omit washing step.
- Add 600 µl of 0.5 % trypan blue/0.9 % NaCl.
- Determine cell number and percentage of dead cells under the microscope. Count 10 randomly chosen fields at 400 × magnification using an ocular with a grid. The total cell number is determined by multiplying the mean by the correction factor: f(6-well) = 6038, f(24-well) = 1258). Since trypan blue is cytotoxic process one well at a time to minimize exposure before counting.
- Alternatively, trypsinize the cells and count in a hemocytometer as it is outlined in protocol 5.

2.1.2 MTT Cytotoxicity Assay

In this assay the capacity of mitochondrial dehydrogenase to convert MTT into a water-insoluble dark blue formazane product is measured. It provides a sensitive and quantitative colorimetric assay for viability but also for proliferation. It allows for processing of a large number of samples with a high precision using a multi-well ELISA reader (scanning spectrophotometer). If linked to a computer, data may be analyzed on-line. The assay is very reliable for the determination of cell numbers between 200 and 300,000 cells.

Protocol 2: MTT Cytotoxity Assay

Special material and equipment
- MTT (Sigma M2128) stock solution of 5 mg/ml in PBS, pH 7.5. To sterilize and to remove insoluble particles pass through a 0.22 micron filter.
- PBS without calcium or magnesium
- 3 % SDS
- 0.04 M HCl in isopropanol (add 0.4 ml 10 M HCl to 100 µl isopropanol)
- microELISA reader

- Transfer cells (logarithmic growth) into a 96-well microtiter plate. Add PBS to outer wells to minimize evaporation. Allow cells to adhere and to recover for 24 hours.
- Replace supernatant by adding 100 µl fresh medium containing oligos at different concentrations (see protocol 1). Incubate cells e.g. for 24 hours.

- Add 10 µl MTT solution directly into the supernatant, mix gently with a pipette or by mild agitation. Place for 2h in the incubator for the enzymatic reaction to occur.
- Adherent cells: Draw off supernatant, add 20 µl 3 % SDS and 100 µl 0.04 M HCl in isopropanol, agitate gently for 20 min.
- Cells in suspension: Spin at 800 × g after incubation with MTT in a centrifuge with a micro plate rotor. Then draw off supernatant, add 20 µl 3 % SDS and 100 µl 0.04 M HCl in isopropanol, agitate gently for 20 min.
- Read plate on a microELISA reader at 570 nm and 630 nm (reference wavelength) within 30 min after addition of acidified isopropanol.

2.2   Cellular Uptake

Unmodified DNA oligos and phosphorothioates are hydrophilic polyanionic macromolecules with molecular weights between 3000 and 10000 Da. They cannot diffuse passively through the bilipid layer of the plasma membrane. Still, the uptake of charged oligos has been described as a surprisingly efficient and rapid process [42]. It has been found to be time, temperature and length dependent [16] [8]. A 80 kD protein has been identified as a putative receptor for polyanionic molecules like oligonucleotides, polysulfates or dextran sulfate [46]. Addition of different DNA and RNA leads to competitive uptake with no preference for sequence or length [20]. In contrast, uncharged modified oligos (e.g. Me-ODN) enter the cell in a length-independent and linear manner. They have been described to pass the membrane by adsorptive endocytosis (pinocytosis) or passive diffusion with no receptor mediated uptake [24].

Fluorochrome labelled charged oligos have been shown to be transported rapidly into the nucleus if injected intracellularly [4], whereas administration to the supernatant cell culture medium leads to a punctuate intracellular distribution. It has been postulated that punctuate distribution is caused by entrapment of the oligo in endosomes from where the oligos are slowly released to other cytoplasmatic compartments [16]. Since the release from the endosomal compartments is slow, biostability and acid resistance are of particular importance for oligos. On the other hand, endosomal entrapment prevents rapid exclusion to the extracellular space. This may explain, at least in part, why antisense oligos exert specific inhibition over extended periods of time even if administered only once in cell culture or in laboratory animals [40].

Several attempts have been made to enhance the uptake efficiency either by conjugation with lipophilic molecules or polycations. Conjugation to cholesterol enhances cellular uptake but also toxicity, which has been attributed to an increased influx of calcium [34] [45]. Another approach is to add the polycation poly-L-lysine (PLL) to neutralize the oligo's negative

Figure 2: Uptake studies are best performed on cell culture slides. Mark time points on the opaque strip of the slide. Fluorescein labelled oligo may be added directly into the supernatant.

charge. These conjugates have been shown to be taken up efficiently. In addition, unmodified ODN which are rapidly degraded both intra- and extracellularly may be protected and show a significantly improved inhibitory effect if coupled to PLL [18]. Still, after cleavage, poly-L-lysine acts as a free cytotoxic compound known to inhibit cell growth at concentrations as low as 2 µM [44].

2.2.1 Cellular Uptake – Fluorescein Labelled Oligonucleotides

Uptake studies are best performed on cell culture chamber slides where a glass or synthetic polymer slide is topped with cell culture wells (e.g. Labtek slides from NUNC, Denmark). This allows to grow, treat, fix and examine cells on the same slide. Fluorochrome labelled oligos offer the easiest approach to assess cellular uptake since they can be visualized directly under a fluorescence microscope. Fluorochrome assays are less sensitive than immunogenic methods (see below) but for most uptake studies a strong signal can be achieved. Fluorescein alone is not taken up by cells and is rapidly excluded from both the nucleus and cytoplasm after cleavage from the oligo [19]. Thus, it is a reliable marker for the intracellular localization of oligonucleotides.

Protocol 3: Cellular Uptake – FITC Labelled Oligonucleotides

Special material and equipment:
- Fluorescein labelled oligo
- Cell culture slide
- Fluorescence microscope

• Time points for addition of oligos depend largely on the cell type and on backbone modifications. For polyanionic oligos like unmodified

ODN or S-ODN which are the most widely used compounds the time frame should comprise 48 hours. As a suggestion, monitor uptake at hours 2, 4, 8, 12, 18, 24, and 48. Since all additions of the oligo are performed on the same cell culture slide (see Figure 2), bear in mind that the oligo is added first to the 48 h well and last to the 2 h well. If cells need fresh medium during the experiment, the labelled oligo is added in equimolar concentration.

- Pretreat wells with poly-L-lysine for better adhesion of cells if necessary
- Prewet wells with 200 µl of PBS or minimal medium to minimize electrostatic disturbance
- Plate 2500 – 5000 cells/well in 200 µl growth medium in 16-well slides. Alternatively, use 8-well slides. Multiply the number of cells and volumes by three.
- If adherent cells are used put the slide in an incubator and avoid any disturbance for several hours or overnight to allow adhesion processes to occur.
- Add the oligo directly to the supernatant medium (time point ‚48 h'): mix gently with a yellow tip pipette. The choice of concentration depends on the stability of the compound. For S-ODN 0.5–2 µM is recommended.
- If cationic lipids are used as enhancers for uptake, add 10 µg/ml to serum-free medium, mix with the oligos and preincubate at 37 °C in a humidified incubator to allow for formation of an oligo-lipid complex. Then add the mixture to the supernatant of the cultured cells. Incubate for 4 hours and then replace supernatant with fresh medium containing 5 % FCS.
- Treat other wells accordingly on the two consecutive days.

Adherent Cells:

- To stop incubation pour off the supernatant and wash the wells twice with minimal medium. The wells may be taken off at this stage or at the end of the experiment.
- Fix cells according to standard protocols, e.g. in 4 % paraformaldehyde/minimal medium for 5 min.
- Wash the wells twice with minimal medium for 1 min.
- Dehydrate cells in a graded series of alcohol (70 %–100 %), 1 min each, allow to fully air dry. Finally peel off the silicone rings
- To examine cells under a fluorescence microscope add a non-quenching immersion oil. Take photographs with a high sensitivity film to minimize exposure time. Fluorescent light causes rapid fading of the signal.

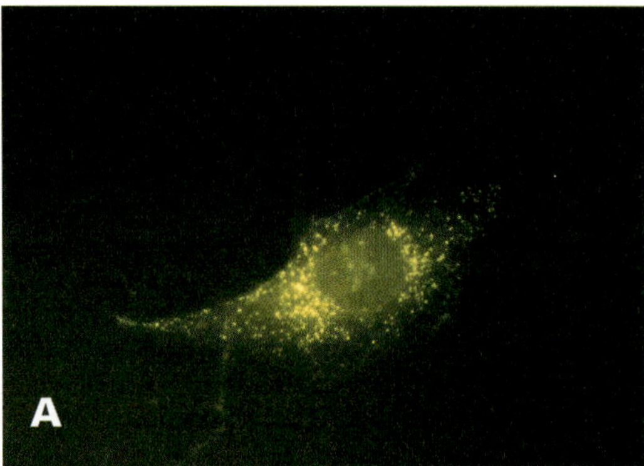

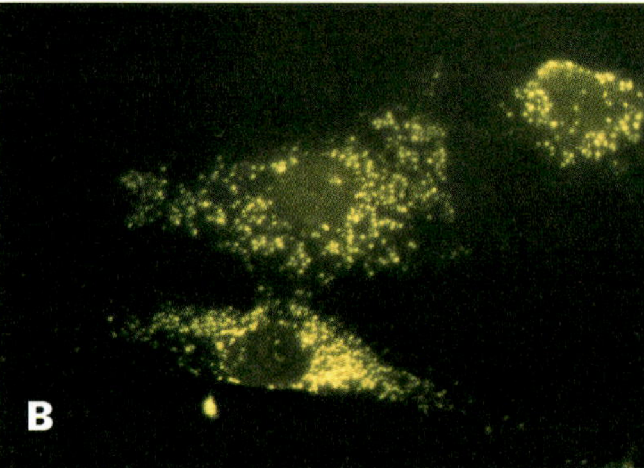

Figure 3: Cellular uptake of fluorescein labelled oligos. NIH 3T3 mouse fibroblasts were incubated with 2 µM of a 14 mer random sequence oligo, 3'-labelled with Fluorescein-isothiocyanate via an aminolinker (Fluoroprime, Pharmacia). After 8 h (**A**) cells were labelled in a punctuate manner in the cytoplasm. In the nucleus the signal was minimal. After 24 h (**B**) the signal pattern was similar and uptake was maximal with no further increase at 48 h (data not shown).

Non-Adherent Cells:

- Stop the incubation by transferring cells into microfuge tubes, spin at 400 × g and discard the supernatant
- Wash the cells in minimal medium, spin at 400 × g and resuspend. Attach the cells to a slide by cytocentrifugation at 1200 × g for 10 min
- Alternatively, resuspend the cells in 20 µl of minimal medium and disperse slightly on a slide previously coated with poly-L-lysine
- Allow to air dry at 37 °C and fix in 100 % methanol (–20 °C) for 5 min or according to an alternative fixation protocol
- Dehydrate the cells and proceed as outlined above

Exemplary results: see Figure 3

2.2.2 Cellular Uptake – Immunodetection

Immunocytochemical assays are more cumbersome but offer greater sensitivity than fluorescence detection. Oligos are labelled with modified nucleotides like bromodeoxyuridine (BrdU) amidites or pendant groups like biotin. These haptens or antigens may be added during automated synthesis and labelled oligos may be purified by reverse phase HPLC.

Protocol 4: Cellular Uptake – Immunodetection

Special material and equipment:
- BrdU labelled oligo. *BrdU is mutagenic! Take adequate safety precautions.*
- Monoclonal (mouse) antibody to BrdU (1st antibody)
- Peroxidase labelled polyclonal antibody to (mouse) immunglobulin (2nd antibody)
- 0.02 % hydrogen peroxidase in Acetone/Methanol (1:1), freshly prepared
- DAB tablets (Sigma)
- Cell culture slides

- Grow the cells on cell culture slides and incubate with BrdU labelled oligo according to recommendations for FITC-labelled oligos (protocol 3).
- To stop the incubation wash the cells with PBS. Remove the wells and silicone rings from the slide
- Fix the cells in 0.02 % hydrogen peroxidase in acetone/methanol for 20 min at –20 °C. Do not allow the cells to become dry after fixation.
- Wash twice in PBS for 2 min
- Incubate the cells in 5 % FCS/PBS for 30 min to block non-specific binding
- Incubate with first antibody according to the manufacturer's recommendations.
- Rinse 3 × 5 min with PBS
- Incubate with the second antibody, (diluted at 1 : 100 usually)
- Rinse 3 × 5 min with PBS
- Prepare DAB solution. Incubate the cells as recommended by the manufacturer.
- The cells may be counterstained. Dehydrate in a graded series of alcohol then transfer to xylene and embed with appropriate medium

Exemplary results: see Figure 4

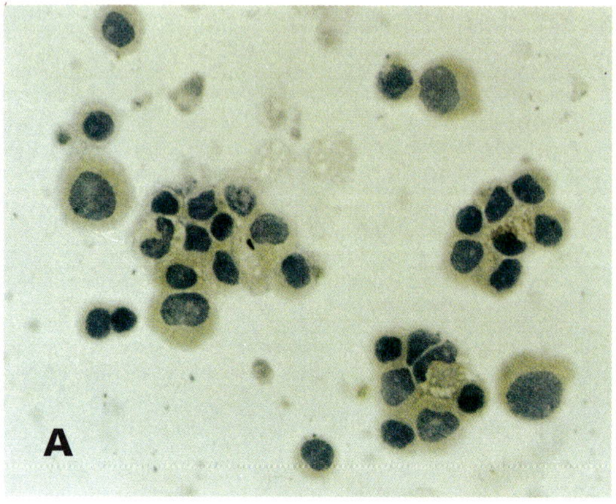

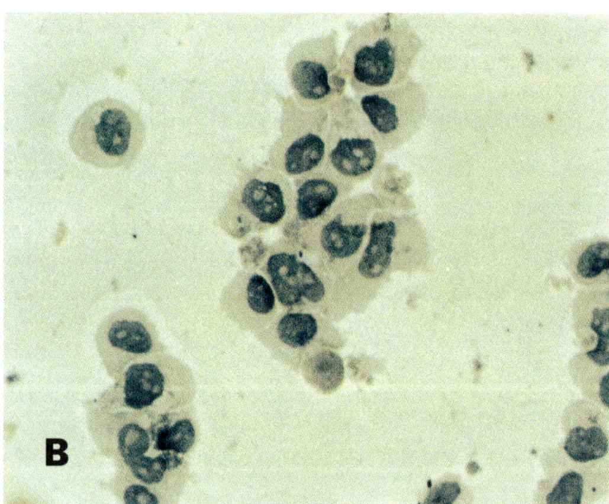

Figure 4: A: Cellular uptake of a 14 mer random sequence oligo, internally labelled with Bromodeoxyuridine amidite (PerSeptive). Peripheral blood mononuclear cells were incubated for 48 h at 1µM concentration and processed for immunocytochemical detection by the biotin-avidin method. The oligo accumulates in the cytoplasm with minimal signal in the nucleus. **B:** Cells treated with unlabelled oligo served as controls. No non-specific background is detectable. (Photographs: courtesy of Piotr Jachimczak).

3   Functional Assays

3.1   Proliferation Assays

Cell proliferation and growth arrest are probably the most studied parameters in antisense research. This is due to the fact that fundamental functional pathways influence cell growth. For example, many cell surface receptors, growth hormones, cytokines, immediate-response early genes, protein kinases and G-proteins interlock in the complex processes involved in cell growth. The search for efficient drugs for cancer therapy demands specific inhibition of aberrant malignant cell growth by selectively blocking pathologically expressed genes. The same applies for cardiovascular disease, where intimal proliferation after angioplasty causes restenosis. In autoimmune diseases like Hashimoto thyroditis or chronic polyarthritis, lymphoid clones grow aberrantly and attack autologous tissues. In fact, examples can be given for most clinical specialties. Consequently, proliferation assays are valuable for both basic science and drug development. On the other hand, cell proliferation assays alone may not distinguish between specific antisense effects, non-antisense effects and non-specific toxicity. Thus, it is mandatory to combine these assays with adequate assays of specificity like Western blot analysis of the targeted gene and related proteins (see section 4).

3.1.1 Direct Assessment of Proliferation – Cell Counting

Cell counting provides the most reliable data on proliferative activity. Cells may grow in size and by number but cell proliferation assays should ideally measure exclusively the latter. Indirect assays measuring enzymatic activity or the incorporation of DNA precursors may not distinguish the two different ways of cell growth and are not as reliable as cell counting (see 3.1.2). On the other hand, cell counting is more cumbersome and time consuming. Consequently, it is less suitable for processing of large numbers of samples.

Protocol 5: Proliferation Assay – Cell Counting

Special material and equipment:
– 96-well, flat bottom cell culture plates
– Neubauer counting chamber (hemocytometer)
– 0.1 % trypsin, 0.01 EDTA in PBS
– Trypan blue solution

• Transfer cells from flasks to flat bottom cell culture plates. 96-well plates are suitable but handling may be more difficult than in 48-well or

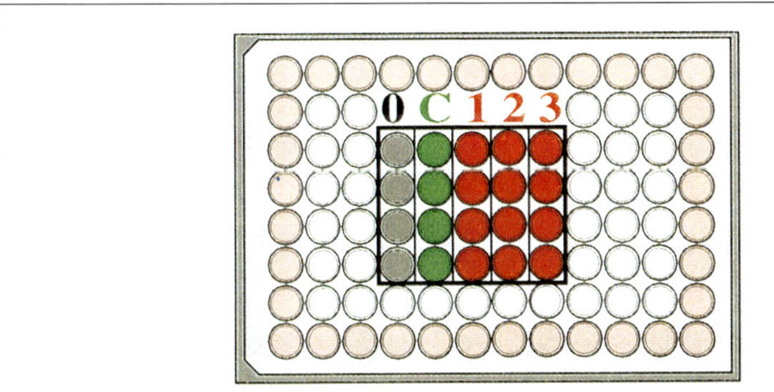

Figure 5: 96-well cell culture plate prepared for cell proliferation studies. Cells are grown in quadruplicate for each condition: 0 = Control cells without antisense oligo, C = Cells treated with control oligo, 1–3 = different antisense oligos. Prepare identical plates for all time points. Draw a frame with a marker pen on the cover and bottom of the plate to facilitate location of the wells. Add PBS to the outer wells to reduce evaporation from the inner wells.

24-well plates. The protocol will be outlined for 96-well plates. Multiply the number of cells and volumes by three or six for 48-well or 24-well plates, respectively.

- Set up as many identical plates as there are time points, e.g. 5 plates. Suggestions for studies of a gene with a rapid turnover are day 1, 2, 3, 4, 7; and if turnover is low, days 1, 3, 6, 9, 14. Take four wells for each condition (see Figure 5).
- To allow for assessment of both an increase or decrease of proliferation rate adapt the FCS concentration, e.g. from 10 % to 5 % as suggested in 1.1.
- Prewash the wells with 1x PBS directly before plating cells to minimize electrostatic disturbance. Otherwise, the cells tend to accumulate in the center of the wells.
- Plate 2500 cells/100µl supplemented medium into each well. Fill the remaining wells with PBS or serum free medium to reduce evaporation.
- Allow the cells to sit for several hours without any further manipulation for the adhesion process to be initiated. At this stage addition of oligos may interfere non-specifically with cell adhesion.
- If crude oligos are used, add antisense and control directly into the supernatant medium of each well. The concentration needed for efficient specific inhibition may vary considerably, depending on the chemistry used, on the purity of the compound, its length and sequence (see above and chapter 1). For highly purified S-ODN a concentration between 0.5 and 5 µM will be appropriate. If the medium has to be changed, add the oligos at identical concentrations each time.
- Check daily under the microscope for morphological changes. Document changes photographically.
- To count the cells, wash them with PBS, replace supernatant medium with 50 µl trypsin solution, incubate at 37 °C. Do not trypsinize more than four wells at a time. When the cells have detached, add 50 µl of 0.5 % trypan blue/0.9 % NaCl into one well at a time and count immediately. Trypan blue is cytotoxic and may cause cell death within minutes. Count the cells in a (Neubauer) hemocytometer or in an automatic cell counter.
- Count each well in duplicate. Determine the percentage of dead cells by trypan blue exclusion. Counting may become difficult if the number of cells per well is too low. No more than 2.5 cells per large square of a hemocytometer will be counted if the 100 µl aliquot contains 2500 cells. Thus, it may be necessary to plate more cells if the antisense oligo inhibits growth completely, leading to a decreasing cell number.

Modification for Cells in Suspension:

- Plate 2500–5000 cells in 100 µl supplemented medium
- In contrast to adherent cells, oligos may be added directly after plating.
- To count cells, add 100 µl trypan blue directly into the well, count immediately in a hemocytometer. Since the total volume is 200 µl, the number of cells must be multiplied by two.

Exemplary results: see Figure 6

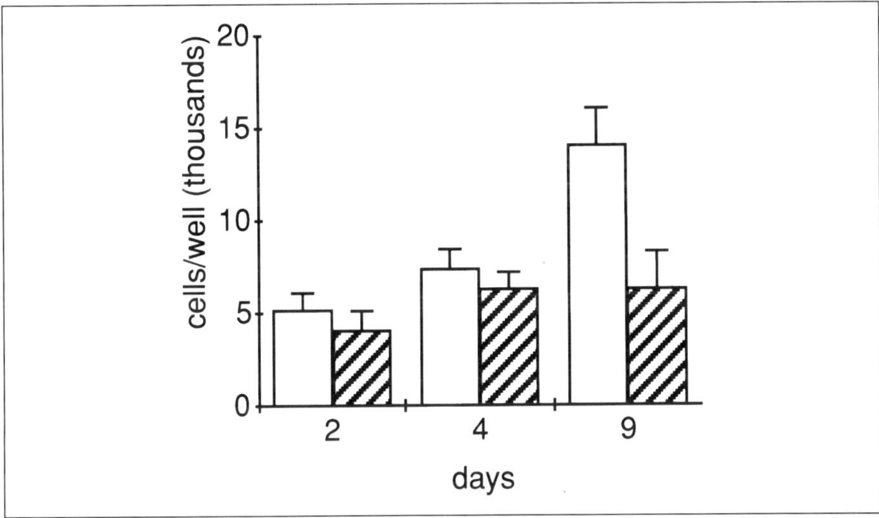

Figure 6: Cellular proliferation assay: Cell counting. SK-Br-3 cells were treated with 2µM of randomized sequence oligo (white bars) or specific anti-c-*erb*B-2 oligo (hatched bars), [1]. Cells were harvested by trypsination and counted in a Neubauer hemocytometer. Antisense treated cells were growth arrested while control oligo treated cells continued growing. Trypan blue labelling revealed that in both groups the percentage of dead cells did not exceed 7 % (data not shown).

3.1.2 Long-term Proliferation Assay

To study the long-term effect of antisense oligos either after single application or after repeated treatment it is necessary to grow cells in parallel for counting and for transfer.

Protocol 6: Long-term Proliferation Assay

- Set up experiments as described in protocol 5 using 96-well plates. Increase the number of wells per condition from four to ten, to allow both counting and transfer of cells (Figure 7).
- The cells should be transferred during log phase growth and before confluency is reached. Since cells for transfer and for counting are kept on the same culture plate it is advisable to transfer cells first, to minimize the risk of contamination.

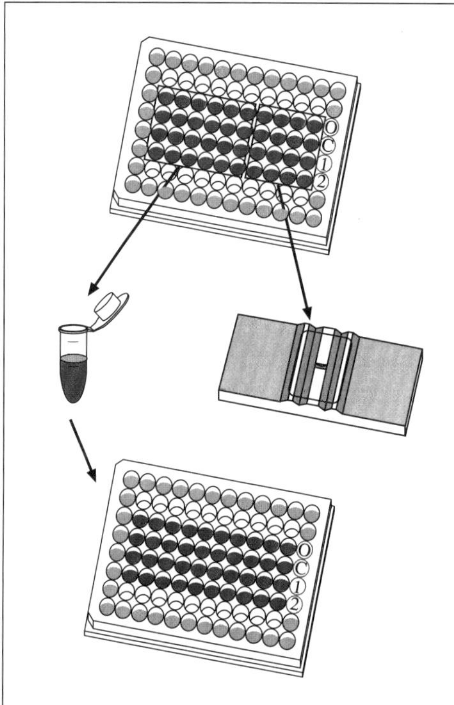

Figure 7: 96-well cell culture plate prepared for long-term proliferation studies. Cells were grown in ten replicates for each condition: 6 wells for transfer and replating (left), 4 wells for cell counting (e.g. in a Neubauer hemocytometer, right). 0 = Control cells without antisense oligo, C = Cells treated with control oligo, 1, 2 = different antisense oligos.

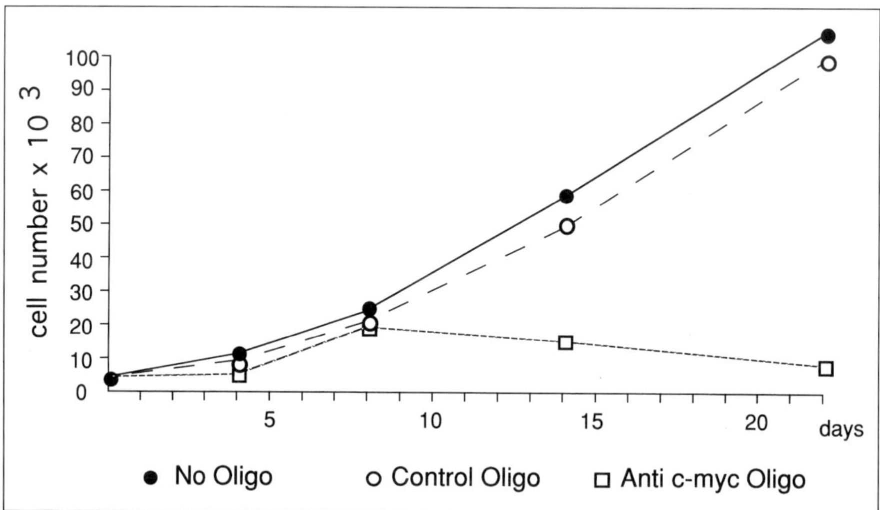

Figure 8: Long term proliferation assay of HTB 187 medulloblastoma cells. For experimental details refer to chapter 8, Neurology, section 3.5. Cells were supplemented with 2 µM of random sequence control oligo, anti-c-*myc* oligo or no oligo at day 0 and twice weekly when changing medium or when transfering cells. After one week no significant difference in cell growth was apparent but after 10 days suppression of c-Myc led to complete growth arrest and a subsequent decrease of cell number.

- To <u>transfer</u> replace medium with 100 µl of trypsin solution. Process all six wells of one condition at a time. Incubate cells at 37 °C until they detach. Pool the contents of the six wells in one microfuge tube containing 600 µl of medium supplemented with 10 % FCS (if cells are grown in 5 % during the experiment).
- Spin the cells at 400 × g and wash in medium/5 % FCS. Resuspend the cells in 600 µl medium/5 % FCS. Count a 50 µl aliquot in a hemocytometer and dilute resuspended cells to the same concentration as at the beginning of the experiment.
- Plate 10 wells of a new cell culture plate with the same amount of cells as before (e.g. 2500 cells in 100 µl supplemented medium). Add the oligos at the same concentration as before.
- To <u>count</u> the remaining four wells proceed as described in protocol 5. Determine the multiplication factor of each population (e. g. 75.000 cells/ml – 7.500 cells/100 µl. The multiplication factor is three where 2500 cells were plated originally).
- In the case of accidental infection of single wells by bacteria or fungi, non-infected wells may be saved to continue the experiment. Add an aqueous non-volatile disinfectant like M + D Aqua-Clear (Baxter) into the infected wells. Do not use alcohol or formalin since evaporation may cause cell death in neighbouring wells.

Exemplary results: see Figure 8

3.1.3 Indirect Assessment of Proliferation – ^{3}H-Thymidine Incorporation[1]

^{3}H-thymidine incorporation assays are routinely used to assess the modulation of cell growth by antisense oligos. Since thymidine is a precursor of DNA but not RNA its incorporation into the genome is an indirect indicator of proliferative activity. Experimental procedures are less time consuming and cumbersome than cell counting. On the other hand, the test is more susceptible to disturbances [10]. False positive results occur when resting cells incorporate the radioactive label during S-phase without further division. In addition, malignant cell clones exhibiting hetero- and polyploidy may increase their genomic content independently from cell growth. False negative results may occur in particular if unmodified ODN are used. Rapid degradation increases the intracellular nucleotide pool, inhibiting competitively the incorporation of the radiolabelled marker [21]. This may be circumvented by applying nuclease resistant modified oligos. In any case, base composition of control oligos should be identical to that of the antisense oligo. Sense controls are often inappropriate, since the base content is normally different to that of the antisense oligo. Con-

---

[1] Alternatively, the MTT assays may be used for indirect assessment of cell growth (see 2.1.2)

trols with identical base content will affect ^{3}H-thymidine incorporation to the same extent as the antisense upon degradation. For design of control oligos refer to chapter 5, Assays Systems and Controls.

In general, ^{3}H-thymidine assays should be sporadically confirmed with other measures of cell proliferation, ideally, cell counting [21].

Protocol 7: Proliferation Assay – ^{3}H-Thymidine Incorporation

Special material and equipment:
– ^{3}H-thymidine (1 mCi/ml, e.g. NEN/Du Pont,)
– scintillation fluid, scintillation tubes
– 5 % trichloric acid
Additional equipment:
– Cell harvester (e.g. Dunn PHD Cell Harvester)
– Glass filter (e.g. grade 934 AH, Dunn)

• Grow and set up cells exactly as for the cell counting assay, plate 2500 (or more depending on cell size and doubling time) in 100 µl complete medium. Reduce the FCS concentration as recommended in 1.1.
• Prepare an identical 96-well plate for each time point (e.g. 24 h, 48 h, 72, 120 h) (see Figure 9). Include at least 6 replicates for each condition.
• Add antisense and control oligos for all time points, e. g. one hour after plating to start the experiment.
• Dilute ^{3}H-thymidine to 50 µCi/ml with serum free medium and add 0.15 µCi (i.e. 3 µl) to each plate 6 or 8 h before the indicated time points. Mix gently with a pipette adjusted to 50 µl. Incubate the cells for 6 hours for cells with short doubling time or for up to 18 h for slowly growing cells.

To harvest cells with an automatic harvester proceed as follows:

• To stop the incubation lyse the cells by freezing the plate at –20 °C for at least 12 h
• After having finished all incubations, heat the plates to 60 °C for 1 h
• Transfer the lysates onto glass-fiber filters with a cell harvester according to the manufacturer's recommendations.
• Wash the filters 5 × with sterile water to eliminate unincorpated ^{3}H-thymidine and cell debris. Wash the filter with 5 % cold TCA, then wash in cold absolute ethanol.
• Either allow the wet glass filters to calibrate at RT for at least 2 h in the scintillation fluid or dry the filters completely prior to counting.
• Record disintegrations per minute in a scintillation counter. Disintegration allows better interexperimental comparisons than counts alone.
• For long-term assays prepare the experiments according to protocol 6.

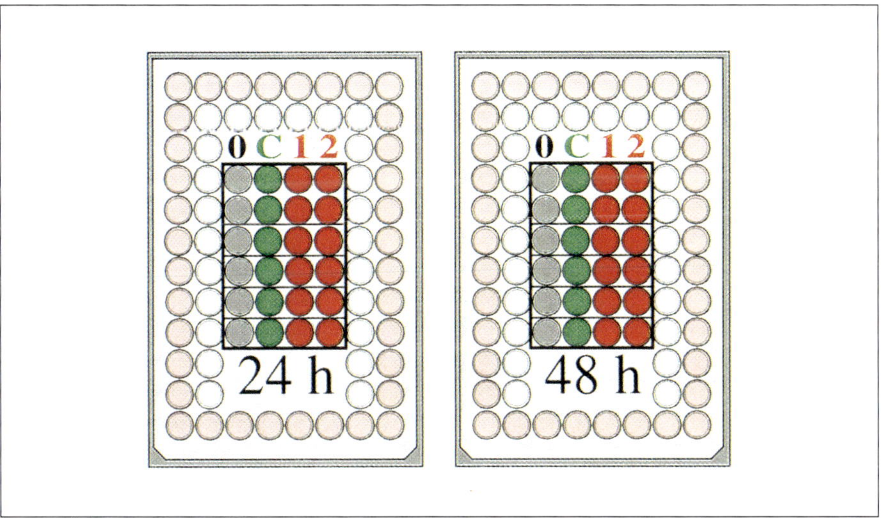

Figure 9: 96-well cell culture plate prepared for ³H-thymidine incorporation assay. Cells are grown in 6 replicates for each condition: 0= Control cells without antisense oligo, C = Cells treated with control oligo, 1, 2 = different antisense oligos. Prepare identical plates for each time point. Draw a frame with a marker pen on the cover and bottom of the plate to facilitate location of the wells. Add PBS to the outer wells to reduce evaporation from the inner wells.

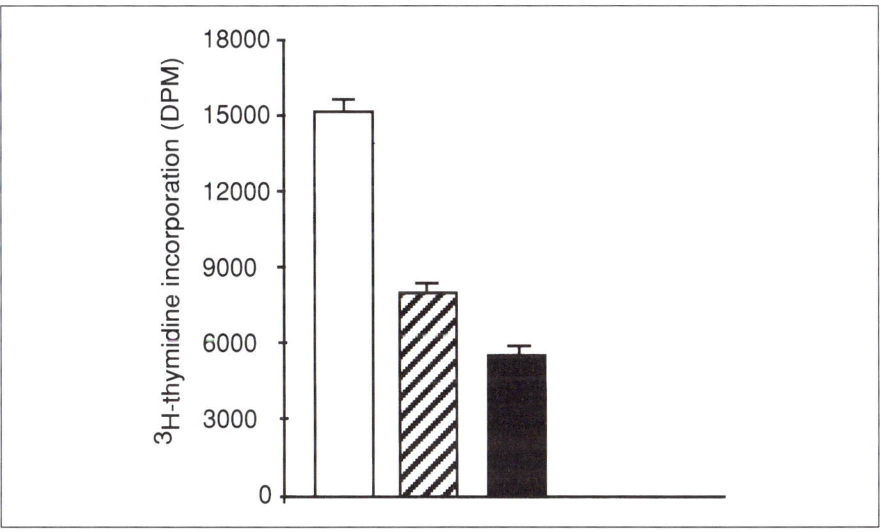

Figure 10: Cellular proliferation assay: ³H-thymidine incorporation assay. Primary rat astrocytes were grown as described before [11] and treated with 2 µM of random sequence control oligo (white bars), anti-c-myc oligo (hatched bars) and specific anti-bFGF oligo (black bars) over 2 days. In comparison to control oligo treated cells proliferation was reduced to 36 % and 50 % by anti-c-myc and anti-bFGF oligo, respectively.

> If no cell harvester is available, proceed as follows:
>
> - To stop the incubation with ^{3}H-thymidine draw off the supernatant and collect radioactive waste appropriately.
> - Wash the cells twice with PBS (containing calcium) and fix the cells twice with 100 µl 100 % methanol for 5 min each. Wash the cells once with water.
> - Incubate once with 5 % trichloric acid for 10 min and then wash three times with water.
> - Add 150 µl 0.3 N NaOH, mix thoroughly, then transfer the lysate into scintillation vials after 15 min. Neutralize alkali with equinormal amount of HCl. Count the samples in a scintillation counter after one hour when chemiluminescense has decayed (see above).

Exemplary results: see Figure 10

3.2 Morphology

Morphological changes are expected in particular when targeting structural proteins like collagens, crystallins, fibrillary proteins or adhesion molecules. In addition, targeting proteins that are involved in central signaling cascades (e.g. protein kinases, growth hormones, hormone receptors, tumor suppressor genes and transcription factors) may lead to both morphological and functional changes.

Generally, undifferentiated cells have a high proliferation rate. Antisense inhibition of genes involved in the mediation of cell growth may lead to growth arrest and differentiation of such cells. If they were rounded with an increased nucleus/cytoplasm ratio they may flatten and grow processes. Also, undifferentiated malignant cells which grow in a disorganized fashion in multiple layers may become contact inhibited and form monolayers. The opposite may occur when blocking tumor suppressor genes or other genes being involved in growth arrest and cellular differentiation [35] [17]. Examples for the modulation of cell growth and cell morphology in transformed and non-transformed cells are given in chapter 12, Oncology and chapter 8, Neurology.

Assessment of Morphological and Functional Changes

> - Changes in the proliferation rate generally precede morphological changes by days. Check cells daily under an inverted microscope over 2-3 weeks after the first addition of antisense oligos.
> - Check for cell size, cell shape (flatness, cellular processes) and nucleus/cytoplasm ratio.
> - Adhesion properties may change significantly. This also becomes apparent during transfer of cells, where detachment of cells during incu-

bation with trypsin may vary between differently treated cells. Directly after plating, the time taken for adhesion and for reshaping may be monitored as well.

- Look for changes in cell to cell communication like contact inhibition: Do cells form homogenous monolayers? If one makes a "wound" in the layer with a cell scraper, does it get rapidly closed? Conversely, do cells grow in dispersed colonies, in a disorganized fashion, one over the other. Cells may still grow in monolayers but in a reticular fashion leaving large "holes".

3.3   Soft Agar Colony Formation

Most non-malignant cells need a solid substrate like the bottom of microtiter plates to attach and to proliferate. Primary cells usually need a surface coated with supportive agents like poly-L-lysine or collagen to improve adhesion and to allow the growth of cellular processes. On the other hand, many malignantly transformed cells may grow anchorage-independently in soft agar to form colonies, while growth of non-transformed cells is suppressed at the same time [23]. Reversal of malignant properties usually also results in reduced colony formation in soft agar [43]. This functional test has been used to compare the colony forming capacity and colony morphology of antisense versus control oligo treated cells [28] [25] [6]. Cells that show reduced colony formation after antisense treatment (e.g. against the protooncogene c-*myc*) may be significantly less tumorigenic if injected in immunodeficient mice [39].

3.3.1

Protocol 8: Soft Agar Colony Formation

Special material and equipment:
- antisense treated cells
- soft agar stock solution
- agar (Difco Laboratories)
- 0.1 % trypsin/0.01 M EDTA in PBS
- 24-well culture plates
- iodonitrotetrazolium violet (0.5 mg/ml sterile water) (Sigma)

- Grow cells in culture flasks and treat with antisense and control oligos. Incubation should exceed the half-life of the protein at least twofold. The cells are then plated without adding the oligos again. Alternatively, cells may be treated directly in the soft agar.
- Prepare 24-well cell culture plates with a basal layer of 0.5 ml 0.6 % agar/minimal medium).
- Trypsinize and wash the cells in PBS, draw gently up and down with a pipette to obtain a single cell suspension. Spin the cells and resuspend

at 10,000 cells/100 μl. Add increasing amount of cells (5,000 to 20,000) into 0.5 ml 0.25 % soft agar/complete medium. (The optimal soft agar concentration usually ranges from 0.2 to 0.4 %)
- Pipette the soft agar/cell suspension onto the basal agar layer and incubate for about 10 to 14 days.
- Count colonies in unstained cultures with the aid of a low power microscope. Alternatively, stain the cells, e. g. with iodonitrotetrazolium violet. Determine the colony forming efficiency (CFE) between treated and untreated cells and pay attention to changes in colony morphology
- Eventually, colonies that can be viewed by eye may be scorable. When first scoring a new cell line, use an inverted microscope to count cells. In general, more than 50 cells are scored as a colony.

Exemplary results: see Figure 11

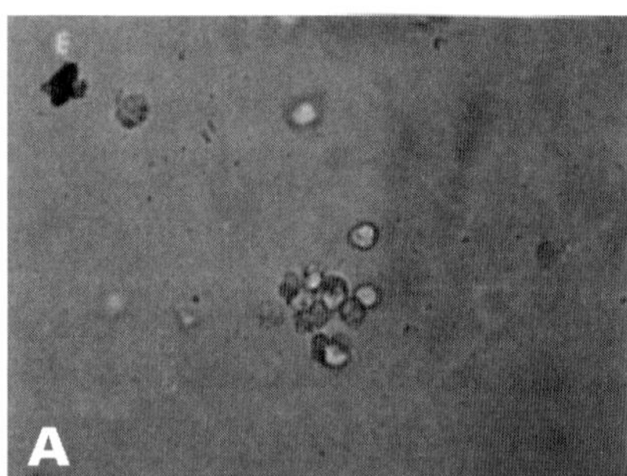
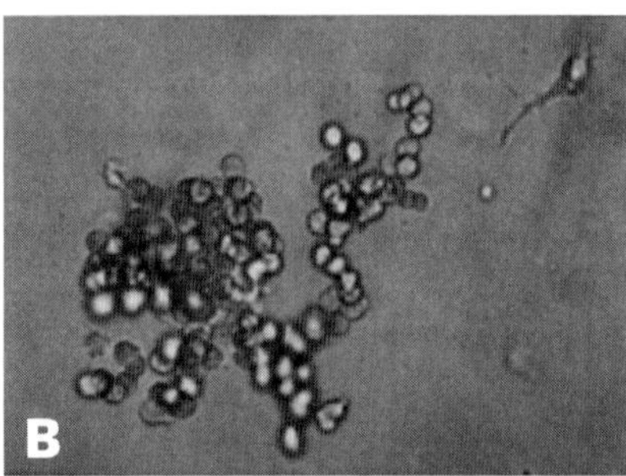
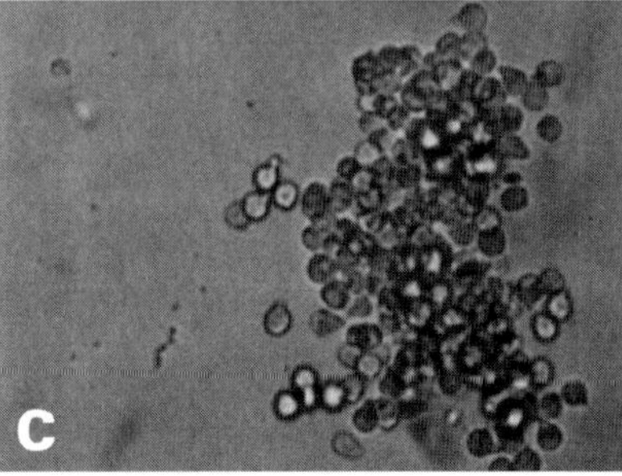

Figure 11: Colony formation assay. Human hematopoietic cells were obtained from healthy donors. Granulocyte-macrophage cells were grown in a plasma clot system as described elsewhere [13]. Cells were treated either with an 18 mer c-*myb* antisense, myeloperoxidase (MPO) antisense or c-*myb* sense. After 12 days colony formation was suppressed by specific anti-c-*myb* oligo **(A)** but not by the corresponding sense oligo **(B)**. The anti-MPO oligo served as a further control since it is not a protooncogene like c-*myb*. As expected anti-MPO does not inhibit colony formation **(C)**. (Photographs: courtesy of Alan M. Gewirtz).

4   Assays for the Evaluation of Antisense Specificity

4.1   Measurement of the Protein Product

Undoubtedly visualization of the protein product by immunodetection offers the most convincing proof of specific antisense activity. Most commonly, protein levels are assayed by Western blot analysis, immunoprecipitation or immunocytochemistry. Comparison of the specific antisense oligo with carefully chosen controls (see chapter 5, Assay Systems and Controls) allows one to distinguish non-specific from specific translation inhibition of the targeted gene.

The effect of antisense inhibition may become apparent a few hours after the oligo has been taken up. This is the case for proteins with a short half-life and a high turnover rate like immediate-early genes [35]. Other proteins like many cell receptors, house keeping genes etc. have a half-life of days rather than hours [2]. Here, significant decrease in signal will become apparent earliest after days when preexisting protein has decomposed, since the antisense oligo suppresses only *de novo* synthesis.

4.1.1

Protocol 9: Setup of Antisense Experiments for Western Blot Detection

For detailed protocols on Western blot procedures please refer to standard lab manuals. In the following, steps important for setup of an antisense experiment will be outlined:

- Principles for experimental set-up and handling of cells are the same as outlined in 1.1 and section 3.
- To obtain sufficient amounts of protein, cells should be grown in 75 ml cell culture flasks. Choose appropriate periods of incubation (e.g. 0 h, 6h, 12 h, 24 h, 48 h for high turnover proteins or 0 h, 12 h, 24 h, 48 h, 92 h or longer for low turnover proteins).
- Reduce the amount of medium during incubation to save oligo. Add the oligos at an effective but non-toxic concentration (previously determined by functional and toxicity assays (see 2.1 and 3)).
- Cells may be harvested by two different methods. Since trypsin digests cell surface proteins it should only be applied if intracellular proteins are investigated. Otherwise, harvest the cells with a cell scraper.
- Lyse the cells in an SDS containing buffer. Omit addition of β-mercaptoethanol at this stage since it interferes with the BCA method (see next step) but add it prior to separation of the protein.
- Determine the total protein concentration of each sample e.g. by the Bicinchoninic Acid (BCA) method as described by Smith et al. [41]. This is particularly important to allow semiquantitative assessment of the Western blots.

> • Proceed with protein separation on an SDS-polyacrylamide gel, immobilization by blotting to membrane and detection according to standard protocols.

Exemplary results: see Figure 12

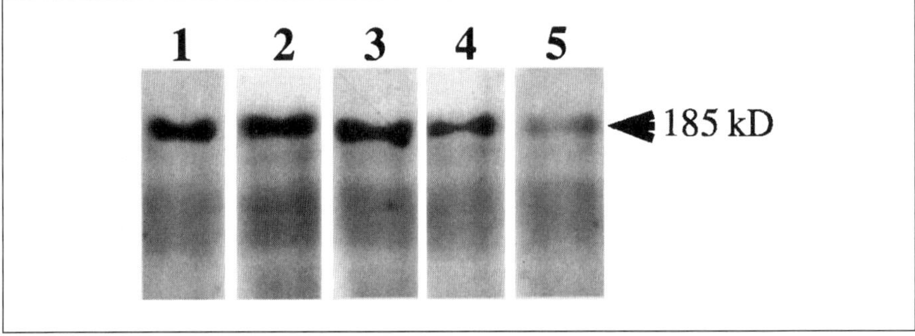

Figure 12: Western blot analysis of SK-Br-3 mammary carcinoma cells. Cells were treated with no oligo (lane 1) or 2 µM of either randomized control 14 mer (lane 2 = 24 h; lane 3 = 48h) or with anti-c-*erb*B-2 oligo (lane 4 = 24 h; lane 5 = 48 h) as described in [2]. Cells were harvested and separated electrophoretically according to standard techniques. The c-ErbB2 protein (p185) was detected with an anti-c-ErbB2 specific antibody. Antisense treatment leads to a reduction at 24 h and an almost complete disappearance of the c-ErbB-2 band at 185 kd at 48 h. In control oligo treated cells the signal of p185 is not reduced as compared to untreated cells.

4.2   Measurement of the Targeted Messenger RNA

Two mechanisms for the mediation of antisense inhibition have been described: a direct hindrance of translation due to a steric block and cleavage of the RNA strand by RNase H (see chapter 1, Introduction). RNase H plays an important role only in some cells or tissues for the mediation of the antisense effect and, additionally, its activation depends on the backbone chemistry of the oligo (see chapter 1). Theoretically, cleavage of the mRNA by RNaseH should enhance the antisense effect since the oligo may bind to several molecules successively. Northern blotting and reverse transcription assays give indirect evidence of this mechanism by semiquantitative comparison of antisense- and control oligo treated versus untreated cells. If the specific signal for the targeted mRNA goes down, an RNase dependent mechanism is likely. Upregulation due to a positive feedback or unaffected levels of the specific mRNA suggest steric hindrance as the predominant mechanism. Generally, these data may only be interpreted as antisense effects if downregulation of the correspondent protein has been documented in parallel experiments.

Protocol 10: Set-up of Antisense Experiments for Northern Blot Detection
Detailed protocols for the complex Northern blot procedure cannot be

outlined here. Please refer to standard lab manuals. The following points should be taken into account:

- To obtain sufficient amounts of mRNA, cells should be grown in 75 ml or 250 ml cell culture flasks. Since the reduction of the mRNA may precede an apparent reduction of the protein product, earlier time points should be chosen as compared to protein detection. To start with, 0 h, 2 h, 6 h, 12 h, 24 h, 48 h may be appropriate for most cell types and targets.
- Reduce the amount of medium during incubation to save oligo. Add the oligos at an effective but non-toxic concentration that has been determined by functional and toxicity assays (see 2.1 and 3)).
- To stop the incubation lyse the cells by adding an SDS containing solution and extract RNA by the acid phenol method [5] or by cesium chloride centrifugation.
- Separate 15–20μg of total RNA on a horizontal 1 % agarose/formaldehyde gel and then blot onto a nylon filter.
- For hybridization and detection, non-radioactive methods using fluorescein or digoxigenin labelled probes are recommended.

5   Preparation of Cells for Tumor Injection

Antisense compounds that have been successfully tested as anti-cancer agents in cell culture should next be studied in animals bearing a tumor. Tumor cells of human origin may be grown in immunodeficient animals. Two breeds of mice with inoculated tumors that have been successfully treated with antisense agents are: nu/nu athymic nude mice [30] [38] and SCID (severe combined immunodeficiency) mice (see chapter 12, Oncology and chapter 13, Hematology and [7] [31] [14]).

Protocol 11: Preparation of Cells for Tumor Injection

Special material and equipment, animals:
- 1 ml tuberculin syringes
- Cannulae for injection, 27G and 20G
- immunodeficient animals, e. g. nude mice or SCID mice

Depending on cell size and proliferation rate, 2×10^6 to 1×10^7 cells should be injected. Too low a number of cells may result either in no tumor growth or in delayed development of tumors. Too high a number of cells requires too big an injection volume and, in addition, may cause multiple adherent tumors being difficult to measure.

- Grow cells with a low number of passages in flasks.
- Treat adequately with antibiotics and mycoplasma removal agents, if necessary. Check daily for morphology and for the absence of infection.

- Do not allow the cells to become confluent. At a maximum of 80 % confluency, cells should be transferred or injected.
- Prior to injection, cells from different flasks may be pooled. Spin down and wash once in appropriate culture medium.
- Resuspend in a defined volume of culture medium (e. g. 10 ml) and determine the cell number in a hemocytometer.
- Spin the cells down, discard the supernatant and redilute to 2/3 of the volume required for injection of all animals. Depending on the cell size and the individual efficiency to form tumors, 1×10^6 to 1×10^7 cells should be injected in a volume of 200 to 300 µl.
- Mix the suspension gently each time before transferring each aliquot into a microfuge tube. Add medium to the final volume.[2]
- Draw the cell suspension into an insulin type syringe using a 20G needle. Replace it with a 27G needle for injection to cause minimal skin damage.
- Cells should be injected within the following 30 to 60 min, e. g. on the upper back of the mice (see Figure 13). Make sure that the volume is injected in one go into one subcutaneous compartment to avoid growth of multiple tumors.
- Check the animals daily for palpable tumors. Measure tumors in all three dimensions as recommended by Morrison [26].
- Antisense experiments may be performed in different ways. The animals may be treated directly after inoculation or after the tumor is palpable. Alternatively, one may treat cells already in culture with the specific antisense oligo for a defined period prior to injection.

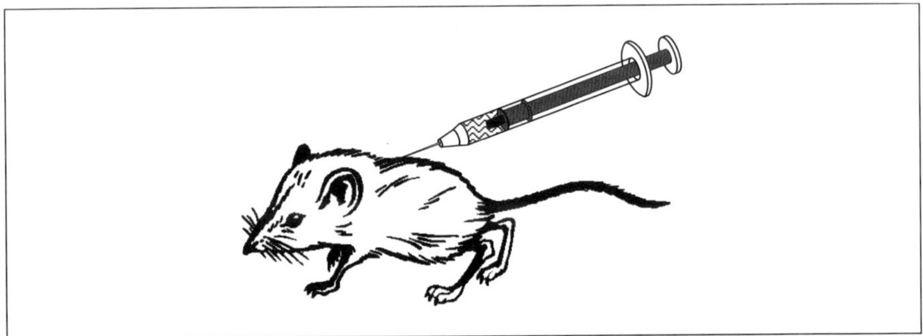

Figure 13: Injection of tumor cells into immunodeficient mice. Cells are suspended in cell culture medium and injected subcutaneously into the back of the animal, ideally, between the scapulae.

---

[2] Take into account that a portion of the cell suspension remains in the syringe. Ideally, the rubber stamp of the syringe should contain a protuberant pin to squeeze out a maximal volume (e.g. 1 ml syringe by Transcoject, Germany). The dead volume may be up to 100 µl. It has to be determined by weighing a syringe when it is empty and when fluid has been drawn up and squeezed out. Increase total cell number and volume accordingly.

6   Mycoplasmal Infection – Importance for Antisense Experiments

Mycoplasmatic organisms are the smallest prokaryotic organisms known. Due to their size of 0.2 to 2 µM and due to their highly variable shape they pass filters with a 0.22 µM pore size. They differ from other prokaryotes by their lack of a cell wall. They are the most common cause for contamination of cultured cells. Mycoplasms often remain undetected because they are invisible under the light microscope and their effects on cells may not be dramatic during routine culturing. However, antisense experiments are affected at least in two ways. First, cell growth of infected cells is inhibited [22]. Interpretation of proliferation assays may thus become impossible. On the other hand, the oligos may not reach the targeted cells since they are taken up and degraded by mycoplasma, which are located in the supernatant and at the cell surface [12]. If tumor cells are grown for injection into immunodeficient mice, infection with such organisms may cause illness of the animal and prevent growth of the tumor. Thus, detection and treatment of cells is mandatory to obtain reliable results. Morphological changes may be discrete, although infected cells may look somewhat fraid. Another indirect sign is accelerated changes in pH, either decreasing (e.g. M. Laidlawi, M. Hyorhinus) or increasing (M. Arginini, M. Orale) before the cells reach confluency. Different detection methods have been developed for mycoplasma in cell culture, e.g. Mycotect by Gibco/BRL or bis-benzimide DNA stain 33258 by Hoechst. More recently PCR techniques have been introduced, where homologous mycoplasmatic ribosome sequences serve as templates for detection of different subspecies (e.g. Mycoplasma Primer Set by Stratagene).

If cells are infected, treat them before starting the antisense experiments to allow for sufficient recovery. Treat cells over three passages. Over the past years many mycoplasms have become resistant to minocycline, thus today quinolone based antibiotics like ciprofloxacin (Bayer) or MRA (Mycoplasma Removal Agent by ICN-Flow) are most suitable [27].

7   Conclusions

This chapter provides an outline of the most common experiments used in antisense research. Careful selection of methods and experimental conditions are of fundamental importance for successful application of antisense oligonucleotides. Toxicological studies are needed to assess the cytotoxic potential of oligos, pendant groups or carriers. Since not all cells take up polyanions at sufficient amounts, uptake studies should be carried out with any new cell line under investigation. For any functional assay it is important to choose the appropriate time frame and to define clear endpoints of the experiment. Any experiment must be carried out with specific antisense and control oligos to rule out non-specific effects. Further validation of the antisense effect has to be provided by measuring the protein level. The experience in our labs and of others has shown that careful design of the experimental setup is as important as selection of the best target sequence, balanced hybridization properties of the oligo and vigorously purified compounds.

References

1. BRYSCH, W., O. D. CREUTZFELDT, K. LUNO, R. SCHLINGENSIEPEN, and K. H. SCHLINGENSIEPEN: Regional and temporal expression of sodium channel messenger RNAs in the rat brain during development. Experimental Brain Research 86(3) (1991) 562.

2. BRYSCH, W., E. MAGAL, J. C. LOUIS, M. KUNST, I. KLINGER, R. SCHLINGENSIEPEN, and K. H. SCHLINGENSIEPEN: Inhibition of p185c-erbB-2 proto-oncogene expression by antisense oligodeoxynucleotides down-regulates p185-associated tyrosine-kinase activity and strongly inhibits mammary tumor-cell proliferation. Cancer Gene Therapy 1(2) (1994) 99.

3. CAZENAVE, C., C. A. STEIN, N. LOREAU, N. T. THUONG, L. M. NECKERS, C. SUBASINGHE, C. HELENE, et al.: Comparative inhibition of rabbit globin mRNA translation by modified antisense oligodeoxynucleotides. Nucl. Acids Res. 17(11) (1989) 4255.

4. CHIN, D. J., G. A. GREEN, G. ZON, F. J. SZOKA, and R. M. STRAUBINGER: Rapid nuclear accumulation of injected oligodeoxyribonucleotides. New Biol. 2(12) (1990) 1091.

5. CHOMCZYNSKI, P. and N. SACCHI: Single-step method of RNA isolation by acid guanidinium thiocyanate-phenol-chloroform extraction. Anal. Biochem. 162(1) (1987) 156.

6. COLLINS, J. F., P. HERMAN, C. SCHUCH, and G. J. BAGBY: c-myc antisense oligonucleotides inhibit the colony-forming capacity of Colo 320 colonic carcinoma cells. J. Clin. Invest. 89(5) (1992) 1523.

7. COTTER, F. E., P. JOHNSON, P. HALL, C. POCOCK, M. N. AL, J. K. COWELL, and G. MORGAN: Antisense oligonucleotides suppress B-cell lymphoma growth in a SCID-hu mouse model. Oncogene 9(10) (1994) 3049.

8. CROOKE, R. M.: In vitro toxicology and pharmacokinetics of antisense oligonucleotides. Anticancer Drug Des. 6(6) (1991) 609.

9. CROOKE, R. M.: In vitro and in vivo toxicology of first-generation analogs. In: S. T. Crooke and B. Lebleu, eds: Antisense Research and Applications. CRC Press Boca Raton (1993) 471.

10. DAVIDSON, P., S. LIU, and M. KARASEK: Limitations in the use of 3H-thymidine incorporation into DNA as indicator of epidermal keratinicyte proliferation *in vitro*. Cell and Tissue Kinetics 12 (1979) 605.

11. GERDES, W., W. BRYSCH, K. H. SCHLINGENSIEPEN, and W. SEIFERT: Antisense bFGF oligodeoxynucleotides inhibit DNA synthesis of rat astrocytes. Neuroreport 3(1) (1992) 43.

12. GESELOWITZ, D. A., L. D. OLSON, and L. M. NECKERS: Incorporation of radiophosphorus from labelled oligodeoxynucleotides into RNA of mycoplasma in cell cultures. Antisense Res. Dev. 2(1) (1992) 41.

13. GEWIRTZ, A. M. and B. CALABRETTA: A c-myb antisense oligodeoxynucleotide inhibits normal human hematopoiesis in vitro. Science 242(4883) (1988) 1303.

14. HIGGINS, K. A., J. R. PEREZ, T. A. COLEMAN, K. DORSHKIND, W. A. McCOMAS, U. M. SARMIENTO, C. A. ROSEN, et al.: Antisense inhibition of the p65 subunit of NF-

kappa B blocks tumorigenicity and causes tumor regression. Proc. Natl. Acad. Sci. USA 90(21) (1993) 9901.

15. ICHIKAWA, M., K. MURAMOTO, K. KOBAYASHI, M. KAWAHARA, and Y. KURODA: Formation and maturation of synapses in primary cultures of rat cerebral cortical cells: an electron microscopic study. Neurosci. Res. 16(2) (1993) 95.

16. IVERSEN, P. L., S. ZHU, A. MEYER, and G. ZON: Cellular uptake and subcellular distribution of phosphorothioate oligonucleotides into cultured cells. Antisense Res. Dev. 2(3) (1992) 211.

17. KEELY, P. J., A. M. FONG, M. M. ZUTTER, and S. A. SANTORO: Alteration of collagen-dependent adhesion, motility, and morphogenesis by the expression of antisense alpha 2 integrin mRNA in mammary cells. Journal of Cell Science (1995).

18. LEONETTI, J. P., G. DEGOLS, and B. LEBLEU: Biological activity of oligonucleotide-poly(L-lysine) conjugates: mechanism of cell uptake. Bioconjug. Chem. 1(2) (1990) 149.

19. LEONETTI, J. P., N. MECHTI, G. DEGOLS, C. GAGNOR, and B. LEBLEU: Intracellular distribution of microinjected antisense oligonucleotides. Proc. Natl. Acad. Sci. USA 88(7) (1991) 2702.

20. LOKE, S. L., C. A. STEIN, X. H. ZHANG, K. MORI, M. NAKANISHI, C. SUBASINGHE, J. S. COHEN, et al.: Characterization of oligonucleotide transport into living cells. Proc. Natl. Acad. Sci. USA 86(10) (1989) 3474.

21. MATSON, S. and A. M. KRIEG: Nonspecific suppression of [3H]thymidine incorporation by "control" oligonucleotides. Antisense Res. Dev. 2(4) (1992) 325.

22. McGARRITY, G. J., D. M. PHILLIPS, and A. B. VAIDYA: Mycoplasmal infection of lymphocyte cell cultures: infection with M. salivarium. In Vitro 16(4) (1980) 346.

23. McPHERSON, I. and L. MONTAGNIER: Agar suspension culture for the selective assay of cells transformed by polyoma virus. Virology 23 (1964) 291.

24. MILLER, P. S. , K. B. McPARLand, K. JAYARAMAN, and P. O. Ts'o: Biochemical and biological effects of nonionic nucleic acid methylphosphonates. Biochemistry 20(7) (1981) 1874.

25. MORRISON, R. S. , S. GIORDANO, F. YAMAGUCHI, S. HENDRICKSON, M. S. BERGER, and K. PALCZEWSKI: Basic fibroblast growth factor expression is required for clonogenic growth of human glioma cells. Journal of Neuroscience Research 34(5) (1993) 502.

26. MORRISON, S. D.: In vivo estimation of size of experimental tumors. Journal of the National Cancer Institute 71(2) (1983) 407.

27. MOWLES, J. M.: The use of ciprofloxacin for the elimination of mycoplasmas from naturally infected cell lines. Cytotechnology 1 (1988) 355.

28. PATINKIN, D., L. E. LEV, H. ZAKUT, F. ECKSTEIN, and H. SOREQ: Antisense inhibition of butyrylcholinesterase gene expression predicts adverse hematopoietic consequences to cholinesterase inhibitors. Cellul. & Molec. Neurobiol. 14(5) (1994) 459.

29. PAULUS, W., I. BAUR, C. HUETTNER, B. SCHMAUSSER, W. ROGGENDORF, K. H. SCHLINGENSIEPEN, and W. BRYSCH: Effects of transforming growth factor-beta 1 on collagen synthesis, integrin expression, adhesion and invasion of glioma cells. Journal of Neuropathology & Experimental Neurology 54(2) (1995) 236.

30. PERLAKY, L., Y. SAIJO, R. K. BUSCH, C. F. BENNETT, C. K. MIRABELLI, S. T. CROOKE, and H. BUSCH: Growth inhibition of human tumor cell lines by antisense oligonucleotides designed to inhibit p120 expression. Anticancer Drug Des. 8(1) (1993) 3.

31. RATAJCZAK, M. Z., J. A. KANT, S. M. LUGER, N. HIJIYA, J. ZHANG, G. ZON, and A. M.

GEWIRTZ: In vivo treatment of human leukemia in a scid mouse model with c-myb antisense oligodeoxynucleotides. Proc. Natl. Acad. Sci. USA 89(24) (1992) 11823.

32. REGELSON, W.: The growth-regulating activity of polyanions: a theoretical discussion of their place in the intercellular environment and their role in cell physiology. [Review] Advances in Cancer Research 11(223) (1968) 223.

33. RODRIGUEZ, F. J., B. GEIGER, D. SALOMON, and Z. e. A. BEN: Suppression of vinculin expression by antisense transfection confers changes in cell morphology, motility, and anchorage-dependent growth of 3T3 cells. J. Cell Biol. 122(6) (1993) 1285.

34. SAXON, M., I. SCHIEREN, L. M. ZHANG, J. L. TONKINSON, and C. A. STEIN: Stimulation of calcium influx in HL60 cells by cholesteryl-modified homopolymer oligodeoxynucleotides. Antisense Res. Dev. 2(3) (1992) 243.

35. SCHLINGENSIEPEN, K. H., R. SCHLINGENSIEPEN, M. KUNST, I. KLINGER, W. GERDES, W. SEIFERT, and W. BRYSCH: Opposite functions of jun-B and c-jun in growth regulation and neuronal differentiation. Developmental Genetics 14(4) (1993) 305.

36. SCHLINGENSIEPEN, K. H., F. WOLLNIK, M. KUNST, R. SCHLINGENSIEPEN, T. HERDEGEN, and W. BRYSCH: The role of jun transcription factor expression and phosphorylation in neuronal differentiation, neuronal cell death, and plastic adaptations in vivo. Cellul. & Molec. Neurobiol. 14(5) (1994) 487.

37. SCHMIDT, R., S. ROTHER, K. H. SCHLINGENSIEPEN, and W. BRYSCH: Neuronal plasticity depending on a glycoprotein synthesized in goldfish leptomeninx. Prog. Brain Res. 91(7) (1992) 7.

38. SCHWAB, G., C. CHAVANY, I. DUROUX, G. GOUBIN, J. LEBEAU, C. HELENE, and B. T. SAISON: Antisense oligonucleotides adsorbed to polyalkylcyanoacrylate nanoparticles specifically inhibit mutated Ha-ras-mediated cell proliferation and tumorigenicity in nude mice. Proc. Natl. Acad. Sci. USA 91(22) (1994) 10460.

39. SHUMAKER, D. K., M. D. SKLAR, E. V. PROCHOWNIK, and J. VARANI: Increased cell-substrate adhesion accompanies conditional reversion to the normal phenotype in ras-oncogene-transformed NIH-3T3 cells. Experimental Cell Research 214(2) (1994) 440.

40. SIMONS, M., E. R. EDELMAN, J. L. DeKEYSER, R. LANGER, and R. D. ROSENBERG: Antisense c-myb oligonucleotides inhibit intimal arterial smooth muscle cell accumulation in vivo. Nature 359(6390) (1992) 67.

41. SMITH, P. K., R. I. KROHN, G. T. HERMANSON, A. K. MALLIA, F. H. GARTNER, M. D. PROVENZANO, E. K. FUJIMOTO, et al.: Measurement of protein using bicinchoninic acid [published erratum appears in Anal Biochem 1987 May 15;163(1):279] Anal. Biochem. 150(1) (1985) 76.

42. STEIN, C. A. and Y. C. CHENG: Antisense oligonucleotides as therapeutic agents – is the bullet really magical?. [Review] Science 261(5124) (1993) 1004.

43. TESTA, N. G.: Clonal assays for hematopoietic or lymphoid cells in vitro. In: C. S. Potton and J.H. Hendry, eds: Cell clones: Manual of mammalian cell techniques. Churchill Livingstone New York (1985) 27.

44. UHLMANN, E. and A. PEYMAN: Antisense oligonucleotides: A new therapeutic principle. Chemical Reviews 90(4) (1990) 544.

45. VLASSOV, V. V., L. A. BALAKIREVA, and L. A. YAKUBOV: Transport of oligonucleotides across natural and model membranes. Biochim. Biophys. Acta 1197(2) (1994) 95.

46. YAKUBOV, L. A., E. A. DEEVA, V. F. ZARYTOVA, E. M. IVANOVA, A. S. RYTE, L. V. YURCHENKO, and V. V. VLASSOV: Mechanism of oligonucleotide uptake by cells: involvement of specific receptors? Proc. Natl. Acad. Sci. USA 86(17) (1989) 6454.

7 Immunology

Immunological Applications of Antisense Oligonucleotides

Piotr Jachimczak and Ulrich Bogdahn
Neurologische Universitätsklinik im Bezirkskrankenhaus, Regensburg,
Germany

1   Introduction

The knowledge of immunological processes involved in various physiological and pathological conditions has been rapidly increasing over the past few years. Investigation of immune cell function and of the signaling action of various cytokine families, including hematopoietins, the interferons, tumor necrosis factor (TNF)-related molecules, immunoglobulin superfamily members and the chemokines, are essential for the understanding of pathological disorders like autoimmune disease, infectious disease, cancer, allergy, inflammation or allograft rejection [38]. Immunomodulation offers new perspectives for new strategies of these disorders. Neutralizing antibodies, soluble receptors or inactive analogs of cytokines have been proposed to interfere with the effector molecules alluded to above (for review see [19]). Antisense offers a different approach to both investigation and potential therapy [17]. Instead of interfering with the effector molecule, antisense is designed to inhibit the generation of the targeted molecule (see Figure 1). Any gene involved in immunological processes and which has been sequenced may be targeted. Theoretically, even functionally or structurally highly related genes (e.g. different interleukins) may be inhibited separately (see chapter 1, Introduction and chapter 6, Cell Culture Protocols). The use of antibodies for investigation or therapy is limited in this field, since cytokines may act locally or even intracellularly through autocrine or paracrine mechanisms. Since antisense oligos interfere intracellularly with the generation of proteins they may target an immunomodulator independently from its destination and mode of action.

In this chapter, we review recent work in the field of immunology and provide some currently used experimental methods employing antisense oligonucleotides (oligos).

2   Review

2.1   Modulation of Cytokine Expression by Antisense Oligonucleotides

Cytokines are a family of multifunctional proteins involved in the regulation of various cellular processes including cell proliferation, differentiation and metabolism [38]. Here, we focus on the antisense approach to study the physiological role of cytokines within the immune system and their effects in disease. Cytokines express a wide range of biological effects by binding to specific, high-affinity receptors located on target cells. They may activate extra- or intracellular ligands, therefore it is not always possible to determine their function by use of neutralizing antibodies, as these do not have access to targeted intracellular proteins. Oligos complementary to cytokine-specific-mRNA may selectively inhibit protein synthesis, and therefore provide an excellent tool for studying their effects *in vitro*

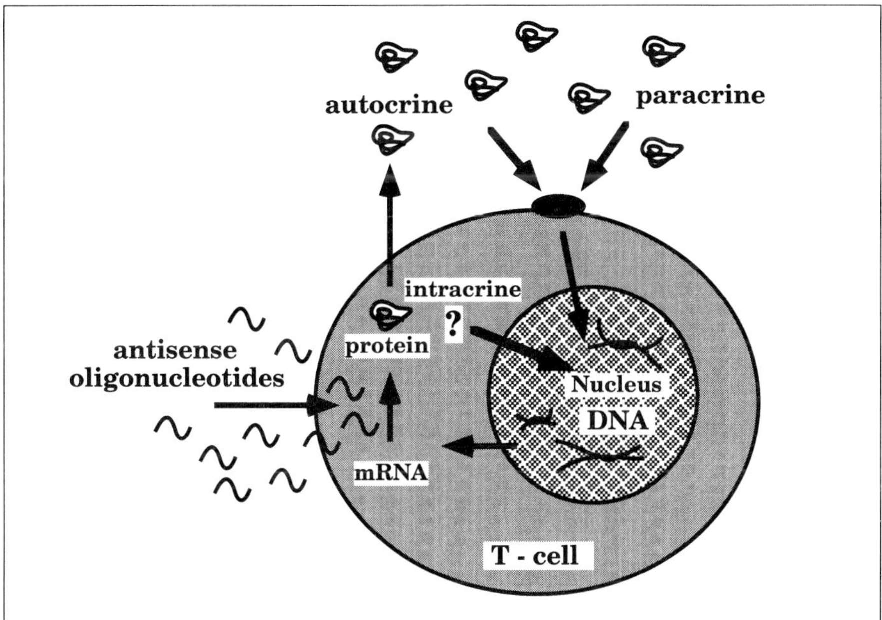

Figure 1: Cytokine-mediated growth modulatory loops in human T-Lymphocytes

and *in vivo* (see Figure 1, [29]). The effects of antisense oligos on the expression and function of various cytokines has been reported by numerous investigators (see Table 1).

Boeve et al. investigated the effects of 12 unmodified antisense oligos (ODN) (15–21 mer) targeted to interferon gamma (IFN-γ) in human T-lymphocyte and neutrophil cultures, stimulated by various mitogens. To obtain inhibitory effects upon IFN-γ production (up to 60 %), the minimum antisense concentrations required were 150 µM. Repeated supplementation of the antisense oligos, to a final concentration of 150–200 µM, were more effective than adding the same concentration in a single dose. Optimal inhibition of IFN-γ may be achieved, when oligos are added to the culture prior to or during the T-lymphocyte activation process. Specific antisense ODN reduce IFN-γ production up to 50 % compared to controls as measured by ELISA. Native cells do not produce detectable amounts of IFN-γ und thus, do not respond to antisense treatment. Addition of serum (FCS) to the cultures abrogated antisense-mediated inhibition of IFN-γ production, an effect, which is associated with the enzymatic cleavage of unmodified phosphate backbone oligos by nucleases [7]. To overcome this problem, various analogs of oligos have been synthesized, which have been found to be more resistant against nucleases and thus, more efficient inhibitors of gene expression [3]. The most widely used modification of oligos is the substitution of a non-bridging oxygen in each internucleotide

Table 1: Effects of Antisense Oligonucleotides on Cytokines

Author	Target	Antisense Oligonucleotides	Effects
Abken et al.	hu IL-1α, IL-6, TNFα and TNFβ, translation initiation region	ODN, 15-mer, (10-60μM, f.c.,) Control: sense ODN Cell uptake: ^{32}P labeled sense ODN	up to 50% (IL-1α-ODN) inhibition of human B-cell clone proliferation (cell counting)
Acha-Orbea et al.	hu and mu perforin, translation initiation region	ODN, 18-mer (human) and 16-mer (murine), (up to 40μM,f.c.,) Control: sense ODN, Cell uptake: ^{32}P-labeled hu ODN	inhibition of hu-PBL (up to 62%) and mu Lymphocyte (up to 69%)-mediated cytotoxicity to P815 cells (^{52}Cr-release assay), inhibition of perforin expression up to 65% (Western blot)
Benbernou et al.	mu IL-4 translation initiation region	ODN and S-ODN, 15-20-mer (up to 20μM, f.c.,) Control: sense ODN	up to 36% / 58% reduction of IgE / IgG2a production by rat spleen cells
Bennett et al.	hu ICAM-1, hu VCAM-1 and E-Selectin translation initiation region and others	S-ODN/ 18-21-mer/ up to 100nM, f.c., S-ODN disolved in DOTMA/DOPE Control: sense and random ODN	inhibition of adhesion molecule expression in HUVEC and HL-60 cells (Northern blot, ELISA, Flow cytometry, Cell adhesion assays)
Birchenall-Roberts et al.	mu CSF-1, translation initiation region	ODN / 15-mer /1-10μM, f.c., Control: sense and random ODN Cell uptake: ^{32}P-labeled-ODN	up to 75% inhibition of monocyte cell line (^{3}H-TdR inc.), inhibition of CSF-1 exprestion (immunofluorescence, bio-assay)
Boeve et al.	hu IFNγ,translation initiation region splice junction sites, a coding region, 5`-and 3`-untranslated region	ODN, 15-21-mer, (50-300μM, f.c.,) Control: random ODN,	up to 60-75%inhibition of hu-IFNγ production in lymphocyte and PBMC cultures (ELISA)
Fonagy et al.	hu P120, splice junction site	ODN, 15-mer, (up to 400μM,f.c.,) Control: random ODN	up to 90% inhibition of hu-PBL proliferation (^{3}H-TdR inc.), inhibition of P120 expression (Western blot, Northern blot)
Fujiwara et al.	hu-IL-1β, translation initiation region	ODN, 18-mer, (up to 20μM,f.c.,) Control: sense ODN	up to 50% inhibition of allogenic hu-LAK cytotoxicity (^{51}Cr-release assay), inhibition of IL-1β production (Western blot)
Gewirtz et al.	hu c-myb, coding region	ODN, 18-mer, (up to 40μg /ml, f.c.,) Control: sense ODN	up to 88% inhibition of T-cell proliferation (^{3}H-TdR inc.), inhibition of c-myb expression (Northern blot, immunofluorescence).

Gewirtz et al.	hu c-myb, coding region	ODN, 18-mer, (up to 40µg /ml, f.c.,) Control: sense ODN	up to 88% inhibition of T-cell proliferation (^{3}H-TdR inc.), inhibition of c-myb expression (Northern blot, immunofluorescence).
Harel-Bellan et al.	mu IL-2 and mu-IL-4, translation initiation region	ODN, 15-mer, (up to 12µg /ml, f.c.,) Control: sense ODN	up to 100% inhibition of T_1 and T_2 cell proliferation (^{3}H-TdR inc.), inhibition of Il-2 and IL-4 expression (Northern blot)
Harel-Bellan et al.	hu c-myc, 5`-second exoon	ODN, 15-mer, (up to 50µM, f.c.,) Control: sense ODN Cell uptake: ^{32}P-labeled ODN	up to 100% inhibition of T-cell prroliferation (^{3}H-TdR inc.), inhibition of protein expression (two-dimensional analysis)
Haruna et al.	mu IL-4, 5`-17-36 nucleot de	S-ODN, 20-mer, (up to 20µg /ml, f.c.,) Control: not done	up to 90% inhibition of IL-4 production in murine B-cells (Bioassay)
Hatzfeld et al.	hu TGF-β1, RB1 and p53 translation initiation region	S-ODN, 21-mer, (up to 100µM, f.c.,) Control: sense and random ODN	up to 1.5-4.5-fold increase in the colonies formation from CD34$^+$ cells (TGF-β1-AS and RB1 AS)
Heikkila et al.	hu c-myc, translation initiation region	ODN, 15-mer, (up to 30µM, f.c.,) Control: sense and random ODN	above 80% inhibition of T-lymphocyte proliferation (^{3}H-TdR inc.), inhibition of c-myc expression (immunocytochemistry, Western blot)
Hikida et al.	mu IL-4, 5`-region (17-36 nucleotide)	S-ODN, 20-mer, (up to 10µM, f.c.,) Control: not done	up to 90% inhhibition of murine Th$_2$-clone (^{3}H-TdR inc.,), inhibition of mRNA expression (Northern blot), inhibition of protein expression (up to 90%)
Ikizawa et al.	hu IL-4-R, translation initiation region	S-ODN, 20-24-mer, (up to 5µM, f.c.,) Control: not done	up to 70% inhibition of IL-4-R-expression on human B-cells, inhibition of IgE production, (RT-PCR, Immunofluorescence)
Jachimczak et al.	hu TGF-β1, translation initiation region	S-ODN, 14-mer, (up to 10µM, f.c.,) Control: random ODN Cellular uptake: BrdU-labeled ODN	up to 50 stimulation of human PBMC proliferation (^{3}H-TdR inc., cell counting), inhibition of TGF-β production (ELISA)
Jun et al.	hu TGF-β1, translation initiation region	S-ODN, 25-mer, (200µg /ml, f.c.,) Control: sense ODN Cellular uptake: 5`-FITC-labeled ODN,	increase IFNγ-induced NO production by peritoneal macrophages, inhibition of TGF-β expression (Northern blot)

Table 1: Effects of Antisense Oligonucleotides on Cytokines (cont.)

Author	Target	Antisense Oligonucleotides	Effects
Kato et al.	hu IL-2, hu c-myc, hu TfR translation initiation region	ODN, 15-mer, (up to 100µM, f.c.,) Control: sense ODN	up to 80% inhibition of PBMC proliferation (^{3}H-TdR inc.), decrease TfR-expression (FACS, Northern blot), G_0/G_1 arrest of PBMC.
Lapidot-Lifson et al.	mu IL-6, 2nd exon	ODN, 15-mer, (up to 20µM, f.c.,) Control: sense ODN	up to 85% inhibition of meyloma cell line proliferation (^{3}H-TdR inc.,),
Levy et al.	hu-IL-6, 2nd exon	ODN, 15-mer, (up to 20µM, f.c.,) Control: sense ODN	up to 95% inhibition of human myeloma cell line proliferation (^{3}H-TdR inc.,),
Louie et al.	mu-IL-4, translation initiation region	ODN, 15-25-mer, (20µM, f.c.,) Control: random ODN 5`-region of cDNA	up to 60% inhibition of B-cell lymphoma clone (CH12.LX) (^{3}H-TdR inc.), up to 50% inhibition of mu-IL-4 concentration (ELISA),
Methia et al.	hu c-mpl, translation initiation region	ODN, 18-mer, (up to 70µg /ml, f.c.,) Control: sense and random ODN (Nucleotides: -3 to +15 and +690 to +708)	up to 81% inhibition of megacaryocytic colony formation by human CD^{34+}-cells, up to 70% inhibition of c-mpl-mRNA expression (RT-PCR)
Mojcik et al.	mu MCF env, translation initiation region	S-ODN, 20-mer, *in vivo*: 500µg i.p., *in vitro*: 0.1-1.0µM Control: random ODN, producing cells, increase MHC class II,	up to 6-10-fold increaese in DNA/RNA synthesis in murine spleen cells in vivo: increase in splenic B cells, and Ig
Morrison et al.	mu- and rat CD4, translation initiation region	S-ODN, 19-mer, (100µM, f.c.) Control: sense and random ODN	up to 45% reduction of CD4 expression on cell membrane (Immunofluorescence-FACS)
Skorski et al.	hu-N-ras, translation initiation region	ODN, 18-mer, (up to 80µg /ml, f.c.,) Control: sense ODN	inhibition of granulcyte, macrophage, ery- throid and megakaryocytic coolony formation, inhibition of N-ras-mRNA in CD34+ cells
Small et al.	hu STK-1 /translation initiation region and up-, and downstream	ODN / 18-mer / up to 100µg /ml, f.c., Control: sense and random ODN	up to 53% inhibition of colony formation by CD34$^+$ cells and T-cell depleted PBMC decrease STK-1-mRNA-expression (RT-PCR)
Sokolowski et al.	p50 and p65 subunit of NF-κB translation initiation region	S-ODN, 21-24-mer, (up to 20µM, f.c.,) Control: sense ODN (RT-PCR)	morphological changes of HL 60 leukemia cells, inhibition of p50 and p65 subunit expression

Tanaka et al.	mu-l-region of γ2b Ig heavy chain gene nucleotides 160 to 177	S-ODN, 18-mer, (up to 1μM, f.c.,) Control: sense and random ODN	inhibition of Ig secretion by murine B-cells, inhibition of Ig-mRNAs but induction of germ line transcript mRNA (γ2b),
Witsell et al.	mu TNF-α, translation initiation region	S-ODN, 17-mer, (up to 10μM, f.c.,) Control: sense ODN	stimulation of murine macrophage progenitor cells, decrease TGF-αRNA expression and TGF-α secretion,
Zheng et al.	T-cell receptor (V_α and V_β) translation initiatin region	ODN, 22-23-mer, (up to 50μM, f.c.,) Control: not done	inhibition of TCR expression (FACS) in T-cell Hybridomas and IL-2 production after antigen stimulation,
Zubiaga et al.	mu-IL-α, translation initiation region	ODN, 15-mer, (up to 9μM, f.c.,) Control: not done	up to 40% inhibition of mu-Th2 cell proliferation (^{3}H-TdR inc.,), inhibition of IL-1α-mRNA expression (Northern Blot),
Zucali et al.	hu-TNFα, hu-IL-6, translation initiation region	ODN, 18-mer, (up to 50μM, f.c.,) Control: sense ODN	inhibition of TNFα and IL-6 secretion in human monocytes (ELISA, Bioassay)

phosphate linkage by a sulfur atom, which results in a phosphorothioate oligos (S-ODN). They are stable to cleavage by nucleases and well solubilized in aqueous solutions. However, in high concentrations (above 10 µM, f.c.) they may produce nonspecific cytotoxic effects on lymphocytes, possibly due to an increased hybridization to mismatch sequences or interaction of sequence motifs with cellular proteins by the charged polyanionic backbone [22, 34, 51] and increased concentrations of toxic synthesis byproducts (see chapter 1).

A large number of investigators targeted T cell derived cytokine, interleukin 4 (IL-4) by antisense oligos. Murine B-cell lymphoma CH12.LX has been found to be stimulated in an autocrine fashion by IL-4. Application of 20-mer length, IL-4-S-ODN, targeted to the translation initiation side, resulted in an inhibition of cell proliferation and reduction of IL-4 production of up to 50 % compared to cells treated with random sequences. B-cell clones which do not produce IL-4, were not affected by addition of IL-4-S-ODN [31]. IL-4 also plays an important role in the development of immunoglobulin secretory B-cells [4]. Benbernou et al. inhibited IgE and IgG2a production (up to 36/58 % resp.) in rat spleen cells by 15-mer IL-4-S-ODN targeted to the sequences surrounding the AUG start codon. Interestingly, prolongation of these oligos up to a 20-mer, reduced the antisense effects on IgE and IgG2a secretion. This observation supported the hypothesis that the affinity of the oligos to its mRNA target not only depends on the chemistry of oligo (e.g. base composition and the chemical modification), but also on its length. Increase in affinity by increasing chain length is beneficial only up to a certain length, since too high an affinity allows binding to mismatched sequences, causing non-sequence-specific effects (see also chapter 1) [35, 43, 47].

CD4$^+$ T-lymphocytes are further divided into two functionally different subsets: Th1 and Th2. Upon mitogen or antigen stimulation, Th1 cells secrete and respond to IL-2 in an autocrine manner whereas the Th2 subpopulation releases IL-4. Harel-Bellan et al. targeted both cytokines in murine T-cell clones by unmodified 15-mer antisense oligos [13]. IL-2-ODN inhibited proliferation of Th1 clones (D11) and had no effects on Th2 clones (D10). Alternatively, IL-4-unmodified ODN blocked only proliferation of Th2 clones. In both systems, a maximum inhibitory effect of 90–100 % (^{3}H-TdR inc.) was achieved employing 5–10 µM oligo concentration. Antisense effects have been reversed by adding either exogenous IL-2 for Th1 clones or IL-4 for Th2 clones. Northern blot analysis of mRNA specific for the two cytokines showed a decrease in the steady-state level of the relevant cytokine mRNA, suggesting that the specific degradation of the mRNA is caused by RNase H-like enzymatic activity. Hikida et al. employed IL-4-S-ODN, to inhibit the same, D10 murine Th2 clone [20]. Here, dose-dependent inhibition of IL-4 secretion up to 90 % has been reached in the presence of more than 5 µg/ml of oligos. In most experiments, however,

5–10 % of IL-4 production remained unaffected in the presense of antisense oligos. The suppression of IL-4 secretion did not result in the accumulation of IL-4 in D10 cells, indicating that oligos inhibited the synthesis but not the secretion of IL-4. Moreover, RT-PCR of IL-4 clearly showed an inhibition of respective mRNA compared to controls [15].

Some additional cytokines seem to be responsible for the growth of CD4$^+$ Th2 cells, as demonstrated by Zubiaga et al. [51]. They targeted the translation initiation region of interleukin 1α (IL-1α) by S-ODN causing inhibition of the Th2-cell proliferation. Thus IL-1 acts as an additional autocrine factor controlling Th2 cell proliferation. Abken et al. investigated the role of endogenously secreted cytokines in human B-cell clones that proliferate under serum-free conditions [1]. All B-cell clones secreted IL-1α, interleukin 6 (IL-6) and tumor necrosis factor (TNF-α and TNF-β). Addition of 15-mer ODN directed to the 5'-end (including the start codons) of the corresponding mRNAs of IL-1α, IL-6 and TNF-α in various concentrations (10–60 µM, f.c.) resulted in growth arrest and cell death, when applied in combination. The addition of single-type antisense oligos reduced cell growth significantly, but the cells survived exhibiting prolonged generation times. This effect could be reversed by exogenous cytokines, which further demonstrates the importance of cytokines in autonomous growth of immortalized B-cells. To monitor uptake and persistence of oligos in lymphoid cells, ^{32}P-labelled oligos have been used. Maximum amounts of the oligos were incorporated within 24 h; SDS-PAGE analysis disclosed intracellular undegraded ODN up to 4 days [1].

Fujiwara and Grimm investigated the role of IL-1β in the generation of human Lymphokine-Activated Killer (LAK) cell activity [11]. They targeted IL-1β by unmodified 18 mer antisense oligos complementary to the translation initiation region. IL-1β-ODN reduced intracellular IL-1β production (Western blot) and inhibited in a dose-dependent fashion LAK cell activity against human Daudi Burkitt lymphoma cells. The maximal inhibition of LAK-cell cytotoxicity (^{52}Cr-release assay) was detected when the effector cells were pre-incubated for 4 h with 20µM IL-1β-ODN (max. up to 60 % at the Effector / Target ratio of 80:1). Application of an IL-1 receptor antagonist or specific neutralizing antibodies did not inhibit IL-1β-mediated LAK-activity as strongly as specific oligos, which indicated the existence of an intracellular autocrine IL-1 circuit.

Interleukin 6 (IL-6) is another immunorelevant cytokine for which an intracellular autocrine mechanism of action was demonstrated employing antisense technology. This cytokine is synthetized by various normal and transformed immune cells as well as by many other cell types including fibroblasts, endothelial cells, astroglia cells [38]. Levy et al. targeted the second exon of the IL-6 gene with 15-mer unmodified ODN in the human myeloma cell lines U 266 and RPMI 8226. IL-6-ODN inhibited in a dose-dependent fashion myeloma cell proliferation (^{3}H-Tdr incorporation). This effect was already significant at an oligo concentration as low as 2 µM (50 % inhibition compared to sense treated controls) and was maximal at 20 µM (85 % inhibition compared to sense treated controls). Antibodies to IL-6 did not alter the proliferation of these myeloma cells, which suggests an intracellular autocrine stimulatory loop [30].

Tumor necrosis factor and (TNF-α/β) are produced by activated macrophages and other cells and have a broad spectrum of biological actions on many different target cells, both immune and non-immune [38]. Witsell et al. investigated the role of autocrine secreted TNF-α by murine macrophage progenitors employing antisense oligos to the initiation region of the mRNA. After 20 h, GM-CSF-stimulated and TNF-α-S-ODN-treated cells (f.c. 5 µM) showed a decrease of TNF-α-mRNA expression (up to 50 %) compared to sense- and untreated cells. The concentration of TNF-α-protein demonstrated by bio-assay reached 75% inhibition compared to 38 % inhibition for sense-treated cells [49]. Zucali et al. targeted TNF-α in LPS-activated human monocytes with 18 mer unmodified ODN. They observed a dose-dependent inhibition of TNF-α secretion up to a maximum of 95 % (50 µM, f.c.) compared to sense treated controls as measured by ELISA and bioassay with L-929 cell line [52].

Autocrine production of cytokines has been suggested as one of the mechanisms responsible for the unregulated growth of hematopoietic cells. Birchenall-Roberts et al. targeted Colony-Stimulating Factor-1 (CSF-1) in murine growth-factor-independent monocyte cell line (FL-*ras*/*myc*) with unmodified 15-mer ODN. CSF- ODN (up to 10 µM, f.c.) inhibited FL-*ras*/*myc* cell proliferation as strongly as neutralizing antibodies to CSF-1 (up to 75 %, ^{3}H-thymidine incorporation). When combined, however, the treatments inhibited cell growth by 82 % and 95 % at 12 and 24 hours, respectively. The efficiency of antisense treatment was further investigated by immunofluorescence and bioassay with the CSF-sensitive BAC1.2F5 cell line. Cell uptake studies with ^{32}P-labelled ODN indicated that 1 to 2 % of the labelled oligo was incorporated into cells after 2 hours and that this level of incorporation remained nearly constant for 10 hours. However, after 10 hours a continuous increase in uptake of radioactivity was evident. This increase, as disclosed by gel-elecrophoresis of cell supernatans, was associated with the incorporation of oligo-degradation products by the cells [6].

Transforming growth factor β (TGF-β) represents a family of structurally related polypeptides with numerous cell-specific effects on proliferation, differentiation and metabolism. TGF-β is secreted by various cells including activated immune cells (T-cells, B-cells, neutrophils and macrophages). To investigate the role of endogenously secreted TGF-β in human peripheral blood mononuclear cells (PBMC), we have cultured the cells with 14 mer TGF-β_1-S-ODN. IL-2- or PHA- mediated activation of PBMC-proliferation was increased after TGF-β_1-S-ODN treatment. Long-term inhibition of endogenous TGF-β_1 production by PBMC did not significantly change the proliferative response to IL-2 compared to random ODN treated cells. The proliferation of IL-2-preactivated PBMC was not significantly altered by TGF-β_1-S-ODN. Antisense dependent growth modulation of PBMCs was associated with decrease of TGF-β_1-protein synthesis (up to 65 %, ELISA), without significant changes in TGF-β_1-mRNA expression. In our experiments we have investigated cellular uptake of antisense oligos employing BrdU- and FITC-labelled S-ODN. Both methods revealed a rapid cellular incorporation of oligos. We were able to detect FITC/BrdU-labelled S-ODN after 30 min. The oligos were mainly localized in the cytoplasm of all PBMCs, irrespective of their individual morphology, indicating no difference in uptake between cell subsets [22] [23]. Jun et al. investigated the role of autocrine secreted TGF-β upon macrophage nitric oxide (NO) production. The endogenous inhibition of TGF-β production by 25 mer TGF-β-S-ODN resulted in stimulation of NO synthesis by IFN-γ treated mouse peritoneal macrophages [25]. Maximal effects have been observed employing 20µg/ml (f.c.) of S-ODN. Higher oligo concentrations decreased IFN-γ-induced NO-synthesis, an effect which might be associated with non-specific cell toxicity. Biological response of activated macrophages to TGF-β-S-ODN was associated with a slight decrease in TGF-β-mRNA expression. The authors performed cellular uptake studies with FITC-labelled S-ODN, which detected antisense oligos after only 5 min of incubation. Hatzfeld et al. targeted the translation initiation region of the TGF-β-mRNA in CD34$^+$ human bone marrow cells. Addition of the TGF-β-S-ODN at concentrations of 5–8 µM resulted in a dose-dependent increase in the formation of hematopoietic (erythroid, granulacyte-monocyte and granulocyte) colonies (1.5 to 2-fold); higher oligo concentrations seemed to be toxic. In contrast, TGF-β-S-ODN had no effect on hematopoietic progenitors [16]. These results further support the hypothesis of autocrine negative control of various cell procesess by TGF-β in immune cells.

2.2 Inhibition of Immunoglobulin Synthesis by Antisense Oligonucleotides

Another interesting antisense strategy has been recently used to examine the role of germline γ2b transcript in Ig class switching [46]. Tanaka et al. targeted a sequence in the intervening I-region of the murine γ2b immunoglobulin heavy chain gene with unmodified (P-ODN) and phosphorothioate (S-ODN) antisense oligos.

Incubation of B-cells with ODN (0.1–33 µM, f.c.) had no effect on the production of IgG2b or any Ig isotype. However, cellular uptake studies with ^{32}P-labelled oligos clearly indicated that unmodified ODN were rapidly degradated in the culture medium (10 % FCS) and therefore did not enter the cell intact, as demonstrated by SDS-PAGE. On the other hand, S-ODN reduced Ig secretion by B cells stimulated with LPS (suppressed subclasses IgG2b, IgM and IgG3), or LPS plus IL 4 (suppressed subclasses IgG1 and IgE) up to 10–50-fold, but did not inhibit the expression of membrane subclasses of IgM, IgG1, IgG2b and IgG3-types. Maximum inhibition of Ig secretion was observed when S-ODN (1 μM, f.c.) were added in conjunction with mitogen-activation. Antisense-mediated inhibition of Ig secretion was not attributed to increased B-cell death, as measured by trypan blue exclusion test or by non-specific inhibition of total protein synthesis (SDS-PAGE analysis of ^{35}S-methionine incorporation). Antisense inhibited mRNA for secretory Ig, but induced the expression of the targeted γ2b germline transcript by 10–20-fold. Other experiments indicate that the observed increase in mRNA is associated with enhanced transcription rate rather than slowing of γ2b transcript turnover. This unexpected increase in mRNA was associated with an increase of DNA-synthesis in B cells. In order to recognize "pure" antisense effects of oligos, the choice of appropriate control sequences is a critical element. The authors employed sense oligos complementary to the antisense sequence and a random mixture of all four nucleotides with the same length and base composition as the antisense sequence. Because of oligo-instability in culture fluid (nucleases), and possible interactions of degradation products with cell metabolism, it seems important to include controls having the same base composition as antisense (e. g. random sequence oligos) [27] [32] [45].

2.3   Inhibition of Adhesion-Molecules, Receptors and other Structural Proteins by Antisense Oligonucleotides

Acha-Orbea et al. provided experimental evidence for a crucial role of perforin, a pore-forming protein, in lymphocyte-mediated cytotoxicity [2]. Cytotoxic activity of various effector cells including human PBMCs and murine spleen lymphocytes was measured by coincubation with ^{51}Cr-labelled P815 mastocytoma target cells. Il-2/anti-CD3+ Ab stimulated effector cells were treated either with 18-mer ODN complementary to a sequence starting at the ATG initiation codon of human perforin or 16-mer unmodified ODN targeted to the nucleotide +3 of the coding region of murine perforine mRNA. As controls, cells were treated with respective sense oligos. The inhibition of lymphocyte mediated cytotoxicity was dose-dependant with maximal effects (up to 5.7-fold) at oligo-concentrations of 40 μM. Decrease of cytotoxicity was associated with an 65 % decrease of perforin protein detected by immunocytochemistry and Western blot. To verify antisense uptake by PBMC, oligos were 5'-labelled with ^{32}P. After 4 hours of incubation, approx. 1 % of perforin- ODN were detected in the cellular pellet.

Migration of immune cells into injured tissue depends on cellular interactions of leukocytes with vascular endothelium, which are mediated by various endothelial-leukocyte adhesion molecules. Bennett et al. investigated a series of 18–21-mer S-ODN targeted to human intercellular adhesion molecule 1 (ICAM-1), vascular cell adhesion molecule 1 (VCAM-1) and E-selectin in human vascular endothelium (HUVEC) [5]. The authors tested a large number of antisense oligos by "gene-walking" procedures, in order to experimentally identify the optimum target sequence for each gene. They found that the optimal cleavage of ICAM-1-mRNA (up to 70 % compared to sense and random oligo controls) was obtained employing oligos that hybridize to the 3'-untranslated sequences. In contrast, oligos hybridizing to the 5'-untranslated region of ICAM-1 did not significantly reduce the level of ICAM-1 mRNA, indicating a mechanism other than RNase H mediated degradation. Interestingly, both 3'- and 5'-targeted antisense specifically inhibited the ICAM-1 protein expression as detected by ELISA and Flow cytometry. In the case of E-selectin expression, again oligos targeted to the 3'-untranslated region (up to 97.7 %) or the 5'-untranslated region (up to 93.4 %) exhibited the greatest activity. The expression of VCAM-1-mRNA was inhibited up to 78.7 % by oligos targeting the 3'-untranslated region. For all 3 genes antisense oligos hybridizing to the AUG translation initiation sites or the 5'-untranslated sequences had less dramatic effects on respective mRNA expression. In these cases, mechanisms of antisense action other than an RNase H mediated mechanism should be looked for. Steric blocking of all essential events associated with maturation of mRNA and its translation into protein could be a possible explanation of this phenomen [3] [43] [44] [47] [48] (see chapter 1). The most apparent finding of these studies is that oligos hybridizing to different regions of the target mRNA can differ markedly in their ability to inhibit gene expression. Targeted sequences usually include translation initiation sites (see Table). As the intracellular targets (mRNA, pre-mRNA or genomic DNA) are protein bound, many theoretically convenient sites for ODN targetting have to be further selected empirically [5]. The study by Bennet et al. demonstrates that the best way to identify optimal antisense sequences is to test as many different oligos as possible [5]. Several methods of oligo delivery have been employed *in vitro* including direct intracellular injection, electroporation, the incorporation of oligos into liposomes or the formation of complexes with positively charged lipids [35]. In the work of Bennett et al., cationic lipids (10 µg/ml solution of DOTMA/DOPE) were used in order to protect the oligos from degradation by serum or lysosomal nucleases and to enhance the binding and membrane penetration of the cationic complex/oligos into the cellular cytosol [5].

Morrison et al. targeted the first 19 nucleotides after the initiation codon of rat- and mouse- T-cell molecule, CD4+-mRNAs. Specific phosphorothioate oligo treatment (final concentration of 100 µM) reduced CD4 membrane surface levels by 45 % (FACS), compared to sense-controls. In order to enhanced cellular uptake of S-ODN the application was performed in water containing 1 % DMSO. CD4+-specific antisense treatment resulted in increased antigen-induced CD4+ T-cell proliferation, which suggested that even such high oligo-concentrations are non toxic. However, strict cell viability controls were lacking in this study [37].

Another approach to study expression of the T-cell surface receptor (TCR, α and β chains) by unmodified ODN was published by Zheng et al. [50]. They synthesized 22–23-mer oligos, which targeted the ATG codon of TCR in T-cell hybridomas (A1.1 and B1.1). Trypsin-treated T-hybridoma cells respond to TCR specific antisense oligos by reduction of T3 expression as assessed by FACS, and failed to produce lymphokine (IL-2) upon exposure to specific antigen. These effects were specific for the TCR, as could be disclosed by comparison of effects on other T-cell surface molecules. However, this study did not employ sense or random-controls to exclude non-sequence related oligo effects.

2.4   Inhibition of other Targets in the Immune System by Antisense Oligonucleotides

The expression of the nuclear proto-oncogene c-*myc* is closely associated with cell proliferation of various cells including human T-lymphocytes. To study its function 15-mer- unmodified ODN to the 5'end of the second exon of c-*myc*, including the translation start codon, were synthesized [14]. Corresponding sense oligos were used as controls. The cellular uptake of ^{32}P-labelled oligos was time-dependent and reached a plateau after 3 hours. The intracellular distribution, as assessed by autoradiograpy, showed submembraneous cytoplasmatic radioactivity; SDS-PAGE analysis of cell lysates showed a significant undegradated fraction of oligos in cytoplasm. When the cells were preincubated for 2 hours with the antisense oligos prior to activation by the mitogen, c-*myc* protein could not be detected on the gels, neither in immunoblotting nor on the autoradiography. T-cell-entry into S-phase, as measured by ^{3}H-TdR incorporation, was blocked by the antisense in a dose-dependant manner compared to sense treated cells. The specificity of the anti-proliferative effects of the unmodified ODN was further assessed by the reversal of the inhibition mediated by an excess of oligo sense strand. The inhibition of c-*myc* by antisense was more efficiently achieved on resting cells compared to IL-2 activated T-cells. The study of Burgess et al., however, has recently shown that a continuous stretch of 4 guanosine residues (see c-*myc*-ODN sequence, [14]) within antisense oligos is potentially responsible for the antiproliferative effects of such oligos, and that the subsequent inhibition of cell proliferation may be due to other non-antisense mechanisms [8].

The cellular proto-oncogene c-*mpl* expressed in CD34^{+} cells, encodes a protein with high homology to the conserved hematopoietin receptor superfamily. Oncogenic version of c-*mpl* (v-*mpl*) has been recently detected in the genome of the highly leukemogenic murine Myeloproliferative Leukemia Virus (MPLV). In order to investigate the role of c-*mpl* expression in the regulation of normal hematopoiesis, Methia et al. used 18-mer oligos targeted to nucleotides –3 to +15 and +690 to 708 of c-*mpl*-mRNA. Antisense treatment of CD34+ cells resulted in c-*mpl*-mRNA degradation (up to 70%), whereas no mRNA cleavage could be detected in sense and random treated controls. Decrease in c-*mpl* mRNA expression, was associated with a significant inhibition (range 54–81%) of *in vitro* megakaryocytic colony formation, which provides evidence that c-*mpl* is involved in human megakaryocytopoesis [33].

To assess the functional significance of N-*ras*, a GTP-binding protein involved in transduction of growth signals inside the normal hematopoeitic progenitor cells via increased amount of the active GTP-bound protein, antisense technology has been employed [39]. Skorski et al. targeted N-*ras* in human T-lymphocyte-depleted mononuclear marrow cells (A-T-MNC) and CD34$^+$-cells by unmodified antisense oligo (18 mer). N-*ras* antisense treatment resulted in a decreased number of granulocyte/macrophage colony-forming units of A-T-MNC/CD34$^+$ induced by IL-3 (up to 48.9 %/73.6 %), GM-CSF (up to 36 %/53.3 %) and M-CSF (up to 67.5 %/61.5 %). Erythroid Colony formation was inhibited by 20.4 % (A-T-MNC) and 40.3 % (CD34$^+$); IL-3, IL-6 and erythropoietin induced megakaryocytic colony formation was strongly inhibited (up to 81.7 % and 66 %) by N-*ras* antisense compared to sense treated cells. The N-*ras* antisense treatment was associated with a decrease of N-*ras*-mRNA in progenitor cells as well as in individual myeloid and erythroid colonies.

NF-κB transcription factor complex is a pleiotropic regulator of various cellular processes implicated in the cellular responce to injury by binding to the distinct regions of the promoter elements of numerous genes including cytokines, growth factor receptors and adhesions molecules. Antisense oligos to the NF-κB subunits p50 and p65 inhibit cellular adhesion and CD11b integrin expression in H-60 leukemia cells (neutrophils). These effects have been associated with specific p50- and p65-mRNA cleavage. Treatment with the p65 antisense did not affect the viability of the cells, as ascertained by trypan blue exclusion test and by demonstrating the reversibility of antisense effects after 4 hours culture without oligos. These findings indicate that p65-antisense can be used to define the role of NF-κB in the activation pathways of neutrophils [41].

To examine the role of acetylcholinesterase (ACHE), an enzyme which regulates the growth of hematopoietic cells, Soreq et al. investigated the effect of specific 15-mer phosphorothioate antisense oligos on the proliferation of mouse hematopoietic cells. ACHE-S-ODN treated progenitors enhanced cell proliferation and colony formation sugesting that the physiological role of ACHE is to reduce proliferation of multipotent stem cells. The proliferative response of ACHE-S-ODN was associated with a decrease of ACHE-mRNA compared to sense treated cells [42].

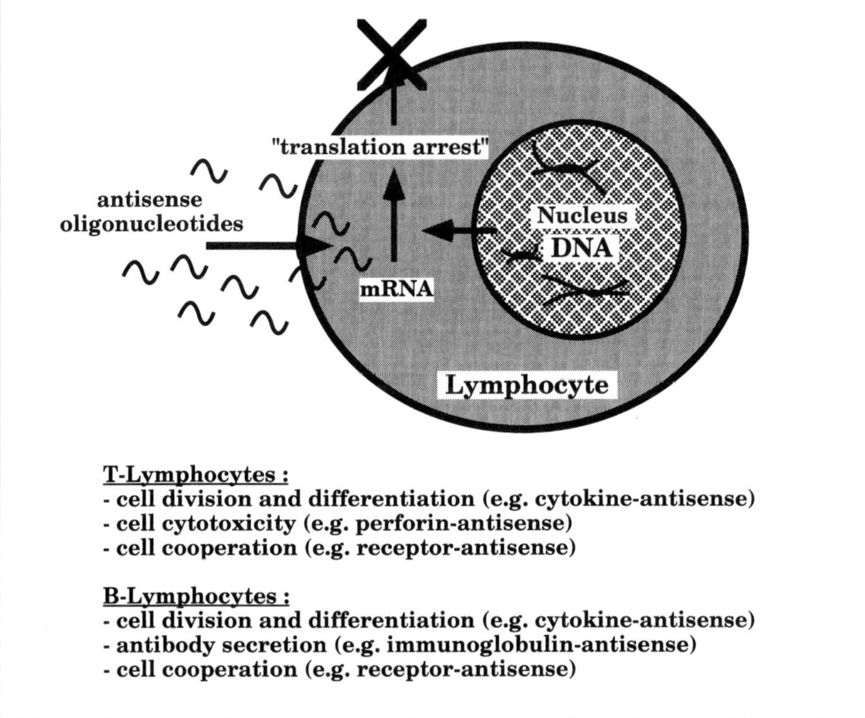

Figure 2: Major targets for antisense therapy in human immune cells.

3 How to Design Antisense Experiments in Immunological Systems

The main steps in preparing an *in vitro* antisense experiment include the following:

3.1 Target Determination

What gene would I like to target? – Targeted gene/gene products should possess a biological function, which may be easily measured (e.g. changes in cellular function resulting in modulation of cell proliferation, cell morphology, motility, cytotoxicity, etc.)

What is the molecular structure of the targeted mRNA/DNA? – Primary and secondary structure of targets should be known in order to design appropriate oligos with maximum affinity, optimal sterical access and hybridization efficiency (Figure 2).

What is the product of targeted gene? – The function of the protein targeted by the chosen oligos should be well known (structure, cellular turnover, location).

3.2   Antisense Determination

Which regions of the chosen gene should be targeted? – Search for sequences with no known homology to any other genes in DNA-Library. Avoid sequences causing known non-specific biological responses (see above 2). To find the best oligo use "gene walking" procedures, if possible, and check all antisense oligos experimentally for their biological activity. The translation initiation region is usually the "classic" approach. However, other target sequences may be preferable for targetting (see chapter 1).

What type of chemical antisense modification should be chosen? – For in vivo or in vitro use, choose analogs which are stable in the extracellular and intracellular milieu (e.g. phosphorothioates). Other chemical, physical and biological properties of oligos like water solubility, hybridization affinity, cellular uptake, bioavailability and pharmacokinetics should be taken into consideration before planning the experiment. In any case, use only highly purified (HPLC) oligos to reduce possible non-specific effects and cellular toxicity by degradation products and chemical impurities (see chapter 1, Introduction).

3.3   Studies of Cellular Uptake

How can the cellular uptake of the antisense be validated? – There is a number of experimental procedures which help to solve this question (e.g. uptake of ^{32}P-labelled, BrdU-labelled and FITC-labelled oligos). The authors of this article recommend the 2 last methods. These control experiments allow assessment of the kinetics of cellular uptake and intracellular distribution by light and fluoroscence microscopy.

3.4   Studies of Antisense Function

How can the specificity and efficacy of "hybridization arrest" by antisense be monitored? – A large number of experimental techniques may help to estimate the effects of antisense oligos. However, two important experiments measuring targeted mRNA and protein should be performed. The most common techniques for determining mRNA include Northern blot analysis and RT PCR. The protein level can be detected by several methods including Western blot analysis, immunocyto-chemistry, FACS and ELISA). The specificity may be determined by cross-checking with control sequences including random- and sense-sequences.

   How can the biological effects of antisense be measured? – Depending on the target try to find the best bio-assay to help answer this question. Do not forget controls like the trypan blue exclusion test, which may disclose any antisense-related cell toxicity and the level of non-specific toxicity. Biological effects also have to be excluded with control oligos.

4   Application of Antisense Oligonucleotides (Protocols)

This section describes experimental application of antisense oligos in immunology. Protocols presented here are based primarily on human peripheral blood mononuclear cells, as this cell population is one of the major targets in clinical antisense application. For those interested in selective targeting of particular immune cell subsets like T-lymphocytes, B-lymphocytes, monocytes and neutrophils the same procedures are generally applicable. However, the basic techniques describing isolation and handling of these cell populations will not be presented here and the readers should refer to a textbook of immunology [9].

4.1   Isolation of Peripheral Blood Mononuclear Cells (PBMCs)

Peripheral blood is the primary source of lymphoid cells for investigations of the human immune system. PBMCs are separated from other blood cells (erythrocytes, neutrophils) by Ficoll-Hypaque density gradient centrifugation.

Protocol 1: PBMCs Isolation

Note: The following procedures are performed aseptically.

Special materials:

- venous blood in heparinized syringes
- Ficoll Hypaque (Pharmacia)
- culture medium (e.g. RMPI 1640) (Gibco/BRL)
- fetal calf serum (FCS) (Gibco/BRL)
- human AB serum (Flow Laboratories Inc.)
- trypan blue (Sigma)

• Collect venous blood in heparinized syringes and centrifuge at 400xg for 10 min at room temperature.
• Decant the supernatant (serum) and mix the cell pellet with an equal volume of PBS room temperature.
• Slowly layer 2 vol. (e.g. 30 ml) of blood onto 1 vol. (e.g. 15 ml) of Ficoll-Hypaque and centrifuge at 900 × g for 30 min at room temperature.
• Collect PBMCs banded at the plasma-Ficoll interface and wash 3 times in complete medium (RMPI 1640 medium supplemented with 10% (v/v) human AB or fetal calf serum and 1µM L-Glutamine).
• Count the cells and determine cell viability by Trypan Blue exclusion test.

4.2   *In vitro* Culture of PBMC

To study growth modulatory effects of antisense oligos on immune cell proliferation a cell proliferation assay can be used.

Note: For each antisense oligos evaluate the optimal dose and time response
curve.

Protocol 2: In vitro Culture of PBMC

Special materials:

- PBMC (see Protocol 1)
- culture medium (e.g. RMPI 1640) (Gibco/BRL)
- fetal calf serum (FCS) (Gibco/BRL)
- IL-2 or PHA (R & D Systems, Genzyme)
- Phosphorothioate Oligos (Biognostik)

- Plate 10^4 to 10^5 cells into each well of a 96-round bottom plate in a volume of 100 µl (complete medium).
- Prepare 0.4µM to 40µM oligo-solutions (*dose response curve*) in complete medium prior to assay.
- Add 50 µl of oligo solution and 50 µl of an appropriate mitogen (e.g. 10^3 IU of IL-2 or 5 mg/ml PHA) in order to obtain a final volume of 200 µl (oligo concentrations from 0.1 to 10 µM). As controls use the same concentrations of "control"-oligos (sense, random, etc.) and oligo-untreated cells.
- Culture the cells for various time periods (e.g. 12–72 hours) in order to determine the optimal incubation time (*time-response curve*).

4.3   Quantification of PBMC Growth

Various methods can be used in order to determine the end-point of cell proliferation. Here, we describe the two most widely used protocols.

Protocol 3: Quantification of PBCM Growth

(for detailed protocols see chapter 6, Cell culture protocols):

- Measurement of DNA synthesis is often taken as representative of cell proliferation. For this, add 20 µl of ^{3}H-TdR (e.g. by Amersham) (1 µCi/well) solution into each well of a 96-well plate 16–18 hours before termination of the culture.
- Alternatively, count the cells stained with trypan blue in a Neubauer chamber or any other cell counter. Using this method you can simultaneously assess the toxic effects of antisense oligos on the cell populations under investigation (dye exclusion test).

Note: Short-term proliferation assays with antisense oligos (up to 72 hours) help to estimate antisense-effects on cell targets with fast turnover such as cytokines. However, they do not allow the investigation of antisense effects on targets (proteins) with long cellullar turnover (e.g. structure proteins, adhesion molecules), neither do they mimic the clinical situation where the cells are exposed to oligos for a prolonged period of time (days or weeks). Such treatment regimes may be simulated in long-term cultures where targeted cells are treated with oligos of interest for a long time. The most important issue of such experiments is to disclose possible non-specific and/or toxic antisense-effects which can be additionally mediated by antisense-degradation products (check by SDS-PAGE). In order to minimize these side effects of oligos, regular medium replacement is mandatory(!). In our long-term experiments with PBMCs (up to 30 days) we changed the culture medium regularly every 48 hours. For medium changes, all PBMC-suspensions were centrifuged, and the medium was replaced with fresh medium supplemented with antisense oligos and cytokines (e.g. IL-2).

4.4   Evaluation of Antisense Effects by Measuring other Biological Responses

4.4.1 Cytotoxicity Assay (Chromium Release Assay)

Cell-mediated cytotoxicity may be modulated by antisense oligos targeting various molecules participating in that process such as perforin, adhesion molecules, cytokines and their receptors. Depending on the target, estimate the optimal antisense-treatment regime (dose and time response curves, see above).

Protocol 4: Chronium Release Assay

Special materials:

- antisense-pretreated target cells (e.g. spleen cells, tumor cells, etc.)
- effector cells (e.g. PBMCs)
- 96-well microtiter plates (Falcon)
- ^{51}Cr solution (Amersham)
- gamma-scintillation counter

- Transfer antisense-pretreated target cells in complete medium into 15 ml conical tubes, wash once, determine the cell count and centrifuge for 10 min at 300–400 × g.
- Add 0.2 ml of 1mCi/ml ^{51}Cr solution and 20 µl FCS, mix gently and incubate for 45 minutes for lymphocytes or for 1–2 hours for tumor cells.

- Prepare effector cells (e.g. PBMCs) in complete medium and dispense 0.1 ml aliquote of cells into the wells of a 96-well microtiter plate with replicates of three wells for each effector cell concentration,
- Wash ^{51}Cr-labelled cells (target cells) 3 times with complete medium and add 0.1 ml of labelled-cells into plates containing effector cells.
- Centrifuge the plates (30 sec) and incubate for 3 to 6 hours.
- Centrifuge the plates again and count ^{51}Cr in γ-scintillation counter.
- Calculate the correct percentage of cell target lysis.

4.4.2 Modified Mixed Lymphocyte-tumor Culture

The biological effects of cytokines on immune cell function may be evaluated by different experimental techniques including cell proliferation assays, cell cytotoxicity assays, colony formation assays and others [9]. As the cytokines may express their function in an autocrine or paracrine fashion, antisense oligos seem to be the appropriate tool to investigate the role of cytokines in cellular interactions. For this purpose, a modified mixed lymphocyte-tumor culture (mMLTC) system has been developed [24] (Figure 3). The cell populations to be investigated are separated by semipermeable membranes (e.g. Nunc or Falcon tissue culture inserts).

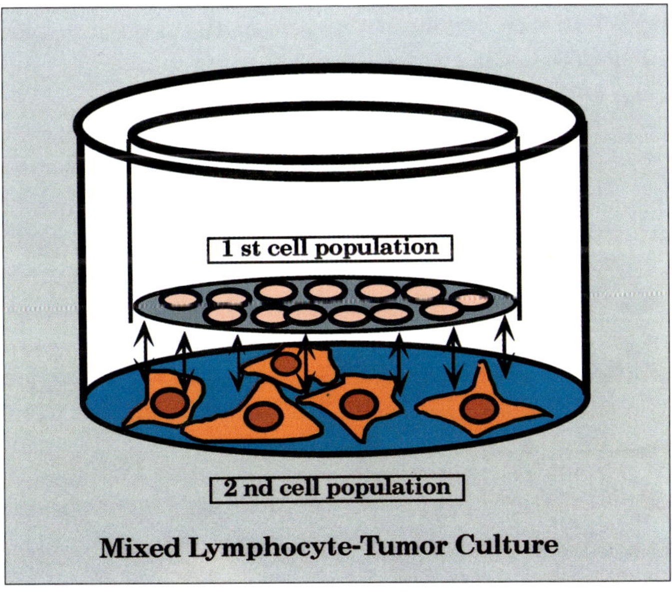

Figure 3: Modified mixed Lymphocyte-Tumor Culture (MLTC)

Protocol 5: Modified Mixed Lymphocyte-tumor Culture

Special materials:

- "1st cell population" (e.g. PBMC)
- "2nd cell population" (e.g. tumor cells)
- 24-well flat-bottom tissue culture plates (Falcon)
- Phosphorothioate oligos (Biognostik)
- Cytokines

- Plate the "2nd cell population" 12 hours before co-culture experiment into 24-well flat-bottom tissue culture plates (5×10^4 cells per well) in a medium supplemented with optimal concentration of antisense (e.g. 1–5 µM, oligos, f.c.).
- Plate the "1st cell population" 12 hours before co-culture into tissue culture, insert (5×10^5 cells per well) in a medium supplemented with optimal concentration of antisense (e.g. 1–5 µM, oligos, f.c.) and cytokines (e.g. 10^3 IU of IL-2).
- After 12 hours, put the tissue culture inserts (e.g. PBMC) into the 24-well plate (tumor cell culture) and co-incubate for an additional 48 hours up to 7 days. Replace complete medium every 48 hours.
- Measure cell proliferation by cell counting or ^{3}H-TdR incorporation
- assay in order to estimate the growth regulatory effects of cytokines targeted by antisense oligos in the cell populations under investigation.
- Collect supernatants for evaluating changes in protein concentrations (controls).

4.5 Control of Antisense Effects by Measuring the Level of Targeted Protein and mRNA.

Direct measurement of the target RNA and protein levels in antisense treated cells compared to "control"-oligo-treated cells, as well as internal control (e.g. β-actin), is a good assay for controlling the antisense-blocking efficiency. Western blot facilitates the identification and quantification of proteins (targets) resolved by SDS-PAGE (see chapter 5). In the case of soluble proteins (antigens) secreted into the culture fluid, an enzyme-linked immunosorbent assay (ELISA) can be applied (see below). Immunocytochemistry as well as flow cytometry (see below) may detect protein qualitatively by the binding of specific antibodies to antigen. Moreover, immunocytochemistry may help to assess the morphological distribution of proteins in the cells (see below).

4.5.1 Immunocytochemistry – Avidin-Biotin Complex Immunoperoxidase Technique

Immunocytochemistry may be used to evaluate the expression of various cell antigens. Morphological distribution of antigen in immune cell subsets can be specifically detected by monoclonal antibodies. This approach can be used to stain cells in smears or on cytocentrifuge preparations.

Protocol 6: Protein Detection – Immunocytochemistry

Special materials:

- Antisense- and control-treated cells
- rabbit / goat serum (Gibco/BRL)
- antigen-specific antibody
- biotin-labelled rat/goat anti-mouse immunoglobulin (Sigma)
- Avidin-biotin complex (ABC) solution (Sigma)
- Diaminobenzidene (DAB) solution (Sigma)
- Methyl Green or Hematoxylin solution (Sigma)
- Solutions as indicated in the protocol

• Prepare cell smears or cytocentrifuge slides with antisense- and control-treated cells.
• Air dry the slides and store in a plastic box overnight at 4 °C before fixation.
• Immerse the slides in acetone 5 min at room temperature.
• To block endogenous peroxidase (if necessary) place the slides in 0.5 % H_2O_2 in methanol for 30 min.
• Wash the slides 3 times in wash buffer (Tris-buffered saline, TBS).
• Incubate the slides for 30 min in 1–5 % rabbit/goat serum in TBS.
• Dilute first antibody in TBS (1: 20 to 1: 200).
• Incubate the slides with first, antigen-specific antibody for 2 hours at room temperature in a humidified chamber.
• Wash the slides 3 times in wash buffer (TBS).
• Incubate the slides with second, biotin-labelled rat/goat anti-mouse immunoglobulin for 30 min at room temperature in a humidified chamber.
• Wash the slides 3 times in wash buffer (TBS).
• Apply Avidin-biotin complex (ABC) solution and incubate for 30 min at room temperature.
• Wash the slides 3 times in wash buffer (TBS).
• Apply Diaminobenzidene (DAB) solution and incubate at room temperature while monitoring the DAB-reaction under a light microscope.

- Stop the reaction by washing in TBS and counterstain the slides in Methyl Green or Hematoxylin solution up to 1 to 2 min.
- Dehydrate the slides by immersing five times in 30 %, 60 %, 90 % and 100 % ethanol and clear by immersing five times in xylene.
- Mount with Eukitt or Permount and apply coverslip.

4.5.2 Enzyme-linked Immunosorbent Assay (ELISA)

To quantitate soluble proteins (antigens) in cell cultures treated with antisense oligos, different ELISA systems can be employed. Here, we describe the most useful and sensitive Antibody-Sandwich ELISA. In this technique plates are coated with specific antibody and followed by incubation with test solutions containing antigen [9]. Depending on the target you can use commercial kits (if available) or you can prepare detection plates yourself as described [9].

Protocol 7: Protein Detection – ELISA

- Add 100 µl of standard, sample (antisense-, sense-, random-oligo-treated cell culture supernatants – check protein concentration before assay) and control supernatants into the appropriate test wells in triplicate.
- Incubate the wells at 37 °C for 60 minutes.
- Aspirate the contents of the wells and wash vigorously 5 times.
- Add 100 µl of specific antibody-AP-conjugate and incubate at 37 °C for 60 min.
- Aspirate the contents of the test wells and wash again 5 times.
- Add 100 µl of substrate solution into each well and incubate the wells at room temperature for 20 min.
- Pipette 100 µl of stop solution into each well.
- Record the absorbance of each well within 30 min after the reaction has been stopped.

4.5.3 Flow Cytometry Analysis Using FACScan

Flow cytometry is a useful method for analyzing the expression of intracellular and surface proteins. This technique measures fluorescence intensity produced by fluorescent-labelled antibodies or ligands that bind specific cell-associated molecules [9].

Protocol 8: Protein Detection – FACScan

- Prepare antisense- and control-oligo-treated cell suspensions in 10 ml of culture medium and determine the cell viability by the trypan blue exclusion test.

- Centrifuge the cell suspension, discard the supernatant and resuspend the cell pellet in 4 °C staining buffer at 2×10^7 cells /ml.
- Pipette 50 µl of cell suspension (10^6 cells) into the tube, add 10 µl of appropriately diluted labelled antibody, mix gently and incubate for 20 min at 4° C.
- Wash the cells with staining buffer 3 times (centrifuge for 6 min at 1000 rpm ($300 \times g$) at 4 °C).
- Resuspend the stained cell pellets in 400 µl cold staining buffer and keep on ice until analyzed by flow cytometry.

Suppliers of special reagents

ELISA kits	R & D Systems Inc., Genzyme
Antibodies (detection of specific antigen)	Becton Dickinson, Dakopatts, R&D Systems Inc., Genzyme
Anti-mouse IgG-biotin conjugate antibodies	Sigma
Avidin-biotin complex	Sigma
Diaminobenzidene (DAB) solution	Sigma

5   Concluding Remarks

One of the most important aspects of each "antisense" experiment is the timing of the first addition of antisense oligos into immune cell cultures. The best times for antisense supplementation should be determined experimentally for each individual cell- and oligo-type. However, we would recommend adding the primary antisense treatment just before (1-2 hours) or during the immune cell activation.

References

1. ABKEN, H., FLUCK, J. and K. WILLECKE. Four cell-secreted cytokines act synergistically to maintain long term proliferation of human B cell lines in vitro. J. Immunol., 149: 2785–2794, 1992.
2. ACHA-ORBEA, H., SCARPELLINO, L., HERTIG, S. , DUPUIS, M. and J. TSCHOPP. Inhibition of lymphocyte mediated cytotoxicity by perforin antisense oligonucleotides. EMBO J., 9: 3815–3819, 1990.
3. AGRAWAL, S. and R.P. IYER. Modified oligonucleotides as therapeutic and diagnostic agents. Current Opinion in Biotechnology, 6: 12–19, 1995.
4. BENBERNOU, N., MATSIOTA-BERNARD, P. and M. GUENOUNOU. Antisense oligonucleotides to interleukin-4 regulate IgE and IgG2a production by spleen cells from *Nippostrongylus brasiliensis*-infected rats. Eur. J. Immunol., 23: 659–663, 1993.
5. BENNETT, C.F., CONDON, T.P., GRIMM, S. , CHAN, H. and M-Y. CHIANG. Inhibition of endothelial cell adhesion molecule expression with antisense oligonucleotides. J. Immunol., 152: 3530–3540, 1994.

6. BIRCHENALL-ROBERTS, M.C., FERRER, C., FERRIS, D., FALK, L.A., KASPER, J., WHITE, G. and F.W. RUSCETTI. Inhibition of murine monocyte proliferation by a colony-stimulating factor-1 antisense oligodeoxynucleotide. Evidence for autocrine regulation. J. Immunol., 145: 3290–3296, 1990.

7. BOEVE, CH. M. A. and M. DE LEY. Inhibition of human interferon-gamma expression by antisense oligodeoxynucleotides. J. Leukocyte Biology, 55: 169–174, 1994.

8. BURGESS, T.L., FISCHER, E.F., ROSS, S. L., BREADY, J.V., QIAN, Y.-X., BAYEWITCH, L.A., COHEN, A.M., HERRERA, CH. J., HU, S. , KRAMER, T.B., LOTT, F.D., MARTIN, F.H., PIERCE, G.F., SIMONET, L. and C.L. FARRELL. The antiproliferative activity of c-myb and c-myc antisense oligonucleotides in smooth muscle cells is caused by a nonantisense mechanism. Proc. Natl. Acad. Sci. USA, 92: 4051–4055, 1995.

9. COLIGAN, J. E., KRUISBEEK, A.M., MARGULIES, D.H., SHEVACH, E.M and W. STROBER. Current Protocols in Immunology. John Wiley & Sons. 1992.

10. FONAGY, A., SWIDERSKI, C., DUNN, M. and J.W. FREEMAN. Antisense-mediated specific inhibition of P120 protein expression prevents G1-to S-Phase Transition. Cancer Res., 52: 5250–5256, 1992.

11. FUJIWARA, T. and E.A. GRIMM. Specific inhibition of interleukin 1 gene expression by an antisense oligonucleotide: Obligatory role of interleukin 1 in the generation of lymphokine-activated killer cells. Cancer Res., 52: 4954–4959, 1992.

12. GEWIRTZ, A.M., ANFOSSI, G., VENTURELLI, D., VALPREDA, S. , SIMS, R. and B. CALABRETTA. G1/S transition in normal human T-lymphocytes required the nuclear protein encoded by c-myb. Science, 245: 180–183, 1989.

13. HAREL-BELLAN, A., DURUM, S. , MUGGE, K., ABBAS, A.K. and W.L. FARRAR. Specific inhibition of lymphokine biosynthesis and autocrine growth using antisense oligonucleotides in Th1 and Th2 helper T cell clones. J. Exp. Med., 168: 2309-2318, 1988.

14. HAREL-BELLAN, A., FERRIS, D.R., VINOCOUR, M., HOLT, J.T. and W.L. FARRAR. Specific inhibition of c-myc protein biosynthesis using an antisense synthetic deoxy-oligonucleotide in human T lymphocytes. J. Immunol., 140: 2431-2435, 1988.

15. HARUNA K-I., HIKIDA, M., OHSUGI, Y. and H OHMORI. The secondary antigen-specific IgE responce in murine Lymphocytes is resistant to blockade by anti-IL-4 antibody and an antisense oligodeoxynucleotide for IL-4 mRNA. Cell. Immunol., 151: 52–64, 1993.

16. HATZFELD, J., LI, M.L., BROWN, E.L., SOOKDEO, H., LEVESQUE, J.P., O'TOOLE, T., GURNEY, C., CLARK, S. C. and A. HATZFELD. Release of early human hematopoietic progenitors from quiescence by antisense transforming growth factor β1 or Rb oligonucleotides. J. Exp. Med., 174: 925-929, 1991.

17. HEIDENREICH, O., KANG, S-H., XU, X. and M. NERENBERG. Application of antisense technology to therapeutics. Molecular Medicine Today, 1: 128–133, 1995.

18. HEIKKILA, R., SCHWAB, G., WICKSTROM, E., LOKE, S. L., PLUZNIK, D.H., WATT, R. and L.M. NECKERS. A c-myc antisense oligodeoxynucleotide inhibits entry into S phase but not progress from G0-G1. Nature, 328: 445–449,1987.

19. HENDERSON B, and S. BLAKE. Therapeutic potential of cytokine manipulation. (Review). Trends Pharmacol. Sci., 13: 145-152, (1992).

20. HIKIDA, M., HARUNA, K.I. and H. OHMORI. Suppression of interleukin 4 production from type 2 helper T cell clone by antisense oligodeoxynucleotide. Immunol. Lett., 34: 297–302, 1992.

21. IKIZAWA, K., KAJIWARA, K., KOSHIO, T., MATSUURA, N. and Y. YANAGIHARA. Inhibition of IL-4 receptor up-regulation on B cells by antisense oligodeoxynucleotide suppresses IL-4-induced human IgE production. Clin. Exp. Immunol., 100: 383–389, 1995.

22. JACHIMCZAK, P., FABEL-SCHULTE, K., HESSDÖRFER, B., BRYSCH, W., SCHLINGENSIEPEN, K-H., BLESCH, A. and U. BOGDAHN. Transforming Growth Factor β-mediated regulation of human peripheral blood mononuclear cell proliferation as detected with phosphorothioate antisense oligodeoxynucleotides. Cell. Immunol., 165: 125–133, 1995.

23. JACHIMCZAK, P., BOGDAHN, U., SCHNEIDER, J., BEHL, CH, MEIXENSBERGER, J., APFEL, R., DÖRRIES, R., SCHLINGENSIEPEN, K-H. and W. BRYSCH. The effect of transforming growth factor β_2-specific phosphorothioate antisense oligodeoxynucleotides in reversing cellular immunosuppression in malignant glioma. J. Neurosurg., 78: 944–951, 1993.

24. JACHIMCZAK, P., SCHWULERA, U. and U. BOGDAHN. *In vitro* studies of cytokine-mediated interactions between malignant glioma and autologous peripheral blood mononuclear cells. J. Neurosurg., 81: 579–586, 1994.

25. JUN, C. D., CHOI, B. M., KIM, S. U., LEE, S. Y., KIM, H. M. and H. T. CHUNG. Down-regulation of transforming growth factor-β gene expression by antisense oligodeoxynucleotides increases recombinant interferon γ-induced nitric oxide synthesis in murine peritoneal macrophages. Immunology, 85: 114-119, 1995.

26. KATO, J., KOHGO, Y., KONDO, H., SASAKI, K. and Y. NIITSU. Antisense oligodeoxynucleotides for IL-2, c-myc and transferrin receptor synchronize mitogen-activated lymphocytes in the G1 Phase. Scand. J. Immunol., 39: 499–504, 1994.

27. KRIEG, A. M., YI, A. K., MATSON, S. , WALDSCHMIDT, T. J., BISHOP, G. A., TEASDALE, R., KORETZKY, G. A. and D. M. KLINMAN. CpG motifs in bacterial DNA trigger B-cell activation. Nature, 374: 546–549, 1995.

28. LAPIDOT-LIFSON, Y.., PATINKIN, D., PRODY, C.A., EHRLICH, G., SEIDMAN, S., BEN-AZIZ, R., BENSELER, F., ECKSTEIN, F., ZAKUT, H. and H. SOREQ. Cloning and antisense oligodeoxynucleotide inhibition of a human homolog of cdc2 required in hematopoiesis. Proc. Natl. Acad. Sci. USA, 89: 579–583, 1992.

29. LEFEBVRE D'HELLENCOURT, C., DIAW, L. and M. GUENOUNOU. Immunomodulation by cytokine antisense oligonucleotides. Review. Eur. Cytokine Netw., 6: 7-19, 1995.

30. LEVY, Y., TSAPIS, A. and J. C. BROUET. Interleukin-6 antisense oligonucleotides inhibits the growth of human myeloma cell lines. J. Clin. Invest., 88: 696–699, 1991.

31. LOUIE, W. S., RAMIREZ, L. M., KRIEG, A. M., MALISZEWSKI, CH. R. and G. A. BISHOP. Endogenous secretion of IL-4 maintains growth and Thy-1 expression of a transformed B cell clone. J. Immunol., 150: 399–406, 19993.

32. MATSON, S. and A. M. KRIEG. Nonspecific suppression of ^{3}H-thymidine incorporation by "control" oligonucleotides. Antisense Res. Dev., 2: 325–326, 1992.

33. METHIA, N., LOUACHE, F., VAINCHENKER, W. and F. WENDLING. Oligonucleotides antisense to the proto-oncogene c-*mbl* specifically inhibit *in vitro* megakaryocytopoiesis. Blood, 82: 1395- 1401, 1993.

34. McIntyre, K. W., Lombard-Gillooly, K., Perez, J. R., Kunsch, C., Sarmiento, U. M., Larigan, J. D., Landreth, K. T. and R. Narayanan. A sense Phosphorothioate Oligonucleotide Directed to the Initiation Codon of Transcription Factor NF-κB p65 Causes Sequence-Specific Immune Stimulation. Antisense Research and Development, 3: 309–322, 1993.

35. Mercola, D. and J. S. Cohen. Antisense approaches to cancer gene therapy. Cancer Gene Therapy, 2: 47–59, 1995.

36. Mojcik, Ch. F., Gourley, M. F., Klinman, D. M., Krieg, A. M., Gmelig-Meyling, F. and A. D. Steinberg. Administration of a phosphorothioate oligonucleotide antisense to murine endogenous retroviral MCV env causes immune effects in vitro in a sequence-specific manner. Clin. Immun. Immunopath., 67: 130-136, 1993.

37. Morrison, W. J., Offner, H. and A. A. Vandenbark. Enhanced T-Helper Cell Function Following CD4 Modulation. Cell. Immunol., 153: 392–400, 1994.

38. Paul, W. E. (editor). Fundamental Immunology. Raven Press. 1989

39. Skorski, T., Szczyklik, C., Ratajczak, M. Z., Malaguarnera, L., Gewirtz, A. M. and B. Calabretta. Growth-Factor-dependent Inhibition of Normal Hematopoiesis by N-ras Antisense Oligonucleotides. J. Exp. Med., 175: 743–750, 1992.

40. Small, D., Levenstein, M., Kim, E., Carow, C., Amin, S., Rockwell, P., Witte, L., Burrow, Ch., Ratajczak, M. Z., Gewirtz, A. M. and C. I. Civin. STK-1, the human homolog of Flk-2/Flt-3, is selectively expressed in CD34$^+$ human bone marrow cells and is involved in the proliferation of early progenitor/stem cells. Proc. Natl. Acad. Sci., 91: 459–463, 1994.

41. Sokolowski, J. A., Sartorelli, A. C., Rosen, C. A. and R. Narayanan. Antisense oligonucleotides to the p65 subunit of NF-κB block CD11b expression and alter adhesion properties of differentiated HL-60 Granulocytes. Blood, 82: 625–632, 1993.

42. Soreq, H., Patinkin, D., Lev-Lehman, E., Grifman, M., Ginzberg, D., Eckstein, F. and H. Zakut. Antisense oligonucleotide inhibition of acetylocholinesterase gene expression induced progenitor cell expansion and suppresses hematopoietic apoptosis ex vivo. Proc. Natl. Acad. Sci. USA, 91: 7907–7911, 1994.

43. Stein, C. A. and Y.-C. Cheng. Antisense Oligonucleotides as Therapeutic Agents – is the bullet really magical. Science, 261: 1004-1012, 1993.

44. Stein, C. A. and R. Narayanan. Antisense oligonucleotides. Current Opinion in Oncology, 6: 587–594, 1994.

45. Stein, C. A. and A. M. Krieg. Problems in interpretation of data derived from in vitro and in vivo use of antisense oligodeoxynucleotides. Antisense Res. Dev., 4: 67–69, 1994.

46. Tanaka, T., Chu, C. C. and W. E. Paul. An antisense oligonucleotide complementary to a sequence in Iγ2b increases γ2b germline transcripts, stimulated B Cell DNA Synthesis, and Inhibits Immunoglobulin Secretion. J. Exp. Med., 175: 597–607, 1992.

47. Tseng, B. Y. and K. D. Brown. Antisense oligonucleotide technology in the development of cancer therapeutics. Cancer Gene Therapy, 1: 65–71, 1994.

48. Wagner, R. W. Gene inhibition using antisense oligodeoxynucleotides. Nature, 372: 333–335, 1994.

49. WITSELL, A. L. and L. B. SCHOOK. Tumor necrosis factor a is an autocrine growth regulator during macrophage differentiation. Proc. Natl. Acad. Sci. USA, 89: 4754–4758, 1992.

50. ZHENG, H., SAHAI, B. M., KILGANNON, P., FOTEDAR, A. and D. R. Green. Specific inhibition of cell-surface T-cell receptor expression by antisense oligodeoxynucleotides and its effects on the production of an antigen-specific regulatory T-cell factor. Proc. Natl. Acad. Sci. USA, 86: 3758–3762, 1989.

51. ZUBIAGA, A. M., MUNOZ, E. and B. T. HUBER. Production of IL-1α by activated Th Type 2 cells. Its Role as an autocrine growth factor. J. Immunol., 146: 3849–3856, 1991.

52. ZUCALI, J. R. and J. MOREB. Use of antisense oligonucleotides to inhibit monocyte derived cytokines. Molecular and Cellular Biology of Cytokines, pages 315–320, Wiley-Liss, Inc., 1990.

8 Neurobiology

Gene Function Analysis and Therapeutic Prospects in Neurobiology

Karl-Hermann Schlingensiepen[1] and Markus Heilig[2]
[1]Max-Planck-Institut für Biophysikalische Chemie, Göttingen, Germany
[2]University of Göteborg, Dep. of Psychiatry and Neurochemistry, Mölndal, Sweden

1   Introduction

During the last three years, antisense oligodeoxynucleotides (hereafter called oligos) have become an increasingly successful tool for selective gene suppression in the nervous system.

Inhibition of protein synthesis with this novel tool has allowed neurobiologists to study the function of individual gene products in complex processes such as light induced resetting of the circadian clock, acquisition and retention of learning tasks, feeding behavior and tissue damage after cerebral ischemia.

Furthermore, selective gene suppression with this powerful technique has allowed the reversal of pathophysiological processes [4] [5] [29] [60]. Thus, antisense oligos are being developed as highly selective pharmacological tools and as a novel class of drugs. The use of antisense oligos as neuropharmacological agents [25] may open the opportunity to tackle hitherto untreatable diseases like neurodegenerative disorders and brain tumors, a class of tumors with rapidly increasing incidence over the last 40 years. In addition, they are becoming an increasingly popular tool for drug target validation in the pharmaceutical industry.

If antisense sequences are selected with the appropriate care, (*e. g.* avoiding non-specific sequence motifs and choosing length and base composition so as to avoid non-specific hybridization to non-target mRNAs) selectivity of gene suppression can be achieved not only in cell culture experiments (see *e. g.* Figure 7, chapter 1), but also *in vivo e. g.* after intracerebroventricular injection of antisense oligos (Figure 1).

As with any other technique (*e.g.* the use of protein kinase inhibitors), careful controls have to be performed to ensure selectivity of gene suppression effects and their functional consequences (see chapter 5, Assay Systems and Controls).

The novel property of antisense oligos lies in the fact that based on oligonucleotide synthesis, an almost indefinite number of molecules can be rationally designed, which differ only in the sequence of the four basic building blocks, the bases A, C, G and T. However, choice of the right target sequence, length of oligonucleotide sequence and base composition are crucial to ensure effectivity and in particular the selectivity of gene suppression [8].

Using antisense oligos, synthesis of such different protein classes as receptors, transcription factors, neuropeptides and growth factors have successfully been suppressed. Examples of antisense inhibition for each of these classes of proteins will be discussed below.

In this chapter we first review antisense studies in neurobiology and in the second part give practical experimental protocols for use of the antisense technology to inhibit expression of different classes of proteins in neuronal cell culture and in rat brain *in vivo*. We describe experimental protocols which have emerged as particularly useful *e.g.* for receptor

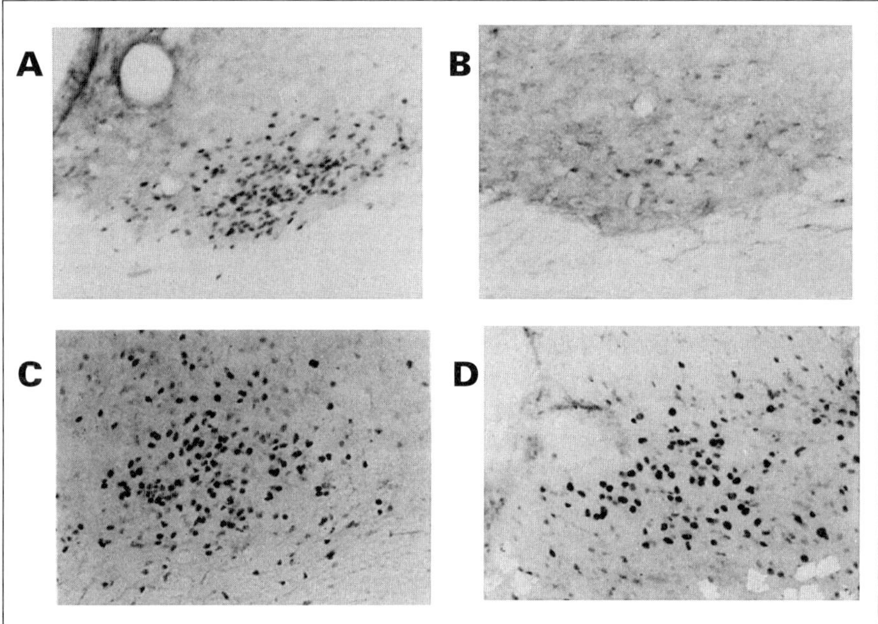

Figure 1: Selectivity of JunB protein suppression *in vivo* in the suprachiasmatic nucleus (SCN) of rat brain. Rats were injected into the third ventricle of the brain either with control oligo (randomized sequence, sections **A** and **C**) or with an anti *jun*B oligo (sections **B** and **D**). In the upper 2 sections (**A** and **B**) immunohistochemistry with a specific JunB antibody was used to detect JunB protein expression. The control animal (**A**) shows marked expression of the JunB protein in the ventral part of the SCN. In contrast, injection of the anti *jun*B oligo markedly reduces the number of JunB expressing neurons and the intensity of JunB protein expression (**B**). Specificity of the results is confirmed by immunohistochemistry detecting the homologous c-Jun protein (sections **C** and **D**). No reduction in c-Jun expression is seen in the anti *jun*B injected rat (**D**) compared to the control animal (**C**). For experimental details see protocol 4. Reproduced from [50] with permission.

knock-down or inhibition of inducible transcription factors. The attenuation of gene expression in the experiments described below modulated neuronal differentiation, neuronal survival and, in *in vivo* experiments, the behavior of animals. In addition, we describe experiments on the inhibition of brain tumor cell growth using antisense oligos.

Before describing these individual experimental protocols we will give a brief overview of some recent discoveries in neurobiology that have become possible through the use of the antisense technology.

2   Review

2.1   Receptors

Conventional receptor antagonists are only available for a minority of receptor proteins expressed in the brain, and often do not distinguish between receptor subtypes.

It is therefore of interest that blocking the expression of a membrane associated receptor can offer a functional equivalent to blocking the receptor. This was originally demonstrated for the T-cell receptor (TCR), [68]. When the expression of appropriate TCR variable-chain (V)-genes was suppressed in cultured T-cell hybridomas using antisense oligos, the functional response to antigen was specifically ablated. Similar results were obtained in culture for the interleukin-1 (IL-1) receptor by Burch and Mahan [9], who could also show that subcutaneous injection with the anti-IL-1 receptor oligo markedly inhibited the infiltration of neutrophils in response to subsequent injection of IL-1.

These principles were adapted for use in the CNS in a line of studies on the functional role of central neuropeptide Y (NPY). Employing agonist administration, evidence had been obtained that NPY acts as an endogenous anti-anxiety signal by activating NPY-Y1 receptors in the amygdala [23]. However, useful receptor antagonists were not available, preventing attempts to establish whether the anti-anxiety actions of exogenous NPY and its analogs reflect a physiological role of the endogenous peptide. When the Y1 receptor was cloned [17] [26] [33], the sequence information necessary for antisense oligo design became available. A combined *in vitro/in vivo* approach could then be utilized [24, 61]: In cortical neurons from rat embryos, the density of Y1-receptors was decreased by approx. 60 % using antisense treatment. The decrease in Y1-receptor density was accompanied by a comparable decrease in the second messenger response to NPY (*i.e.* inhibition of forskolin induced cAMP accumulation). With the efficiency of the oligo thus documented, *in vivo* studies could be performed. Repeated i.c.v. administration of the oligo for two days produced a selective loss of cortical Y1-binding density by approx. 60 %. The loss of Y1 receptors *in vivo* was followed by a marked increase in anxiety as tested in the elevated plus-maze, a model where central NPY administration has the opposite effect.

In a follow-up study [21], this approach has been used to help resolve whether food intake stimulation, another prominent effect of central NPY [58], is mediated by Y1 receptors. Since Y1 antisense treatment not only did not suppress but even increased food intake, it seems likely that another NPY receptor subtype is responsible for the latter action of the peptide. Thus, the utilization of antisense oligos has helped to clarify a heterogeneity which might be beneficial for future drug design efforts.

Antisense receptor knock-down as a strategy has been validated in systems where conventional pharmacological tools are available. An effect similar to that seen with the non-competitive NMDA-receptor antagonist MK-801, *i.e.* a cytoprotective action following experimental cerebral ischemia, was achieved by inhibiting NMDA-R1 protein expression with a combination of two phosphodiester oligos [60]. Similar evidence has been obtained by two different laboratories for dopaminergic D2 receptors.

Creese and colleagues [67] found that i.c.v. infusion of a phosphorothioate oligo corresponding to the rat D2 receptor mRNA reduced rat striatal D2 receptors by approx. 50 %, while D1, muscarinic, and serotonin 5-HT2 receptors were unaffected. D2 receptor autoradiography indicated a D2 receptor down-regulation of about 50 % throughout the striatum and over 70 % in the nucleus accumbens; reversibility after 5 days was also demonstrated. A random control ODN was inactive. The antisense treatment inhibited locomotion induced by the D2 receptor agonist quinpirole, without altering behavior induced by a D1 receptor agonist. Antisense treatment also elicited catalepsy and reduced spontaneous locomotor activity. Thus, the profile of D2 antisense action was virtually identical to that observed with classical competitive D2 antagonists. Similar data have been obtained in mice by Weiss and collaborators, although the reductions in receptor density was lower in these studies [63] [69]. A 20-mer phosphorothioate D2 oligo, administered i.c.v. to mice with unilateral 6-hydroxydopamine lesions of the corpus striatum inhibited rotations induced by a D2 dopamine receptor agonists but did not block rotations induced by D1 or muscarinic cholinergic receptor stimulation. This behavioral effect was seen within 1 day of repeated injections of D2 antisense, and almost complete inhibition was seen after 6 days of treatment. The effect of D2 antisense was reversible: Behavioral recovery occurred by 2 days after cessation of antisense treatment. Repeated anti D2 oligo administration specifically reduced the levels of D2 dopamine receptors and D2 dopamine receptor mRNA; D1 receptors were unaffected.

Subtypes of DA receptors have been cloned over the recent years. The DA-D3 receptor subtype [55] has attracted particular interest due to its largely limbic distribution, and its possible involvement in mediating the abuse potential of central stimulants [10]. The lack of selective, high affinity D3 antagonists has, however, prevented the latter hypothesis from being examined. Recently, we developed a phosphorothioate 15-mer targeted at the D3-transcript which seems to inhibit D3 receptor expression both *in vitro* (in stably transfected CHO-cells) and *in vivo* in the rat striatum [42]. The latter was accompanied by profound effects on DA synthesis. The experimental protocols for these experiments are given below.

Other applications of antisense mediated receptor knock-down in the brain have been developed. For example, the expression of angiotensin-1 (AT-1) receptors has been blocked both *in vitro* and *in vivo* using i.c.v. administration of phosphorothioate oligo [48] [19]. As expected, this treatment attenuated drinking induced by i.c.v. administration of angiotensin II [48]. Perhaps more importantly, a long lasting decrease in blood pressure was produced by AT-1 antisense in spontaneously hypertensive rats [19]. In addition, the expression of progesterone receptors has been consistently blocked with antisense oligos by several groups, resulting in profound effects on sexual behavior [38] [43] [46]. Numerous other reports of antisense expression blockade of neurotransmitter receptors have been presented, including opioid (for a review, see [44]) and serotoninergic receptor subtypes [7]; for the latter, this strategy is becoming the method of choice to bring order into the physiology of 17 cloned receptor species, revealing highly specific behavioral changes upon inhibition of individual receptor subtypes with antisense oligos [7].

In summary, extensive evidence is available to support the idea that neurotransmitter receptor expression can be consistently inhibited by relatively similar treatment schedules of antisense oligos, The dosages and administration schedules required are emerging, and are given in protocol 5.

2.2   Transcription Factors

The role of transcription factor expression in complex brain functions like learning, circadian rhythm changes and locomotion remained a matter of speculation until knock-down of their expression was made possible with antisense oligos. In contrast to neurotransmitter receptors, there are no small molecule agonists or antagonists for most transcription factors. Gene knock-out experiments, for example, of members of the Jun family, may lead to lethal phenotypes [30] [27], or as with c-*fos* knock-outs, the animals may show considerable malformations [62]. While these studies point to the importance of the knocked-out genes for development, they do not allow study of their function in complex behavioral situations in postnatal animals. Transcription factors like Jun, Fos and EGR family members are rapidly and strongly induced during various physiological and pathological changes in the brain, but, until recently their function remained unclear [40] [65]. Transcription factors bind to regulatory sites in the DNA and activate or inhibit expression of downstream target genes. Thus, they may be key regulators for changes in cellular gene expression. It could therefore be hypothesized that long-term changes like plastic adaptation in the brain might require induction of certain transcription factors. Recently, the use of antisense oligos has opened the route for *in vivo* studies in animal brains which could not be performed with any other technique so far.

In particular, new properties of the rapidly inducible Jun and Fos transcription factors have been elucidated with the antisense technique ranging from cell culture experiments to *in vivo* studies.

2.2.1 *Jun Family Transcription Factors*

Using the antisense technique it was discovered that different members of the Jun transcription factor family may have opposite effects on physiological parameters including cell proliferation and neuronal differentiation. In *in vivo* experiments, inhibition of the two Jun transcription factors, c-Jun and JunB, had pronounced behavioral effects.

Both the c-Jun and JunB transcription factors were shown to play crucial, but antagonistic roles in neuronal differentiation as well as in neuronal survival or programmed cell death [49] [51]. Since these properties of the Jun proteins were first discovered using antisense technology and were recently confirmed by two different methods *i.e.* antibody injection and transfection with transdominant negative mutants [16], [20], exemplary protocols of antisense experiments used to elucidate the function of these gene products are outlined below (see sections 3.1–3.3.2).

Furthermore, the functional importance of Jun transcription factor expression could also be demonstrated in *in vivo* experiments after intracerebral or intrathecal injection. In one model system the resetting of the circadian clock by light stimulation was chosen. A phase shift of the circadian rhythm can be induced in rats housed in constant darkness by light stimulation during the subjective night phase. This light stimulation that evokes a resetting of the circadian clock, induces expression of the JunB and c-Fos transcription factors in the suprachiasmatic nucleus (SCN) of rat brain, the pacemaker that sets the circadian rhythm [32]. Antisense oligos against *jun*B and c-*fos* injected into the third ventricle selectively suppressed the c-Fos and JunB transcription factors in rat SCN. This inhibition of the two transcription factors completely prevented light induced phase shifts of the circadian rhythm, while injection of two randomized control oligos had no effect [51] [66].

In a further *in vivo* model, an antisense c-*jun*-oligo, injected into the rat hippocampus, specifically inhibited acquisition and retention of a brightness discrimination learning task in rats. Surprisingly, after injections of antisense c-*jun* oligo before the first training session, the animals did not only perform poorly during this first training session, in which they were impaired in acquiring the learning task, but also they started the retraining sessions on the following three days like naive animals, as though they had not been trained before. Thus, on day 4 after having undergone three full training sessions on the previous three days they did not perform better during their first six runs than they did on the first day, despite the fact that the antisense c-*jun* oligo had only been injected prior to the first day of training. This long-term effect suggests that suppression of the c-Jun transcription factor may have prevented induction and/or silencing of various downstream target genes whose transcription would normally be regulated by c-Jun. Two different control oligos, a mismatch-c-*jun* sequence and the specific *jun*B-antisense oligo had no effect in this behavioral model [59].

2.2.2 Fos Transcription Factor

C-fos and related transcription factors are expressed in the brain in response to a wide range of stimuli, and c-*fos* induction has been extensively used as a marker of neuronal activation. However, again the possible functional role of this phenomenon in the living brain remained poorly understood, largely due to a lack of tools which would make it possible to block it [40]. For example, it has been established that c-*fos* expression is induced in the striatum by cocaine and amphetamine through an activation of dopaminergic D1-receptors [39], and in the central nucleus of amygdala (CNA) by various stressors [11] [2]; yet the functional relevance of these phenomena for reward and anxiety has remained unknown. The situation has been similar in most other systems where c-*fos* induction occurs, such as in cerebral ischemia.

The feasibility of time- and dose-dependently blocking c-*fos* expression in the brain by a single localized injection of a phosphorothioate antisense oligo directly into the target area was demonstrated by Robertson and colleagues [12] [14]. This group reported a sequence specific 70–80 % reduction in the number of Fos-positive nuclei when c-*fos* induction with amphetamine was preceded by antisense oligo administration. Using the same methodology, we could show that inhibiting c-*fos* expression acutely (10h post injection) blocked locomotion induced by cocaine [22]. This somewhat surprising effect of blocking the expression of a transcription factor has been consistently confirmed in subsequent studies with amphetamine and rotational activation [14] [56]. Hence, we were less surprised to then find a reduction in experimental anxiety when stress-induced c-*fos* expression in the rat amygdala was blocked with an antisense oligo [22]. Interesting results following the blocking ischemia-induced c-*fos* expression in the hippocampus have also been published, although the functional consequences were not addressed in that study [36].

In summary, rapidly inducible transcription factors can consistently be targeted with the antisense approach. Their rapid turnover allows fast inhibition of their expression following short term application of antisense oligos. (This is in contrast to some other proteins *e.g.* neurotransmitter receptors which often require application of antisense oligos for several days; see section 2.1 and also protocol 5 below).

2.3 Neuropeptides and Other Secreted Peptide Molecules

Successful inhibition of interleukin-6 (IL-6) expression obtained in culture [35] suggested a potential also for targeting the ‚presynaptic‘ component of peptidergic systems. Accordingly, inhibition of neuropeptide Y (NPY) expression has been obtained in the rat hypothalamus, with an expected decrease in feeding as a result [1]. Expression blockade of corticotropin-releasing hormone (CRH) has also been reported, and produced behavioral effects similar to those seen with a CRH antagonist [53]. In another study, the synthesis of the POMC-derived peptides adrenocorticotropin (ACTH) and -endorphin (-END) was markedly reduced by an unmodified oligo complementary to a region of -END mRNA both in AtT-20 cells and *in vivo* [57]. We have observed a dose- and sequence-dependent inhibition of pro-dynorphin synthesis after intrastriatal administration of a phosphorothioate oligo [18]. In addition, using intraventricular rather than localized administration, inhibition of vasopressin expression has been reported, and was accompanied by a transient diabetes insipidus [54]. Highly sequence specific effects of oxytocin antisense infusion into the hypothalamus on suckling have been recently reported. However, this was observed within 4–5 hours post infusion, and in the absence of a detectable decrease in the tissue content of the peptide. These findings may reflect a depletion of a rapidly releasable peptide pool, which comprises too small a fraction of the total tissue content of peptide to be detected [41].

Different functional properties of newly synthesized versus pre-existing secreted forms of a peptide have been demonstrated in the case of ependymin. Ependymin molecules are constitutively present at high concentration in goldfish brain fluid constituting about 15 % of soluble extracellular protein (reviewed in [52]). However, inactivation of small subfractions of this ependymin pool with antibodies against different epitopes were sufficient to prevent memory consolidation after active avoidance conditioning. Antisense oligo injection experiments then revealed that the ependymin portion required for memory consolidation consists of the newly synthesized ependymin molecules [52][1].

Secreted brain peptides can thus be targeted with antisense oligo, but in view of the above cited study on oxytocin suppression [41] and our own experience it can be more difficult and may require more preliminary experiments to determine the right dosing and the best time window for optimal inhibition of secreted peptides. The efficiency as detected by determination of the amount of total protein may in some cases seem lower than that seen in numerous receptor systems. This may be due to differences between pre-existing and newly synthesized protein. Thus, 96 % of the ependymin present in goldfish brain fluid consists of disulfide bridged dimers. Yet, inactivation of the rest of the 4 % of monomeric ependymin by antibodies prevents memory formation after active avoidance training and so does inhibition of *de novo* synthesis of ependymin with antisense oligos [52]. Presumably the monomeric portion of ependymin consists mainly of newly synthesized protein. This relevance of *de novo* synthesis was also indicated by strong induction of ependymin mRNA after active avoidance conditioning [47].

In summary, the sometimes complex results with peptide antisense experiments may be related to turnover kinetics and posttranslational modi-

---

[1] Antisense oligos prevented recall of an active avoidance task, while randomized control oligos had no effect. Non-specific toxicity of the antisense sequence was ruled out in further control experiments with well trained fish. Their performance after 10 training sessions was not impaired by antisense oligo injection.

fication of the peptide in question. Protocols for inhibiting neuropeptide expression may thus be more varied than those for neurotransmitter receptor or transcription factor knock-down. They are easiest where *de novo* synthesis of a peptide occurs in the absence of high pre-existing levels. An example of this is the *de novo* synthesis of NPY preceding the progesterone induced LH-surge in female rats. Under these conditions, brief i.c.v. administration of unprotected AS-ODN was highly effective, allowing the clarification of a causal relation between the two events [31].

2.4   Cholinesterases

The enzymes acetylcholinesterase (ACHE) and butyrylcholinesterase (BCHE) control the termination of cholinergic signaling by hydrolyzing the neurotransmitter acetylcholine. Malfunction of cholinesterases is of great pathophysiological importance for a variety of diseases. They are targets for various drugs including tacrine for Alzheimer's disease treatment and pyridostigmine used in large-scale prophylaxis against potential nerve gas attacks during the 1991 Persian Gulf War [34] [37]. While their importance for nerve cell function and malfunction is well recognized, the important role of ACHE and BCHE in hematopoietic cell proliferation and differentiation was recently shown using antisense oligos.

Antisense oligos inhibiting ACHE expression increase hematopoietic cell proliferation, reflecting the expansion of progenitors, and divert hematopoiesis towards production of primitive blasts and macrophages. These findings may explain the tumorigenic association of perturbations in ACHE gene expression with leukemia.

Antisense oligos selectively inhibiting the other important cholinesterase BCHE increase myelogenesis and inhibit production of megakaryocytes, suggesting requirement of the BCHE protein for megakaryopoiesis. *In vivo*, injection of BCHE antisense oligos into mice reduced BCHE mRNA levels in megakaryocytes and their progenitors. Furthermore, bone marrow cultures from antisense injected mice displayed a strong reduction in megakaryocyte colony formation [45].

Thus, inhibition of the two cholinesterases with antisense oligos reveals different functions for ACHE and BCHE in hematopoiesis.

Therapeutically, the common adverse effects of anti-cholinesterase drugs make selective antisense oligos important drug candidates for the treatment of Alzheimer's disease, especially since it was recently shown that transgenic expression of human acetyl cholinesterase in mice induces progressive cognitive deterioration [3].

3   Practical Approaches to Antisense Experiments in Neurobiology

3.1   Modulation of Neuronal Differentiation

Neuronal differentiation of proliferating stem cells starts with the commitment stage. During the commitment stage dividing stem cells first have to leave the proliferative cycle. This requires a drastic switch in cellular gene expression pattern. Antisense technology offers an ideal tool to study the role of individual genes during this dramatic change in cell function and structure: First, the dividing cells' gene expression pattern has to be reprogrammed so as to allow irreversible proliferation arrest. Neurons can not re-enter the proliferative stage once they have passed this stage of differentiation (in contrast to some other differentiated cells *e.g.* liver cells).[2] Thus, the commitment stage is a step of fundamental biological and pathophysiological importance. Using antisense oligos in neuronally differentiating PC-12 cells we analyzed the functional role of genes that are rapidly expressed following the induction of differentiation.

3.1.1 *Neuronally Differentiating PC-12 Cells as a Model System*

An important experimental model for early stages of neuronal differentiation is the use of rat PC-12 cells. The PC-12 cell line was derived from a tumor of the adrenal gland, a phaeochromocytoma. This immortal tumor cell line can theoretically be kept in culture indefinitely. If stimulated with nerve growth factor (NGF), however, PC-12 cells stop proliferating and differentiate into neuronal-like cells. Thus, PC-12 cells were used to study the switch in gene expression pattern that reprograms proliferating cells to enter neuronal differentiation.

In NGF-stimulated PC-12 cells, antisense oligos can be used to study the functional significance of particular gene products for the differentiation process. If expression of a particular gene is required for neuronal differentiation, its suppression with the appropriate antisense oligo will alter the differentiation process. Suppression of genes required for commitment, the earliest step of neuronal differentiation, should prevent the differentiation process altogether.

Two proteins rapidly induced upon induction of neuronal differentiation of PC-12 cell are the transcription factors c-Jun and JunB [49]. The functional significance of their expression could be determined with specific antisense oligos using the protocols outlined below.

---

[2] As a clinical correlate of this basic biological phenomenon there are no neuronally derived tumors in adult patients. Instead, brain tumors of adults are derived from non-neuronal, mostly glial cells. In contrast, childhood tumors can be derived from maldifferentiated cells with neuronal characteristics.

Protocol 1A: Morphological Assessment

Special materials:

- Anti-c-*jun* S-ODN, anti-*jun*B S-ODN (Biognostik, Göttingen)
- PC-12 cells (can be obtained from American Type Culture Collections (ATCC)).
- RPMI or DMEM cell culture medium (Gibco/BRL)
- poly-L-lysine (Sigma) for coating of culture plates to allow neurite outgrowth
- NGF 2.5 S (Sigma)
- penicillin, streptomycin (Gibco/BRL)

- Grow the cells in the medium they have been adapted to. We used PC-12 cells adapted to both RPMI or DMEM medium.
- Grow the cells in the medium supplemented with 100 U/ml penicillin, 100 µg/ml streptomycin and 5–10 % Fetal calf serum (FCS). (We found addition of horse serum plus FCS, as frequently described for PC-12 cells, unnecessary in our experiments).
- Coat 96-well flat-bottom microtiter plates with poly-L-lysine to allow for better adherence of cells and outgrowth of neurites: Incubate wells with a sterile 0.1 % solution of poly-L-lysine overnight. Remove poly-L-lysine (may be recycled for next use by sterile ultrafiltration), wash the wells twice with sterile water and once with sterile PBS or culture medium.
- Plate 2500 cells/well in RPMI or DMEM medium with 5 % FCS into the poly-L-lysine coated wells and add phosphorothioate oligos at 2 µM concentration.
- 2–24 h after plating of the cells add 10 ng/ml of NGF.
- Depending on the PC-12 cell clone neurite outgrowth takes different lengths of time, ranging from a few days to more than a week.
- Observe and photograph the cells daily under the microscope. We observed the cells for up to 12 days without changing the medium and without further addition of antisense oligos.

In our antisense experiments, inhibition of *jun*B expression almost completely prevented morphological differentiation of NGF treated PC-12 cells. In contrast, inhibition of c-*jun* expression led to a marked increase in neurite outgrowth (Figure 2)[3]. Thus, use of antisense technology revealed

---

[3] The randomized control S-ODN (Fig. 3) and two specific control antisense S-ODNs directed against *p53* and *jun*-D (for the data of these control experiments see Schlingensiepen *et al.*, 1994) did not inhibit morphological differentiation. The use of several control oligos, randomized, as well as those directed against other genes served as a stringent control to confirm the specificity of the effects.

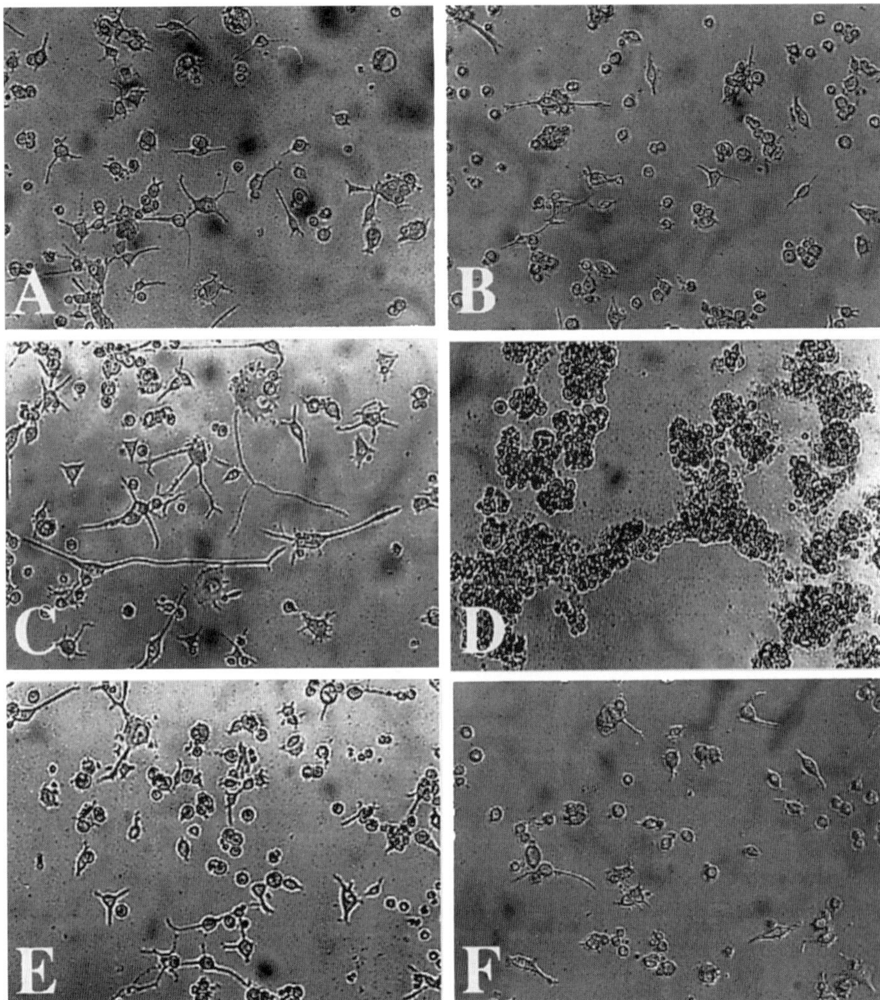

Figure 2: Neuronal differentiation of PC-12 cells. Differentiation of PC-12 cells after stimulation with NGF and inhibition of either JunB, c-Jun, JunD or p53 expression. **A:** Control cells, not treated with oligo. **B-F:** Morphological differentiation after treatment with 2 μM of the following oligos. **B:** Randomized control S-ODN. **C:** Anti-c-*jun* S-ODN. **D:** Anti-*jun*B S-ODN. **E:** Anti-*jun*D S-ODN. **F:** Anti-*p53* S-ODN. Oligos were applied at 2 μM concentration. Inhibition of expression of the two proteins JunD and p53 did not show significant differences to control conditions. Morphological differentiation was completely prevented by anti-*jun*B S-ODN. In contrast, enhancement of neurite outgrowth was seen after treatment with anti-c-*jun* S-ODN. Reproduced from [50] with permission.

opposite functions of c-Jun and JunB during neuronal differentiation of PC-12 cells. The opposite effects of the two different antisense oligos which have equal length and equal C,G content further highlights the selectivity of the technique.

To determine whether specific inhibition of expression of a particular gene (here *jun*B) prevents not only morphological differentiation, but also the proliferation arrest required for neuronal differentiation, thymidine incorporation and cell number can both be determined:

Protocol 1B: Thymidine Incorporation

- Plate 2500 cells/well into poly-L-lysine coated 96-well plates (see protocol 1A).
- Add phosphorothioate oligos at 2 µM concentration 6 h after plating.
- Add 10 ng/ml of NGF 24 h after plating of the cells.
- On day nine of the experiment incubate the cells with 0.15 µCi ^{3}H-thymidine/well for 6 h before harvesting.
- Lyse cells by storing the microtiter plates at -20 °C overnight.
- Harvest the cells as described in protocol 7, chapter 6, Cell Culture Protocols.
- Determine the amount of incorporated tritium by liquid scintillation counting.

Protocol 1C: Cell Counting

- Plate 2500 PC-12 cells/well into poly-L-lysine coated 96-well plates.
- Incubate with antisense oligos and NGF as described above (protocol 1B).
- Trypsinize, harvest and centrifuge the cells.
- Determine the cell numbers either by using a Neubauer counting chamber and trypan blue dye exclusion to determine the cell viability or by using an automatic cell counter (see also protocol 1, chapter 6, Cell Culture Protocols).

In our experiments on *jun* gene function, PC-12 cells treated with control oligo or anti-c-*jun* oligo both approached basal levels of ^{3}H-thymidine incorporation after induction of differentiation with NGF. In contrast, treatment with anti *jun*B oligo resulted in 6-fold higher rates (Figure 3). These levels were equal to those in proliferating cells that were not stimulated with NGF. Cell counting demonstrated a two-fold higher cell number in NGF-stimulated PC-12 cells after treatment with anti-*jun*B oligo (Figure 3).

In summary, the antisense experiments demonstrated that NGF is unable to induce PC-12 cell proliferation arrest if Jun-B protein expression is inhibited with antisense oligo, revealing a crucial role of this gene product in the commitment to neuronal differentiation.

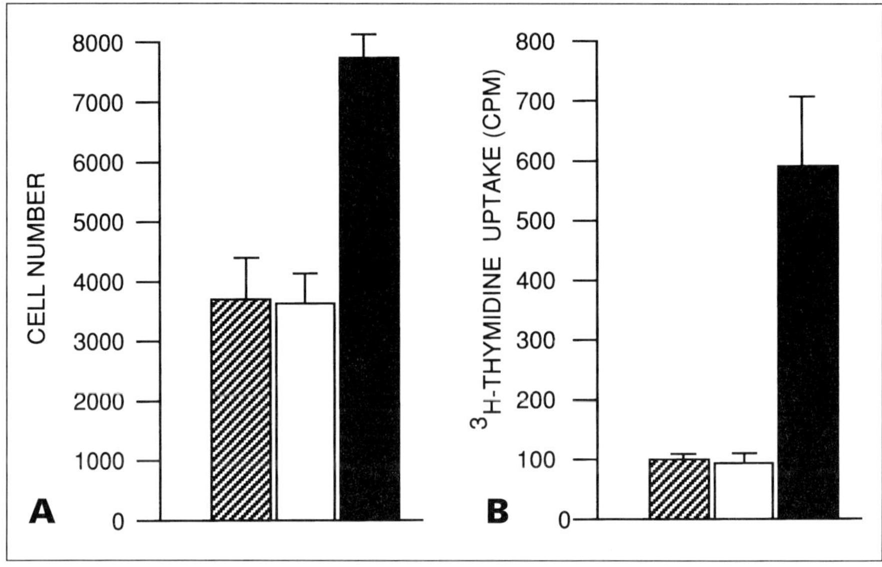

Figure 3: Cell growth arrest in NGF treated PC-12 cells is prevented by anti-*jun*B oligo. **A:** PC-12 cell number after 8 days of NGF treatment. Bars represent the mean of 4 values. **B:** ^{3}H-thymidine incorporation into PC-12 cells, treated with NGF for 48 h. Bars represent the mean of three values. Cells were either incubated with: Hatched bars: 2 µM control oligo; White bars: 2 µM anti-c-*jun* oligo; Black bars: 2 µM anti-*jun*B oligo. Error bars correspond to 1 SD. Reproduced from [49] with permission.

3.1.2 Differentiation of Primary Neuronal Cell Cultures

While PC-12 cells are an excellent system to study the early stages of neuronal differentiation, the later stages including synapse formation can be studied more adequately in primary neuronal cultures from fetal rat brain. Neurons can be cultured best if explanted about two days after they have become postmitotic. Thus, they are not suitable to study the commitment stage of differentiation, but they offer an elegant system to analyze changes in morphological differentiation and, with respect to pathophysiology, to analyze gene function in programmed neuronal cell death.

Protocol 2: Hippocampal Neuronal Cell cultures

Special materials:

– Highly purified FITC-labelled and unlabelled phosphorothioate oligos (Biognostik, Göttingen)
– HM-medium prepared according to [28]
– N2-medium prepared according to [6]

- Remove brain tissue from embryonal day 18 (E18) rats.
- Dissect out both hippocampi, dissociate the tissue by mild trypsinization to disperse the brain tissue into single cells.
- Disperse the cells in serum-free HM/N2 medium (1:1, v:v).
- Plate the cells into 24-well flat bottom plates at a density of 4×10^5 cells per ml in a mixture of 50 % of unconditioned HM/N2-medium with 50 % of preconditioned HM/N2-medium (*i.e.* conditioned by primary rat astrocytes).
- Add antisense oligos with the first change of medium 1 h after plating by mixing oligos into the replacement medium at a concentration of 1 µM.
- If the medium is changed *e.g.* by replacing 50 % of the volume every 3 days, supplement the replacement medium with oligos at the desired final concentration (here 1µM).
- Examine and photograph the cells every day under the microscope.
- To determine oligonucleotide uptake FITC-labelled oligo are particularly useful (see also protocol 7). Use the same protocol as above except for replacing unlabelled phosphorothioate oligos by FITC-labelled phosphorothioate oligos.

For an example see Figure 4. Here, FITC-labelled S-ODNs were added to hippocampal cultures five days after plating of the neurons. Uptake into the cultured neurons was very efficient. Strong labelling of neurons was observed after 24 h of FITC-S-ODN incubation and even more intense labelling after 72 h. FITC-labelling was not confined to neuronal cell bodies, but was also seen in neurites (Figure 4).

In our antisense experiments examining the role of the c-*jun* and *jun*B genes during morphological differentiation of primary neurons the following results were obtained:

Marked alterations in the morphology of the differentiating hippocampal neurons were observed after 5 days in culture. Knock-down of JunB expression led to a strong reduction in neurite density. Conversely, inhibition of c-Jun expression resulted in a denser neuritic network compared to control conditions. These differences were even more pronounced after 12 days in culture (Figure 5). Thus, the opposite effects of two different antisense oligos provided a further stringent control for the specificity of the antisense effects in addition to the application of randomized control oligo.

Furthermore, striking differences in neuronal survival between neurons deficient in JunB or c-Jun expression compared to control became apparent.

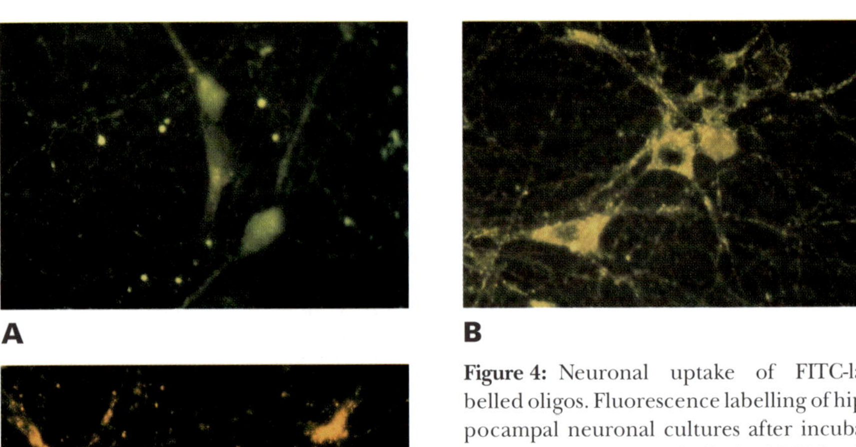

A

B

C

Figure 4: Neuronal uptake of FITC-labelled oligos. Fluorescence labelling of hippocampal neuronal cultures after incubation with 1 µM FITC-labelled S-ODN for **A:** 2 h ; **B:** 24 h; **C:** 72 h. A progressive increase in the intensity of fluorescence labelling is seen with time. Oligo uptake can be visualized in neuronal cell bodies as well as in neurites.

3.2   Modulation of Neuronal Survival

To quantify the changes in neuronal survival a combination of two dyes, one staining dead cells and the other one staining vital cells, is useful:

The assay for neuronal survival described below uses the fluorescein-diacetate/ethidium bromide double staining method. Fluorescein-diacetate selectively stains vital cells, which convert the non-fluorescing FDA to fluorescein, thus yielding intense green fluorescent staining. Ethidium bromide is excluded from vital cells, but stains the nuclei of dead cells which appear red under the fluorescence microscope:

Protocol 3: Semi-Quantitative Analysis of Neuronal Survival

Special materials:

- Anti-c-*jun* S-ODN, anti-*jun*B S-ODN (Biognostik, Göttingen)
- Culture media as above (see protocol 2)
- Ethidium bromide (EtBr) (Sigma)
- Fluorescein-diacetate (FDA) (Sigma)

• Prepare primary hippocampus neuronal cultures as described in protocol 2.
• Incubate the cells for 2–5 min with a solution of 50 mM FDA/50 mM EtBr (final concentration) in culture medium *e.g.* on day 7 and day 15 of the experiment.
• Determine the number of fluorescein-stained cells and of EtBr-stained nuclei, by counting under a fluorescence microscope with a counting grid.

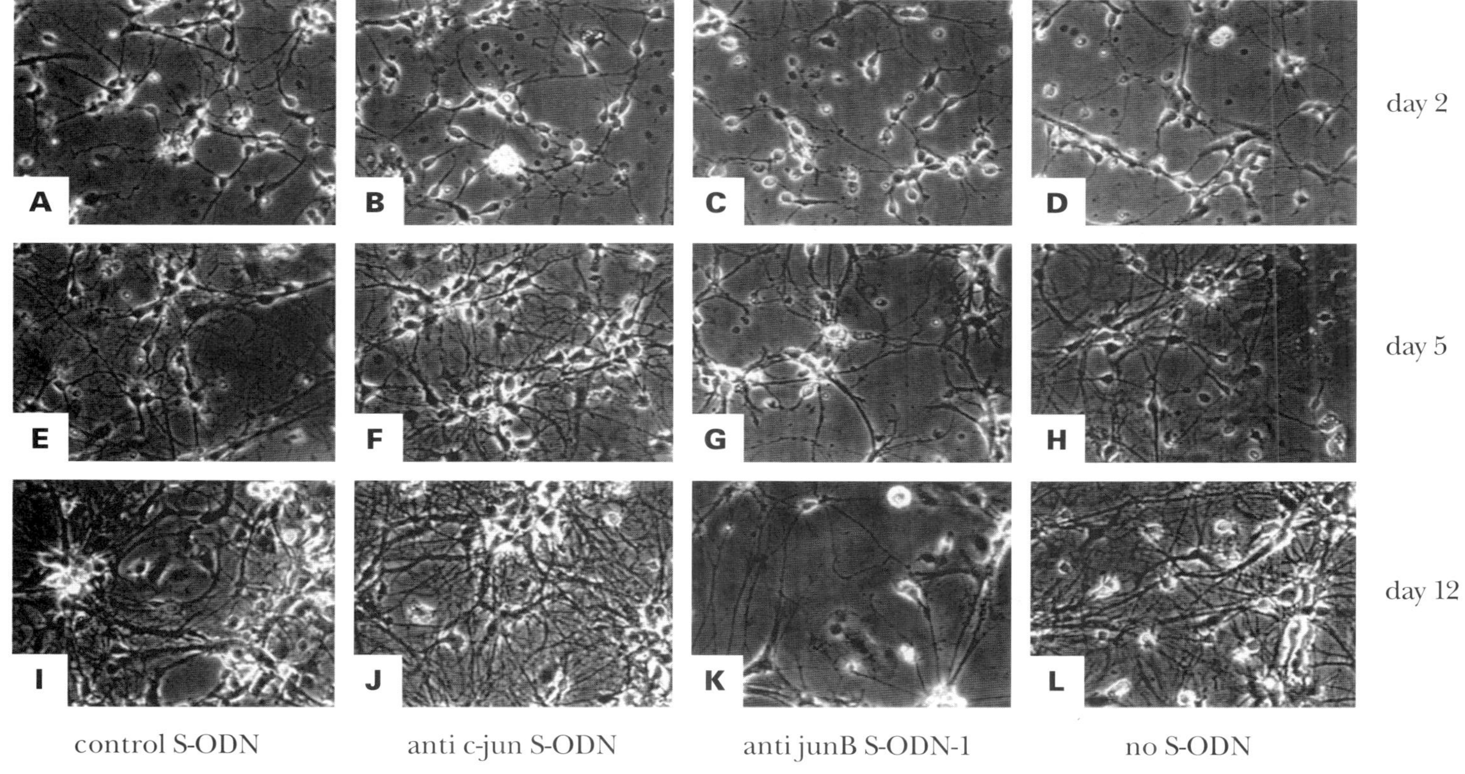

Figure 5: Differentiation and survival of hippocampal neurons. Neurons were photographed on day 2 (**A-D**), day 5 (**E-H**) and day 12 (**I-L**) of neuronal culture. Cells were either treated with control oligo, anti-*jun*B oligo, anti-c-*jun* oligo or no oligo as indicated. Oligos were applied at 1 µM concentration.

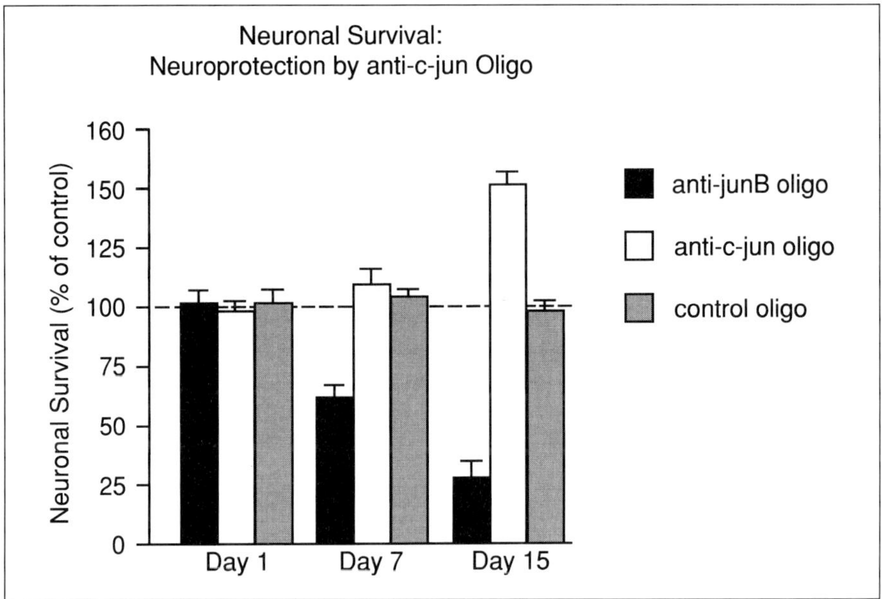

Figure 6: Quantification of neuronal survival. Survival of developing hippocampal neurons after treatment with different antisense oligos. The dotted line marks the value of untreated control neurons, set as 100 %. Neurons were incubated with the following oligos: Black bars: 1 µM anti-*junB* S-ODN; White bars: 1 µM anti-*c-jun* S-ODN; Gray bars: 1 µM control S-ODN. Anti-*junB* oligo strongly increases the rate of cell death. In contrast, treatment with anti-*c-jun* oligo has a protective effect on the neurons. On day 15 their survival is increased by more than 50 %.

In our experiments suppressing c-Jun or JunB expression in hippocampal neurons, we obtained the following results. In response to treatment with anti-*junB* S-ODN, neuronal survival was reduced to 61 % of controls on day 7 and to 27 % on day 15. In contrast, in anti *c-jun* treated cells, neuronal survival was increased by 54 % on day 15 compared to cells treated with randomized control oligonucleotide and by 51 % compared to untreated control neurons (Fig. 6). These results demonstrate the important role of the two transcription factors in programmed neuronal cell death. Furthermore, they indicate the therapeutic potential of antisense oligos to prevent delayed neuronal cell death. Thus, in stroke patients in the so-called penumbra, which surrounds the area of immediate neuronal death, a larger proportion of neurons die after a delay period and this process involves gene expression. Similarly, after prolonged hypoxia *e.g.* after cardiac resuscitation, patients are frequently well and alert in the immediately following 24 h, but subsequently become comatose and suffer irreversible neuronal cell damage. Under these conditions, which involve changes in gene expression prior to the induction of delayed neuronal cell death, selective inhibition of genes mediating programmed cell death like *c-jun* may be of therapeutic value.

3.3   Inhibition of Gene Expression *in vivo*

3.3.1 In vivo Suppression of Rapid Turnover Proteins

For the inhibition of proteins that are rapidly induced after a certain stimulus and have a high turnover rate, such as the immediate early gene products c-Fos, c-Jun or JunB, a single injection (or two injections within a few hours) suffice to block their induction. (In contrast, constitutively expressed genes like many receptors in the nervous system require continuous application over several days; see protocol 5).

Protocol 4: Intraventricular Injection of Antisense Oligodeoxynucleotides

Special materials:

- Stereotaxic frame
- Stainless steel microinjection guide cannula (*e.g.* from Plastics One, Roanoak, VA)
- Specific antibodies to determine suppression of proteins under study
- behavioral paradigm to study functional consequences (for details of one such behavioral paradigm see section 3.3.2).

- Under appropriate anesthesia (*e.g.* ketamine + xylazine or pentobarbitol) place rats in a stereotaxic frame. Make incision and remove periostium from the skull bone. Use bone wax for hemostasis and let the bone dry.
- Drill holes for skull screws and for the cannula holder. Secure the screws to the bone. Place cannula and lower cannula tip in place. With acrylic cement fix the cannula to the anchor screws. After the cement has hardened remove the holder.
- We used microinjection guide cannulae for both, intracerebral injection into the brain tissue next to the dorsolateral hippocampus [59] and i.c.v. injection into the third ventricle to inhibit protein expression in the suprachiasmatic nucleus (SCN) which is located next to the bottom of the third ventricle [51] [66]. (For localization of tissue distribution and cellular uptake of oligos see protocol 7).
- Close the cannula with obturators while not in use.
- Recovery period: Manipulation of the brain tissue during cannula implantation itself leads to induction of immediate early genes. Therefore, the antisense experiments have to be performed after an appropriate delay period following placement of the chronically implanted cannulae. We waited at least 7 days after cannula implantation before antisense oligos were injected.
- Inject 1–2 µl of antisense oligo at 1–2 mM concentration.
- Perform behavioral experiments 6–12 h after oligo injection (*e.g.* subjection of rats to a light pulse that induces phase shifts of the circadian rhythm. For details of this behavioral paradigm see [51] [66].)

- Determine protein suppression by an appropriate assay *i.e.* immuno-histochemistry and/or Western blotting using the standard techniques. In our experiments we used immunohistochemistry to determine inhibition of JunB and c-Fos expression: Under deep anesthesia rats were perfused transcardially with saline followed by 4 % paraformaldehyde, their brains were removed, cryoprotected in 30 % sucrose and sectioned at 35 μm on a cryostat. Immunohistochemistry was performed on free-floating sections using the avidin-biotin protocol for detection [51] [66].

Results showing selective inhibition of JunB protein expression *in vivo* are shown in Figure 1. Inhibition of c-Fos protein expression is shown in Figure 7.

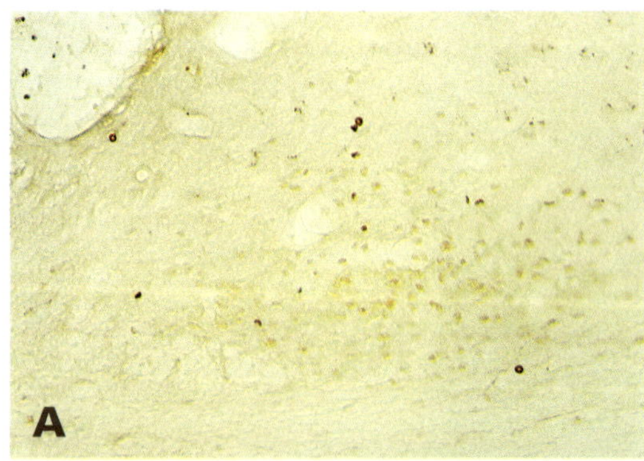
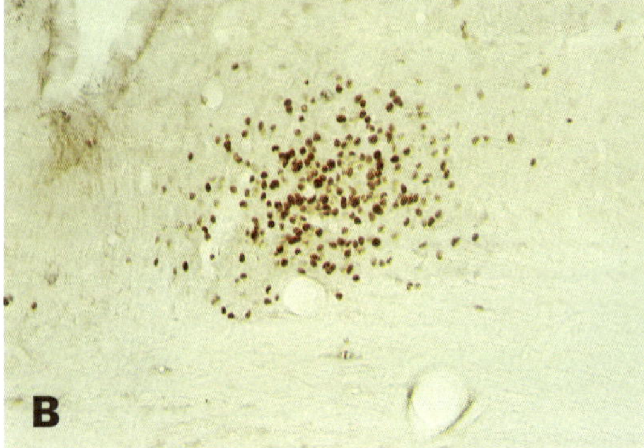

Figure 7: Inhibition of c-Fos expression. Rats were injected into the third ventricle of the brain either with **A:** anti *c-fos* oligo or **B:** control oligo. Immunohistochemistry with a specific c-Fos antibody was used to detect c-Fos protein expression. The control animal (**B**) shows marked expression of the c-Fos protein in the ventral part of the SCN after light stimulation. In contrast, injection of the anti *c-fos* oligo prior to light stimulation markedly reduces c-Fos protein expression (**A**).

3.3.2 Alteration of Animal Behavior by Anti-junB and Anti-c-fos Oligodeoxynucleotides *in vivo*

An example of behavioral alterations in the circadian rhythmic activity of animals modulated by injection of antisense oligos is shown in Figure 8. In this experiment antisense oligos were used to inhibit protein expression in the suprachiasmatic nucleus (SCN) of rat brain. The SCN is the major pacemaker for the generation of the endogenous circadian rhythm.

Rats kept in constant darkness for several weeks show a regular circadian pattern of activity (Fig. 8A). Stimulation with a 1 h light pulse during the subjective night phase of the animals shifts the beginning of circadian rhythmic activity by about 2 h. JunB and c-Fos are the predominantly expressed AP-1 transcription factor components in the SCN after light stimulation that induces resetting of the circadian clock.

To determine the functional role of JunB/c-Fos co-expression in the SCN antisense oligos were used under the following experimental conditions:

Protocol 4B: Modulation of Behavior – Circadian Rhythm

Special materials:
- See protocol 4A

- Implant rats stereotaxically with a microinjection guide cannula as described in protocol 4A (targeted to the bottom of the III. ventricle at the level of the SCN).
- Two weeks later, place the animals in cages equipped with a running wheel connected to an on-line monitoring computer system [51, 66].
- House the rats under constant darkness for 16 weeks.
- Every four weeks expose the animals to a light pulse (300 lux for 1 h) at subjective circadian time 15 h (CT 15). At this time point of the circadian cycle, a light pulse induces maximal phase shift of the circadian activity rhythm.
- Inject either a mixture of anti-c-*fos*/anti-*jun*B antisense S-ODN or a mixture of two randomized control S-ODN (one control oligo for each antisense oligo) 6 h prior to light exposure (2 µl of each oligo, diluted at 4 mM in physiological saline).

Figure 8B shows activity plots from an animal injected with antisense oligos (ASO) against c-*fos* and *jun*B before the stimulation with a light pulse (LP). The phase shift normally induced by light is completely prevented by the antisense oligos. In contrast, injection of control random "nonsense oligos" (NSO) 4 weeks later did not prevent the light induced phase shift in circadian rhythmic activity.

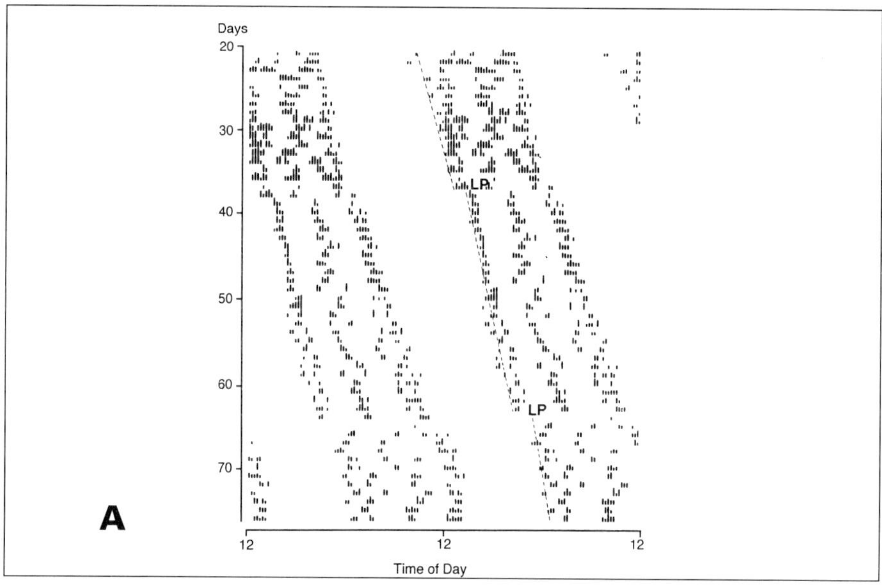

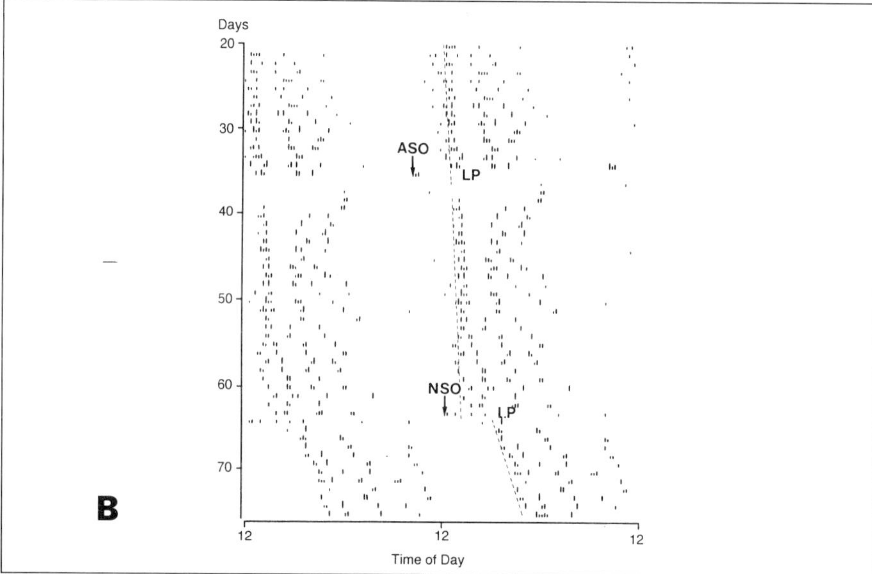

Figure 8: Antisense modulation of animal behavior: Running wheel activity. **A:** Circadian rhythm shift after light pulse treatment. The left half of the figure shows the level of activity of an individual rat as measured by on-line monitoring. Each bar represents 20 min of running wheel activity. The right side of the figure shows the same activity plot with a dotted line connecting the beginning of significant activity. Also, on the right side the times of light pulse treatment are indicated (LP). **B:** Activity plot from a rat injected with a combination of anti-*jun*B and anti c-*fos* antisense oligos (ASO) prior to light pulse (LP) treatment on day 36. No phase shift is induced. A control experiment in the same animal on day 64 shows that injection of a combination of two control nonsense oligos (NSO) prior to light stimulation leads to a normal physiological phase shift as in the untreated control rat (**A**). Modified reproduction with permission from [50].

3.3.3 *Suppression of G-Protein Coupled Receptor Expression* in vivo

The range of targets in this category is so wide that a discussion of their biological relevance is beyond the scope of this text. This section will instead focus on some practical considerations. As mentioned above, inhibiting the expression of G-protein coupled receptors has been one of the most consistently successful applications of antisense methodology in the nervous system. No indications are available to suggest that some receptor systems are per se more suitable targets than others. Not surprisingly, however, it appears that a functionally meaningful degree of expression blockade is more easily achieved with sparsely expressed receptors than with those present at high densities. In addition, receptors with a long half-life require more prolonged application of antisense oligos compared to those with a short half-life.

Surprisingly consistent protocols have emerged from reports of antisense receptor knock-downs (see introduction/review section). They all require longer oligo application times compared to rapid turnover proteins like c-Jun, c-Fos and JunB (see protocol 4). Our own variant for such a protocol follows here and is derived from recent work with the DA-D3 receptor [42]. While earlier work has used repeated intermittent oligo injections, we have lately turned to continuous infusion described in this protocol, since we find it to be more effective and reliable. Our experience with continuous infusion of oligos targeted to G-coupled receptor proteins is in agreement with the report that superior brain tissue distribution and concentrations are achieved in this manner [64].

Protocol 5: Continuous Oligo Infusion into the Rat Brain

Special materials:

- Osmotic minipump mod. 2001 (Alza Corp., Palo Alto, CA)
- Brain infusion kit (Alza Corp.)[4]
- Skull screws (Plastic One (0–80 × 1/16))
- Acrylic cement (Plastic One)
- Cannula holder (Plastic One)

- Day 1: Dissolve oligo in sterile Ringer's solution. We and others have found i.c.v. doses between 30 and 40 nmol/day effective (1.7 nmol/h in [42]). The pumping rate of the pump is 1 µl/h; therefore, these doses translate into concentrations of 1.25–1.7 nmol/µl, and are in agreement with the finding by Whitesell et al. [64] that delivery of phosphorothioate oligo at 1.5 nmol/h for 7 days produces good distribution throughout the brain, and is well tolerated.

---

[4] (L-shaped cannula of the kit can also be purchased cut to custom length directly from: Plastic One, Roanoak, VA).

- Cut tubing of the brain infusion kit to the appropriate length to avoid bending when pump assembly is put in place as described below. Load the pump and brain infusion kit separately; only then connect the two to avoid air bubble formation. Put the pump in saline overnight in order to prime it.
- Day 2: Under anesthesia of choice (we use ketamine + xylazine), place the rat in stereotactic frame; we use tooth-bar placement of -3.3 mm (i.e. 3.3 mm below interaural line).
- Make an incision of sufficient length to later allow inter-scapular placement of the pump. Carefully scrape away the periostium from the skull bone, and take care to achieve good hemostasis (with bone wax, Surgicel (R) or similar). Allow the bone to dry.
- Manually drill holes for 3–4 skull screws, and secure these to the bone. Place the cannula of the brain infusion kit into the cannula holder, take bregma readings, and calculate final coordinates, with the coordinates for the lateral ventricle being 0.8 mm posterior and 1.4 mm lateral to bregma, 4.4 mm ventral to skull surface. Drill a hole for the cannula, and lower the cannula tip in place.
- Prepare a working mixture of cement, and cement the cannula to the skull screws. Wait until the cement has hardened (or suffer the consequences!). Gently separate the claws of the cannula holder with a spatula, and remove the holder.
- With a hemostat, bluntly dissect a subcutaneous cavity from the caudal limit of the incision and down between the scapulae. Place the pump in the cavity, taking care not to bend the tubing. Close the incision with sutures or Wachenfelt-clips, taking care not to puncture the tubing.
- Postoperatively, give antibiotics systemically. We use benzylpenicillin, 200 000 U.
- Days 5–7: Perform functional experiments on day 5 through 7. Then sacrifice the animals, dissect out the brain regions of interest and verify receptor knock-down in a binding assay and/or direct protein assay e.g. by immunoprecipitation and/or Western blotting.

Protocol 5B: Binding Studies, Biochemistry, Behavior

- Densities of DA-D2/D3 receptors are determined by [3H]-spiperone binding in membrane preparations from limbic forebrain or caudate-putamen.
- Homogenates (obtained in a glass-Teflon homogenizer and binding buffer, see protocol 6 below) are centrifuged ($30.000 \times g$, 15 min $\times$ 2) and the final pellet is resuspended in buffer @ 30 mg protein/ml.
- Suspensions are incubated in triplicate (20 min @ 37 °C) with eight concentrations (17–574 pM) of ^{3}H-spiperone (97 mCi/mmol), dissolved in 0.1 % ascorbic acid and 0.01 BSA.

- Mianserin (4.2 µM) is added to mask serotoninergic binding sites. Non-specific binding is determined in the presence of 10 mM 6,7-ADTN. Incubations are terminated by filtration over Whatman GF/B filters coated with 0.1 % polyethylenamine, followed by washing with ice-cold buffer (3 × 4 ml Tris-HCl).
- Bound radioactivity is determined by liquid scintillation counting, and is analyzed using non-linear least-square curve fitting (LIGAND, Elsevier Biosoft).
- We found that density of spiperone sites is unaffected in caudate putamen, where D3 receptors are not expressed. In limbic forebrain, however, where according to various estimates, D3 receptors account for 20–40 % of spiperone binding, binding density is significantly reduced (Bmax 68.2 + 3.8 vs. 84.0 + 4.9, $p < 0.05$).

Biochemistry:

- Using standard reversed-phase HPLC with electrochemical detection, the synthesis rate of DA is determined as the accumulation of DOPA after inhibition of DOPA-decarboxylase by administration of NSD 1015 (115 mg/kg s.c., 30 min prior to killing subjects).
- No effects on DA synthesis were seen in caudate-putamen. In limbic forebrain, DOPAC accumulation was increased by 28 + 2.5 % ($p < 0.01$), indicating an inhibitory role for D3 receptors in regulation of DA synthesis [42].

Behavior:

- Locomotor activity is monitored for 60 min on day 6, in computerized infrared activity boxes (Motron, Stockholm). Exploratory activity, known to be driven by dopaminergic transmission in the basal ganglia, declines over time in normal animals, a phenomenon called habituation. Activity of D3-antisense treated subjects increased. Secondly, the decline in exploratory locomotor activity was slower than in random-oligo treated control animals (Figure 9).

A particular problem is likely to arise in antisense receptor studies. The rationale behind such experiments will most often be that conventional antagonists are lacking for the receptor type/subtype of interest, a limitation conveniently circumvented by the use of antisense. As stated elsewhere in this volume (Chapter 5, Assay Systems and Control), however, if antisense knock-downs are to be used for functional studies, it is mandatory that their effectivity is demonstrated at the protein level. For receptors, this is most appropriately done by assaying binding density. Yet, for the same reason which provides the rationale for targeting a receptor, the efficiency of knock-down will be difficult to demonstrate in a binding assay, since a (suf-

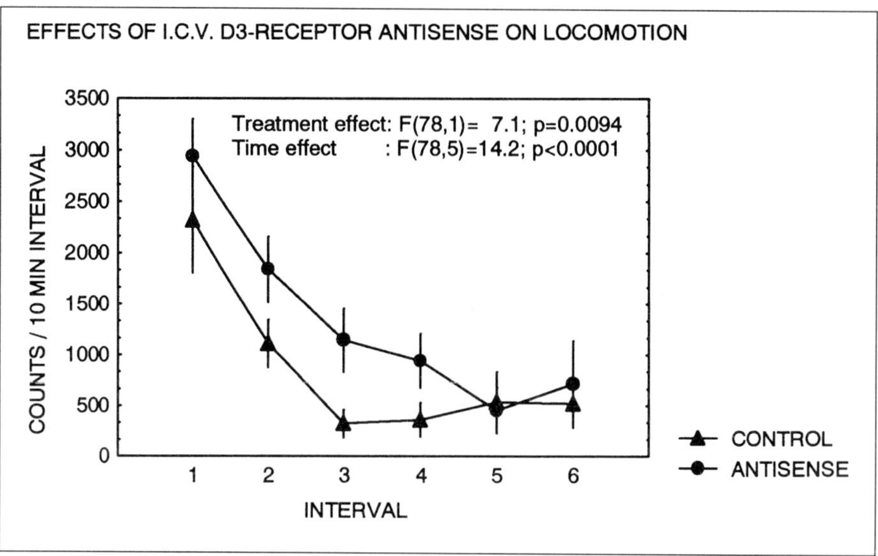

Figure 9: Antisense modulation of animal behavior: Exploratory behavior. Spontaneous locomotor activity of rats treated with anti-D3 oligo (dots) or random control oligo (triangles) treated subjects. Activity is measured over a 1h test session in a novel environment (infrared detector locomotor activity chamber). A highly significant activity increase is seen in the anti-D3 oligo treated group, characterized by a slowing of the normal activity decline over time. This behavioral finding is consistent with the biochemically detected increase in dopaminergic synthesis (see text).

ficiently selective) ligand will also be likely to be lacking. Alternative approaches, *e.g.* methods based on the use of antibodies directed toward epitopes within the receptor protein may be attempted, but the availability of such antibodies is limited, and quantitation of functional receptor sites is not trivial. Furthermore, as described above in the case of secreted peptides (see chapter 2.3) the amount of total protein present may not correlate with the amount of functionally available protein, due to inactivating posttranslational modifications.

Combining *in vitro* and *in vivo* approaches may provide an attractive solution in this type of situation, illustrated by our recent work with the DA-D3 receptor subtype [42] [15]. Established ligands, such as spiperone, label both D2 and D3 sites. Infusion of an antisense oligo targeting the D3 transcript dramatically affected DA synthesis, but it was difficult to establish on the basis of these data alone that this effect was caused by a selective knock-down of D3 receptors. Although spiperone binding showed a significant decrease of combined D2 + D3 binding density in ventral striatum, this decrease was small (approx. 20 %), and effects on D2 type sites could theoretically be involved. However, using transfected CHO cells expressing either D2 or D3 receptors, it could be demonstrated that our anti D3 oligo suppresses the expression of its target, while leaving the closely related D2 subtype unaffected. The protocol used for these *in vitro* studies is given below.

3.3.4 In vitro Blockade of G-Protein Coupled-Receptor Expression in Stably Transfected Non-neuronal Cells

Protocol 6: Determination of Oligodeoxynucleotide Selectivity for D3 versus D2 Dopamine Receptor

Special materials:

- Cells: Chines Hamster Ovary (CHO) cells, stably expressing D3 or D2 receptors, see [13])
- Medium: Alpha-MEM (Biochrom, Berlin, Germany), supplemented with 10 % FCS, 100 U/ml Penicillin G, 100 µg/ml streptomycin and 2 mM L-glutamine
- Binding buffer: 50 mM Tris-HCl, 4 mM MgCl2, 1.5 mM CaCl2, 5 mM KCl, 1 mM EDTA, 129 mM NaCl, 0.05 mg/ml BSA, 0.1 mg/ml ascorbic acid
- Ligand: ^{3}H-spiperone (95 Ci/mmol, Amersham; labels both D3 and D2 receptors)
- Plasticware: Petri dish 10 cm (Seromed (R), Biochrom); 6-well plate (Nunclon (R), Nunc, Roskilde, Denmark)

- Day 1: Seed out 2×10^3 stably D3-transfected CHO cells in 10 ml medium/dish in 10 cm Petri dishes.
- Day 2: Continue culture.
- Day 3 : With the cells growing adherent in dishes, discard old medium, and replace with equal volume of fresh medium containing oligos (antisense, random or mismatched) at a concentration of 4 µM.
- Day 4: Repeat as on day 3.
- Day 5: Discard the medium. Wash the cells in PBS (2×2.0 ml), and trypsinize (0.5 % trypsin). Using fresh medium containing oligos (at the same 4 µM concentration), plate 1.5×10^3 cells into 6-well plates in 2.0 ml/well.
- Day 6: With the cells adherent in the wells, replace with 2.0 ml fresh medium containing oligos.
- Day 7: Repeat as on day 6.
- Day 8: With the cells still adherent in the wells, wash in PBS (2×2.0 ml). Then add ligand, in 1.0 ml binding buffer, at 9 different concentrations (0.078–5.2 nM final; triplicates).
- Incubate 60 min at room temperature. Stop the incubation by washing in 50 mM Tris-HCl (5×5.0 ml).
- To dissolve the cells, add 500 µl glacial acetic acid to each well. Count bound radioactivity in triplicate 100 µl samples from each well.
- For each concentration, non-specific binding is determined in separate wells in the presence of 2 µM (+)-butaclamol.
- To control for effects on cell viability and/or proliferation, parallel wells (triplicates) for each oligo treatment are used to determine cells numbers and trypan blue exclusion.

> • In separate experiments, use the same protocol as above except using D2-transfected CHO cells to demonstrate the specificity of the anti-D3 oligo (which should not, and does not affect spiperone binding in the D2 cells).

Transfected cell lines offer many practical advantages, but express receptors at levels exceeding those of *e.g.* mature neurons. Transfectants therefore require a highly efficacious antisense treatment for successful expression blockade. We found a 30 % reduction of binding *in vitro*, correlated with good activity *in vivo* [42]. A cell line constitutively expressing the receptor of interest may alternatively be used if available, but will require other control experiments. Primary neuronal culture may finally offer an attractive but technically more demanding alternative.

3.4   Analysis of Tissue Distribution and Cellular Uptake *in vivo*

Tissue distribution of oligos and their cellular uptake after i.c.v. of intracerebral injection can be monitored using labelled oligos (for labelling techniques see chapter 6, Cell Culture and chapter 3, Labelling). We found that ^{35}S radiolabelled and FITC-labelled oligos show very similar distribution patterns[5] [59]. Since handling of FITC-oligos is more convenient than that of radiolabelled probes especially for *in vivo* application, the protocols given below describe the use of FITC-labelled probes.

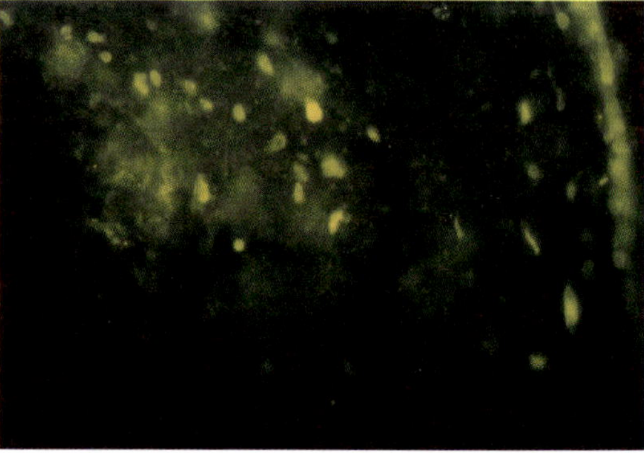

Figure 10: Uptake of FITC-oligo after i.c.v. injection. Detection of fluorescence labelling in a section of rat brain after intraventricular injection of 2 µl FITC-labelled oligo (4 mM concentration) into the third ventricle. The rim of the third ventricle is seen in the upper right corner. Animals were sacrificed 6 h after oligo injection. The ependymal cells lining the ventricle are clearly labelled. Furthermore, oligo uptake is also seen in the adjacent brain tissue indicating that the oligo freely diffuses from the cerebrospinal fluid into brain tissue.

---

[5] Virtually the same results were also obtained using BrdU-labelled oligos and anti-BrdU antibodies for detection. However, this method is more time-consuming compared to direct visualisation of the FITC-labelled oligos. Unless immunohistological detection of BrdU labelled probes is routinely done in a laboratory we recommend the use of FITC-labelled probes, which can easily and reliably be detected with a fluorescence microscope.

Protocol 7: Injection of Labelled Oligos into the Brain

Special materials:

- Highly purified FITC labelled phosphorothioate oligo (Biognostik)
- Skull screws (Plastic One (0–80 × 1/16))
- Acrylic cement (Plastic One)
- Cannula holder (Plastic One)
- Stainless steel guide cannula
- *Continuous i.c.v. infusion into the lateral ventricle:* Use the protocol for osmotic minipump implantation as described in protocol 5 with the following modification: Fill the pump and tubing with FITC-labelled oligo. Preferably more than 90 % of the oligo should be FITC-labelled. This labelling efficiency requires appropriate HPLC purification steps to separate FITC-containing from unlabelled oligo. When determining the amount of oligo by UV-spectrometry use fluorescein-quenching buffer to avoid false quantification of oligo concentration through fluorescein-mediated UV-absorption.
- *Histological examination:* Remove the brains *e.g.* after 5–7 days, at the same time as the functional studies are performed in the antisense experiments and prepare tissue sections for fluorescence microscopy. Compare sections from uninjected animals to determine background fluorescence.
- *Single i.c.v. injections for suppression of proteins with rapid turnover:* Implant a microinjection guide-cannula (see protocol 4): or under deep anesthesia inject with a Hamilton microsyringe. Inject 2–4 µl of a 1–4 mM FITC-oligo solution. Remove the brains after 2–24 h, section the tissue *e.g.* on a cryostat and examine under a fluorescence microscope. For results see Figure 10.
- *Direct injection into brain tissue:* For single or few repeated injections into brain tissue implant a microinjection guide-cannula under anesthesia or directly apply the oligo with a Hamilton microsyringe[6]. To target *e.g.* the dorsolateral hippocampus in adult rats implant the cannula 2.2 mm lateral and 2.8 mm posterior to the bregma with the tip of the cannula placed 3.2 mm ventral of the skull surface. Inject 1–2 µl FITC-oligo at 2–4 mM concentration. Process the tissue as above for fluorescence microscopy. A typical example is shown in Figure 11, showing strong fluorescence labelling of neuronal cell bodies as well as neurites.

---

[6] Induction of immediate early type transcription factors by insertion of the needle is not of equal concern as it is for the antisense experiments to suppress such genes. Using both the microinjection guide cannula and Hamilton microsyringe injection, we found little difference with respect to oligo uptake. Yet, to ensure that uptake studies are performed under exactly the same conditions as the functional studies, we usually prefer application with the microinjection guide cannula.

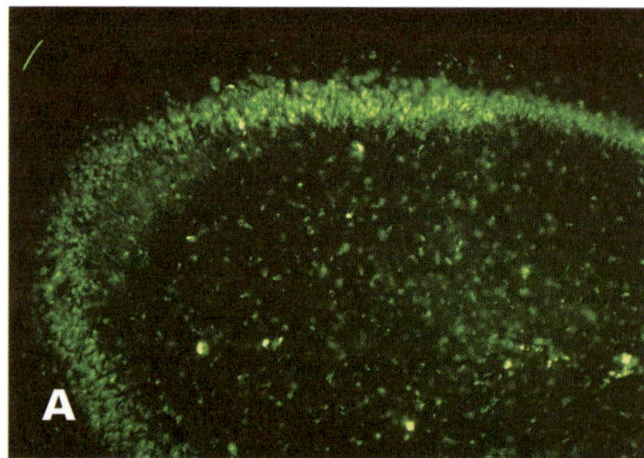

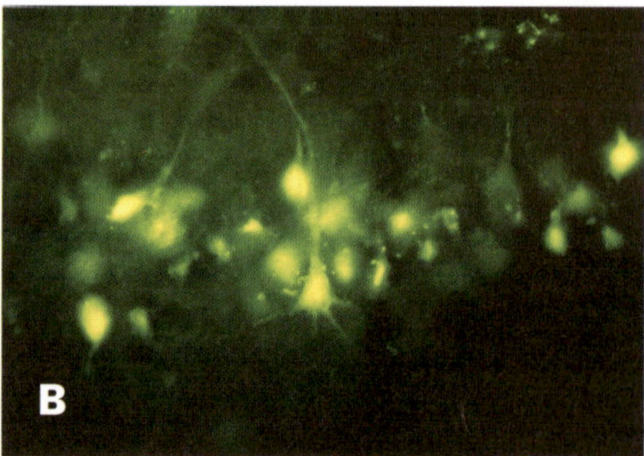

Figure 11: Uptake of FITC-oligo after intracerebral injection. Detection of fluorescence labelling in rat brain after injection of 2 µl FITC-labelled oligo (4 mM concentration) into rat brain tissue of the hippocampal formation. **A**: Overview over the hippocampal region. The pyramidal cell layer is particularly strongly labelled. **B**: Higher magnification of pyramidal neurons in the hippocampal formation. Labelling is seen in the neuronal cell bodies as well as in apical dendrites.

3.5   Antisense Oligodeoxynucleotides in Neurooncological Disorders

The relevance of gene expression for brain tumor cell proliferation is important to identify drug target molecules. Antisense oligos themselves are promising candidates for the treatment of malignant brain tumors. For a protocol on inhibition of TGF-β2 expression in human glioblastoma cells see chapter 12, Oncology, section 5.

Here we will describe antisense experiments in human medulloblastoma cells. Medulloblastoma is the most common childhood brain tumor. Unfortunately, in the majority of children with medulloblastoma relapse occurs after surgical and/or radiotherapy treatment. Development of new therapeutic agents is thus urgently needed. In the following we will describe antisense experiments with medulloblastoma cells overexpressing a well studied oncogene c-*myc*.

Protocol 8: Inhibition of Gene Expression in Brain Tumor Cells

Special materials:

- Human medulloblastoma cells HTB 187 (ATCC, non-adherent suspension cells).
- Eagle's MEM medium (Gibco/BRL)

- Sodium pyruvate (Gibco/BRL)
- Non-essential amino acids (NEAA) (Gibco/BRL)
- Trypan blue (Sigma)
- 96-well microtiter plates (Nunc)
- Sterile 1.5 ml microfuge tubes

- Grow the cells in Eagle's MEM supplemented with 1 mM sodium pyruvate, 1 % NEAA and 5 % FCS without antibiotics.
- Plate 10000 cells per well in 100 µl medium into 96-well microtiter plates. Use 16–20 wells per condition *i.e.* cells treated with antisense oligo, cells treated with control oligo and untreated control cells.
- Add oligos at 2 µM final concentration.
- Resuspend the cells twice weekly in fresh medium: Pipette the cells into sterile microfuge tube. Centrifuge the cells, remove the supernatant and discard. Resuspend the cells in 100 µl of Eagle's MEM/1 mM sodium pyruvate/1 % NEAA/5 % FCS supplemented with 2µM oligo for the antisense treated and control oligo treated cells and transfer the cells into a 96-well microtiter plate.
- Count the cells from 2 wells per condition every week using trypan blue to determine the cell viability. At a density of 50000 cells per well dilute the cells to a density of 10.000 per well to maintain similar proliferation rates.

Results from our experiments with anti-c-*myc* oligo are shown in Figure 12.

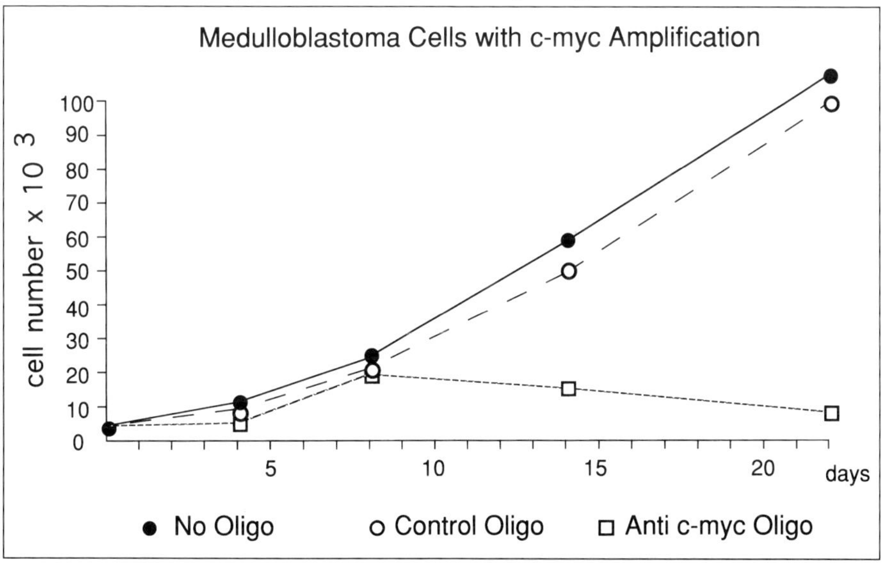

Figure 12: Medulloblastoma cell proliferation. Medulloblastoma cell proliferation after treatment with control oligo (open circles) no oligo (filled circles) or anti-c-*myc* oligo (open squares). After day 8 of oligo treatment a decrease in cell proliferation is seen in the anti-c-*myc* treated tumor cells.

4   Conclusions

The use of antisense oligos is a technology which is rapidly gaining momentum in neurobiology since it allows to determine the function of proteins by selective knock-down of their expression.

Antisense oligos are becoming increasingly important tools to resolve the function of receptor subtypes, transcription factors, enzymes and secreted proteins, to name just a few. The above protocols which proved useful to suppress certain proteins in different experimental systems, *i.e.* cell lines, primary neuronal cultures and in the central nervous system may serve as guide lines for antisense experiments in neurobiology, which can then be fine tuned for each individual protein and experimental system.

Protein half-life, basal expression levels and time course of induction are important variables that determine the optimal application mode and duration of antisense treatment. With a few preliminary experiments the optimal time window for antisense studies can usually be determined.

Today, antisense oligos are successfully applied in three different fields:

1. In basic science they serve as selective tools for gene function analysis in neurobiology.
2. Drug finding is considerably speeded up using antisense oligos for drug target validation. Thus, antisense technology is now being used in most major pharmaceutical companies to rationalize drug-development. This is particularly useful in CNS diseases. With the increasing knowledge on the multitude of different families of receptors, signal transduction proteins and transcription factors, analysis of their function requires new tools. Conventional agonists and antagonists usually lack specificity for different receptor subtypes and are not available for many other proteins like transcription factors. Antisense technology is therefore becoming the method of choice for drug target validation in neuropharmacology.
3. Antisense oligos themselves are being developed as highly selective drugs. For most neurological diseases, including neurodegenerative disorders, currently available therapeutic options are poor. Thus, development of new therapeutic agents for these often severely disabling diseases is required. The specificity of antisense mediated modulation of gene expression is an intriguing approach for the development of novel therapeutic agents.

References

1. AKABAYASHI, A., C. WAHLESTEDT, J. T. ALEXANDER, AND S. F. LEIBOWITZ: Specific inhibition of endogenous neuropeptide Y synthesis in arcuate nucleus by antisense oligonucleotides suppresses feeding behavior and insulin secretion. Brain Research. Molec. Brain Res. 21 (1–2) (1994) 55.
2. ARNOLD, F. J., DE, LUCAS, BUENO, M, H. SHIERS, D. C. HANCOCK, *et al.*: Expression of c-fos in regions of the basal limbic forebrain following intracerebroventric-

ular corticotropin-releasing factor in unstressed or stressed male rats. Neuroscience 51(2) (1992) 377.

3. BEERI, R., C. ANDRES, E. LEV-LEHMAN, R. TIMBERG, T. HUBERMAN, M. SHANI, AND H. SOREQ: Transgenic expression of human acetylcholinesterase induces progressive cognitive deterioration in mice. Current Biology 5(1063–1071) (1995).

4. BEHL, C., U. BOGDAHN, J. WINKLER, R. APFEL, W. BRYSCH, AND K. H. SCHLINGENSIEPEN: Autoinduction of platelet derived growth factor (PDGF) A-chain mRNA expression in a human malignant melanoma cell line and growth inhibitory effects of PDGF-A-chain mRNA-specific antisense molecules. Biochem. & Biophys. Res. Com. 193(2) (1993) 744.

5. BEHL, C., J. WINKLER, U. BOGDAHN, J. MEIXENSBERGER, K. H. SCHLINGENSIEPEN, AND W. BRYSCH: Autocrine growth regulation in neuroectodermal tumors as detected with oligodeoxynucleotide antisense molecules. Neurosurgery 33(4) (1993) 679.

6. BOTTENSTEIN, J. E. AND G. H. SATO: Growth of a rat neuroblastoma cell line in serum-free supplemented medium. Proc. Natl. Acad. Sci. USA 76(1) (1979) 514.

7. BOURSON, A., E. BORRONI, R. H. AUSTIN, F. J. MONSMA, AND A. J. SLEIGHT: Determination of the role of the 5-ht6 receptor in the rat brain: a study using antisense oligonucleotides. J. Pharmacol. & Exp. Ther. 274(1) (1995) 173.

8. BRYSCH, W. AND K. H. SCHLINGENSIEPEN: Design and application of antisense oligonucleotides in cell culture, in vivo, and as therapeutic agents. [Review] Cellular & Molecular Neurobiology 14(5) (1994) 557.

9. BURCH, R. M. AND L. C. MAHAN: Oligonucleotides antisense to the interleukin 1 receptor mRNA block the effects of interleukin 1 in cultured murine and human fibroblasts and in mice. J. Clin. Invest. 88(4) (1991) 1190.

10. CAINE, S. B. AND G. F. KOOB: Modulation of cocaine self-administration in the rat through D-3 dopamine receptors. Science 260(5115) (1993) 1814.

11. CAMPEAU, S. , M. D. HAYWARD, B. T. HOPE, J. B. ROSEN, E. J. NESTLER, AND M. DAVIS: Induction of the c-fos proto-oncogene in rat amygdala during unconditioned and conditioned fear. Brain Research 565(2) (1991) 349.

12. CHIASSON, B. J., M. L. HOOPER, P. R. MURPHY, AND H. A. ROBERTSON: Antisense oligonucleotide eliminates in vivo expression of c-fos in mammalian brain. Eur. J. Pharmacol. 227(4) (1992) 451.

13. CHIO, C. L., M. E. LAJINESS, AND R. M. HUFF: Activation of heterologously expressed D3 dopamine receptors: comparison with D2 dopamine receptors. Mol. Pharmacol. 45(1) (1994) 51.

14. DRAGUNOW, M., P. LAWLOR, B. CHIASSON, AND H. ROBERTSON: c-fos antisense generates apomorphine and amphetamine-induced rotation. Neuroreport 5(3) (1993) 305.

15. EKMAN: (submitted).

16. ESTUS, S. , W. J. ZAKS, R. S. FREEMAN, M. GRUDA, R. BRAVO, AND E. J. JOHNSON: Altered gene expression in neurons during programmed cell death: identification of c-jun as necessary for neuronal apoptosis. J. Cell Biol. (1994).

17. EVA, C., K. KEINANEN, H. MONYER, P. SEEBURG, AND R. SPRENGEL: Molecular cloning of a novel G protein-coupled receptor that may belong to the neuropeptide receptor family. Febs Letters 271(1–2) (1990) 81.

18. GEORGIEVA, J., M. HEILIG, I. NYLANDER, M. HERRERAMARSCHITZ, AND L. TERENIUS:

In vivo antisense inhibition of prodynorphin expression in rat striatum – dose-dependence and sequence specificity. Neurosci. Lett. 192(1) (1995) 69.

19. GYURKO, R., D. WIELBO, AND M. I. PHILLIPS: Antisense inhibition of AT1 receptor mRNA and angiotensinogen mRNA in the brain of spontaneously hypertensive rats reduces hypertension of neurogenic origin. Regul. Pept. 49(2) (1993) 167.

20. HAM, J., C. BABIJ, J. WHITFIELD, C. M. PFARR, D. LALLEMAND, M. YANIV, AND L. L. RUBIN: A c-Jun dominant negative mutant protects sympathetic neurons against programmed cell death. Neuron 14(5) (1995) 927.

21. HEILIG, M.: Antisense inhibition of neuropeptide Y (NPY) Y1 receptor expression blocks the anxiolytic-like action of NPY in amygdala and paradolically increases feeding. Regul. Pept. (in press).

22. HEILIG, M., J. A. ENGEL, AND B. SODERPALM: C-fos antisense in the nucleus accumbens blocks the locomotor stimulant action of cocaine. Europ. J. Pharmacol. 236(2) (1993) 339.

23. HEILIG, M., G. F. KOOB, R. EKMAN, AND K. T. BRITTON: Corticotropin-releasing factor and neuropeptide Y: role in emotional integration. [Review] Trends in Neurosciences 17(2) (1994) 80.

24. HEILIG, M., E. M. PICH, G. F. KOOB, F. YEE, AND C. WAHLESTEDT: In vivo down-regulation of neuropeptide Y (NPY) Y1 receptors by i.c.v. antisense oligodeoxynucleotide administration is associated with signs of anxiety in rats. Soc. Neurosci. Abs. (1992) 642.18.

25. HEILIG, M. AND K. H. SCHLINGENSIEPEN: Antisense oligodeoxynucleotides as novel neuropharmacological tools for selective expression blockade in the brain. In: D.S. Latchman, ed. Genetic manipulation of the nervous system. Academic Press London (1996) 249.

26. HERZOG, H., Y. J. HORT, H. J. BALL, G. HAYES, J. SHINE, AND L. A. SELBIE: Cloned human neuropeptide Y receptor couples to two different second messenger systems. Proc. Natl. Acad. Sci. USA 89(13) (1992) 5794.

27. HILBERG, F., A. AGUZZI, N. HOWELLS, AND E. F. WAGNEr: c-jun is essential for normal mouse development and hepatogenesis [published erratum appears in Nature 1993 Nov 25; 366(6453):368]. Nature 365(6442) (1993) 179.

28. HONEGGER, P., D. LENOIR, AND P. FAVROD: Growth and differentiation of aggregating fetal brain cells in a serum-free defined medium. Nature 282(5736) (1979) 305.

29. JACHIMCZAK, P., U. BOGDAHN, J. SCHNEIDER, C. BEHL, J. MEIXENSBERGER, R. APFEL, R. DORRIES, et al.: The effect of transforming growth factor-beta 2-specific phosphorothioate-anti-sense oligodeoxynucleotides in reversing cellular immunosuppression in malignant glioma. J. Neurosurg. 78(6) (1993) 944.

30. JOHNSON, R. S. , L. B. van, V. E. PAPAIOANNOU, AND B. M. SPIEGELMAN: A null mutation at the c-jun locus causes embryonic lethality and retarded cell growth in culture. Genes & Development (1993).

31. KALRA, P. S. , J. J. BONAVERA, M. G. DUBE, W. R. CROWLEY, AND S. P. KALRA: Inhibition of endogenous neuropeptide Y (NPY) synthesis by antisense oligodeoxynucleotide administration suppresses the progesterone-induced LH surge. 24th Annual Meeting of the Society for Neuroscience, Miami Beach, Florida, Usa, November 20(1–2) (1994).

32. KORNHAUSER, J. M., D. E. NELSON, K. E. MAYO, AND J. S. TAKAHASHI: Regulation of

jun-B messenger RNA and AP-1 activity by light and a circadian clock. Science 255(5051) (1992) 1581.

33. LARHAMMAR, D., A. G. BLOMQVIST, F. YEE, E. JAZIN, H. YOO, AND C. WAHLESTEDT: Cloning and functional expression of a human neuropeptide Y/peptide YY receptor of the Y1 type. J. Biol. Chem. 267(16) (1992) 10935.

34. LEV-LEHMANN, E., G. HORNREICH, D. GINZBERG, A. GNATT, A. MESHORER, F. ECHSTEIN, H. SOREQ, *et al.*: Antisense inhibition of acetylcholinesterase gene expression causes transient hematopoietic alterations *in-vivo*. Gene Therapy 2 (1994) 127.

35. LEVY, Y., A. TSAPIS, AND J. C. BROUET: Interleukin-6 antisense oligonucleotides inhibit the growth of human myeloma cell lines. J. Clin. Invest. 88(2) (1991) 696.

36. LIU, P. K., A. SALMINEN, Y. Y. HE, M. H. JIANG, J. J. XUE, J. S. LIU, AND C. Y. HSU: Suppression of ischemia-induced fos expression and AP-1 activity by an antisense oligodeoxynucleotide to c-fos mRNA [see comments] Ann. Neurol. 36(4) (1994) 566.

37. LOEWENSTEINLICHTENSTEIN, Y., M. SCHWARZ, D. GLICK, B. NORGAARDPEDERSEN, H. ZAKUT, AND H. SOREQ: Genetic predisposition to adverse consequences of anticholinesterases in atypical bche carriers Nature Medicine 1(10) (1995) 1082.

38. MANI, S. K., J. D. BLAUSTEIN, J. M. ALLEN, S. W. LAW, B. W. O'MALLEY, AND J. H. CLARK: Inhibition of rat sexual behavior by antisense oligonucleotides to the progesterone receptor. Endocrinology 135(4) (1994) 1409.

39. MORATALLA, R., E. A. VICKERS, H. A. ROBERTSON, B. H. COCHRAN, AND A. M. GRAYBIEL: Coordinate expression of c-fos and jun B is induced in the rat striatum by cocaine. J. Neurosci. 13(2) (1993) 423.

40. MORGAN, J. I. AND T. CURRAN: Stimulus-transcription coupling in the nervous system: involvement of the inducible proto-oncogenes fos and jun. [Review] Ann. Rev. Neurosci. 14(421) (1991) 421.

41. NEUMANN, I., D. W. PORTER, R. LANDGRAF, AND Q. J. PITTMAN: Rapid effect on suckling of an oxytocin antisense oligonucleotide administered into rat supraoptic nucleus. Am. J. Physiol. (1994).

42. NISSBRANDT, H., A. EKMAN, E. ERIKSSON, AND M. HEILIG: Dopamine D3 receptor antisense influences dopamine synthesis in rat brain. Neuroreport 6(3) (1995) 573.

43. OGAWA, S. , U. E. OLAZABAL, I. S. PARHAR, AND D. W. PFAFF: Effects of intrahypothalamic administration of antisense DNA for progesterone receptor mRNA on reproductive behavior and progesterone receptor immunoreactivity in female rat. J. Neurosci. (1994).

44. PASTERNAK, G. W. AND K. M. STANDIFER: Mapping of opioid receptors using antisense oligodeoxynucleotides – correlating their molecular biology and pharmacology [Review] Trends in Pharmacological Sciences 16(10) (1995) 344.

45. PATINKIN, D., L. E. LEV, H. ZAKUT, F. ECKSTEIN, AND H. SOREQ: Antisense inhibition of butyrylcholinesterase gene expression predicts adverse hematopoietic consequences to cholinesterase inhibitors. Cellular & Molecular Neurobiology 14(5) (1994) 459.

46. POLLIO, G., P. XUE, M. ZANISI, A. NICOLIN, AND A. MAGGI: Antisense oligonucleotide blocks progesterone-induced lordosis behavior in ovariectomized rats. Brain Res. Mol. Brain Res. 19(1–2) (1993) 135.

47. ROTHER, S. , R. SCHMIDT, W. BRYSCH, AND K. H. SCHLINGENSIEPEN: Learning-in-

duced expression of meningeal ependymin mrna and demonstration of ependymin in neurons and glial cells J. Neurochem. 65(4) (1995) 1456.

48. SAKAI, R. R., P. F. HE, X. D. YANG, L. Y. MA, Y. F. GUO, J. J. REILLY, C. N. MOGA, *et al.*: Intracerebroventricular administration of AT1 receptor antisense oligonucleotides inhibits the behavioral actions of angiotensin II. J. Neurochem. 62(5) (1994) 2053.

49. SCHLINGENSIEPEN, K. H., R. SCHLINGENSIEPEN, M. KUNST, I. KLINGER, W. GERDES, W. SEIFERT, AND W. BRYSCH: Opposite functions of jun-B and c-jun in growth regulation and neuronal differentiation. Developmental Genetics 14(4) (1993) 305.

50. SCHLINGENSIEPEN, K. H., F. WOLLNIK, M. KUNST, R. SCHLINGENSIEPEN, T. HERDEGEN, AND W. BRYSCH: The role of jun transcription factor expression and phosphorylation in neuronal differentiation, neuronal cell death, and plastic adaptations in vivo. Cellular & Molecular Neurobiology 14(5) (1994) 487.

51. SCHLINGENSIEPEN, K. H., F. WOLLNIK, M. KUNST, R. SCHLINGENSIEPEN, T. HERDEGEN, AND W. BRYSCH: The role of Jun transcription factor expression and phosphorylation in neuronal differentiation, neuronal cell death, and plastic adaptations in vivo. Cellular & Molecular Neurobiology 14(5) (1994) 487.

52. SCHMIDT, R., W. BRYSCH, S. ROTHER, AND K. H. SCHLINGENSIEPEN: Inhibition of memory consolidation after active avoidance conditioning by antisense intervention with ependymin gene expression J. Neurochem. 65(4) (1995) 1465.

53. SKUTELLA, T., H. CRISWELL, S. MOY, J. C. PROBST, G. R. BREESE, G. F. JIRIKOWSKI, AND F. HOLSBOER: Corticotropin-releasing hormone (CRH) antisense oligodeoxynucleotide induces anxiolytic effects in rat. Neuroreport 5(16) (1994) 2181.

54. SKUTELLA, T., J. C. PROBST, M. ENGELMANN, C. T. WOTJAK, R. LANDGRAF, AND G. F. JIRIKOWSKI: Vasopressin antisense oligonucleotide induces temporary diabetes insipidus in rats. J. Neuroendocrinol. 6(2) (1994) 121.

55. SOKOLOFF, P., B. GIROS, M. P. MARTRES, M. L. BOUTHENET, AND J. C. SCHWARTZ: Molecular cloning and characterization of a novel dopamine receptor (D3) as a target for neuroleptics. Nature 347(6289) (1990) 146.

56. SOMMER, W., B. BJELKE, D. GANTEN, AND K. FUXE: Antisense oligonucleotide to c-fos induces ipsilateral rotational behaviour to d-amphetamine. Neuroreport 5(3) (1993) 277.

57. SPAMPINATO, S. , M. CANOSSA, L. CARBONI, G. CAMPANA, G. LEANZA, AND S. FERRI: Inhibition of proopiomelanocortin expression by an oligodeoxynucleotide complementary to beta-endorphin mRNA. Proc. Natl. Acad. Sci. USA 91(17) (1994) 8072.

58. STANLEY, B. G.: Neuropeptide y in multiple hypothalamic sites controls eating behavior endocrine and autonomic systems for body energy balance. Colmers, W. F. And C. Wahlestedt (Ed.). Contemporary Neuroscience: The Biology Of Neuropeptide Y And Related Peptides. Xvi+564p. Humana Press Inc.: Totowa, New Jersey, Usa. Isbn (1993).

59. TISCHMEYER, W., R. GRIMM, H. SCHICKNICK, W. BRYSCH, AND K. H. SCHLINGENSIEPEN: Sequence-specific impairment of learning by c-jun antisense oligonucleotides. Neuroreport 5(12) (1994) 1501.

60. WAHLESTEDT, C., E. GOLANOV, S. YAMAMOTO, F. YEE, H. ERICSON, H. YOO, C. E. INTURRISI, *et al.*: Antisense oligodeoxynucleotides to NMDA-R1 receptor channel protect cortical neurons from excitotoxicity and reduce focal ischaemic infarctions. Nature 363(6426) (1993) 260.

61. WAHLESTEDT, C., E. M. PICH, G. F. KOOB, F. YEE, AND M. HEILIG: Modulation of anxiety and neuropeptide Y-Y1 receptors by antisense oligodeoxynucleotides. Science 259(5094) (1993) 528.

62. WANG, Z. Q., C. OVITT, A. E. GRIGORIADIS, S. U. MOHLE, U. RUTHER, AND E. F. WAGNER: Bone and haematopoietic defects in mice lacking c-fos. Nature 360(6406) (1992) 741.

63. WEISS, B., L. W. ZHOU, S. P. ZHANG, AND Z. H. QIN: Antisense oligodeoxynucleotide inhibits D2 dopamine receptor-mediated behavior and D2 messenger RNA. Neuroscience 55(3) (1993) 607.

64. WHITESELL, L., D. GESELOWITZ, C. CHAVANY, B. FAHMY, S. WALBRIDGE, J. R. ALGER, AND L. M. NECKERS: Stability, clearance, and disposition of intraventricularly administered oligodeoxynucleotides: implications for therapeutic application within the central nervous system. Proc. Natl. Acad. Sci. USA 90(10) (1993) 4665.

65. WISDEN, W., M. L. ERRINGTON, S. WILLIAMS, S. B. DUNNETT, C. WATERS, D. HITCHCOCK, G. EVAN, et al.: Differential expression of immediate early genes in the hippocampus and spinal cord. Neuron 4(4) (1990) 603.

66. WOLLNIK, F., W. BRYSCH, E. UHLMANN, F. GILLARDON, R. BRAVO, M. ZIMMERMANN, K. H. SCHLINGENSIEPEN, et al.: Block of c-Fos and JunB expression by antisense oligonucleotides inhibits light-induced phase shifts of the mammalian circadian clock. Europ. J. Neurosci. 7(3) (1995) 388.

67. ZHANG, M. AND I. CREESE: Antisense oligodeoxynucleotide reduces brain dopamine D2 receptors: behavioral correlates. Neurosci. Lett. 161(2) (1993) 223.

68. ZHENG, H., B. M. SAHAI, P. KILGANNON, A. FOTEDAR, AND D. R. GREEN: Specific inhibition of cell-surface T-cell receptor expression by antisense oligodeoxynucleotides and its effect on the production of an antigen-specific regulatory T-cell factor. Proc. Natl. Acad. Sci. USA 86(10) (1989) 3758.

69. ZHOU, L. W., S. P. ZHANG, Z. H. QIN, AND B. WEISS: In vivo administration of an oligodeoxynucleotide antisense to the D2 dopamine receptor messenger RNA inhibits D2 dopamine receptor-mediated behavior and the expression of D2 dopamine receptors in mouse striatum. J. Pharmacol. Exp. Ther. 268(2) (1994) 1015.

9 Virology

Oligonucleotides as Antiviral Agents

Sudhir Agrawal and Jamal Temsamani
Hybridon Inc., Worcester, USA

1   Introduction

A critical obstacle to the development of successful antiviral drugs is the problem of specificity. Most antiviral agents that interrupt the virus replication cycle also affect the replication of cells in the host, causing significant problems. It is important, therefore, to identify characteristics of the virus that distinguish it from its host. One such characteristic is the nucleic acid sequence itself. An antiviral agent that would destroy the viral nucleic acid while leaving the host unaffected might be an effective antiviral therapy that would have minimal side-effects.

The ability of antisense oligonucleotides (oligos) to selectively inhibit gene expression led to the suggestion that they would be useful therapeutic agents. Oligos are small segments of DNA that can be synthesized chemically. Gene expression is inhibited by hybridization of the oligo to sequences in the DNA or messenger RNA (mRNA) target in the host by Watson-Crick base pairing. By binding specifically to the target mRNA sequence, oligos form a hybrid DNA-RNA duplex and prevent the translation of that gene into protein. Because oligos act by simple base-pairing, they can be designed to target any gene of a known sequence. Achieving inhibition at the gene or mRNA level is believed to be a much more efficient intervention in the disease process than inhibition at the protein level. The viral genome is much smaller and different than that of humans, so the choice of a target sequence for oligo therapies is relatively straightforward and specific, and the use of antisense oligos as antiviral agents seems simple.

The first example of specific inhibition of viral gene expression by an oligo was reported by Zamecnik and Stephenson [1, 2], who demonstrated that a short oligo inhibited Rous sarcoma virus replication in cell culture. Since then, the field has progressed rapidly and today there are numerous reports in the literature documenting the effectiveness of various antisense oligos. The recent commencement of clinical trials using antisense oligo therapy for viral illnesses including human papilloma virus (HPV), human immunodeficiency virus (HIV), cytomegalovirus (CMV), and cancer, has heralded a new era in drug design [3–8]. In this chapter, we will review the antiviral effect of antisense oligos, emphasizing HIV-1 as a target.

2   Viral Targets

Antisense oligos should be most effective in inhibiting processes involving single-stranded RNA. The desired therapeutic goals influence which steps in the virus life cycle should be targeted. For example, to protect cells against viral infection, one would want to inhibit viral adsorption and viral genome integration whereas to prevent virus replication within cells already infected, inhibition of viral RNA processing or translation may be preferable. For maximum clinical control of the virus, both approaches may be necessary. If the viral genome is RNA, then an appropriate site within this RNA must be targeted (see below). If the viral genome is DNA, then mRNA must be targeted. In the case of negative-strand RNA viruses, such as influenza virus, the oligo can be targeted to the negative-strand RNA. The viral target should code for an essential function and should be expressed early in the viral replicative cycle, prior to genome replication which can increase mRNA levels. For example, many herpesvirus infections show three relatively discrete temporal pulses of RNA, named immediate-early, early, and late, and the expression of each depends to a large extent on the expression of the previous class. An antisense oligo might therefore best be designed to target the immediate-early class, effective inhibition of which should prevent the expression of the subsequent classes.

Another strategy is to target mRNAs that translate proteins essential for viral replication and survival. For example, the regulatory proteins *Rev* and *Tat* are essential for a productive viral replication cycle and inhibition of these proteins will stop viral replication. One must consider genetic variability in choosing a target sequence, especially when dealing with viruses that are highly mutated, such as HIV. Target regions in which the nucleotide sequences are conserved, such as the initiation site of *Gag* region, are ideal because the same oligo sequence can be used against most of the viral strains.

2.1   Virus Replication Cycle

Antisense oligos can be targeted to different stages of the viral life cycle. Figure 1 shows a simplified version of the HIV-1 replication cycle. The viral protein gp120 recognizes the cell CD4 surface protein and the virus enters the cell via membrane-to-membrane fusion. After HIV-1 enters the CD4 positive cell, reverse transcription occurs in the cytoplasm and the viral RNA is transcribed into DNA. This step is followed by the formation of a pre-integration complex, which carries the viral DNA into the nucleus. The viral genomic RNA therefore represents an early target for antisense oligos. By hybridizing to the viral RNA as it enters the cell, oligos may block reverse transcription, and may also degrade the viral RNA by becoming a substrate for RNase H at the hybridization site. Once the proviral DNA is integrated in the host cell genome, the viral genetic information stays in the double-stranded form and remains in the cell as long as the cell is viable. After the viral DNA is transcribed into viral RNA copies, antisense oligos may bind to appropriate sites on the pre-mRNA and inhibit translocation of pre-mRNA to the cytoplasm. Additionally, in the cytoplasm, antisense oligos may bind to the viral RNA and inhibit its translation into viral proteins by: a) blocking ribosome assembly; b) inhibiting their elongation; c) cleaving the specific RNA by RNAse H at the hybridization site; or d) disrupting the secondary structure of the mRNA (e.g. TAR, RRE of HIV-1) which may be essential for binding of regulatory proteins.

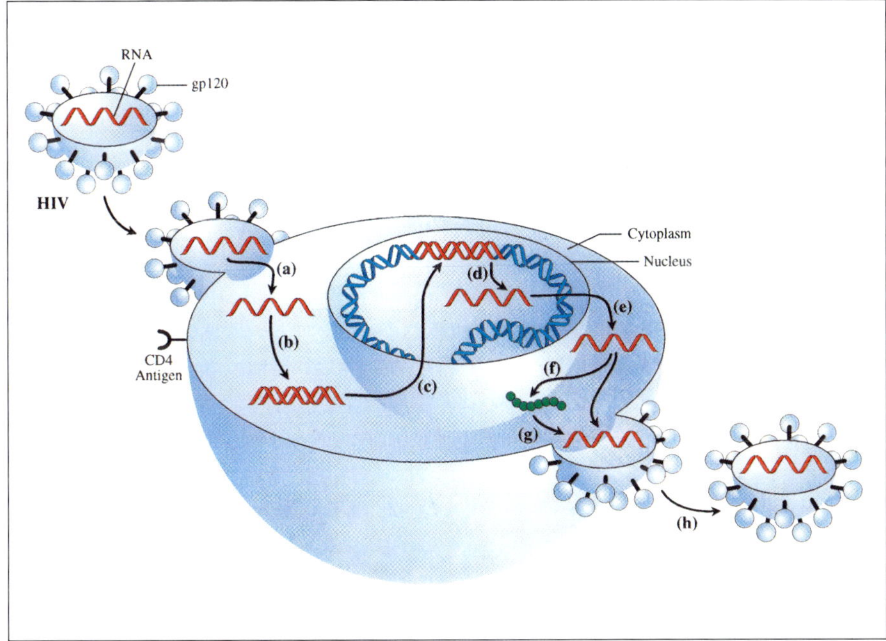

Figure 1: HIV-1 replication cycle. Several possible target sites for phosphorothioate oligos are indicated. (a) adsorption, (b) reverse transcription, (c) integration, (d) transcription, (e) maturation of the RNA, (f) translation, (g) packaging, and (h) release of the virus.

3   Antisense Oligonucleotides

3.1   Synthesis

Advances in oligo synthetic chemistry have contributed significantly to the growing use of antisense oligos. Solid-phase methods using phosphotriester, phosphoramidate, or H-phosphonate chemistry are the most common methods for synthesis of oligos and their various analogs [9, 10]. The synthesis is carried out on a solid support and nucleosides are added sequentially to the growing oligo chain. There are three essential steps in solid-phase synthesis; i) attachment of the first nucleoside to the solid support; ii) assembly of the oligo chain; iii) deprotection and removal of the oligo from the solid support. Oligos are synthesized from the 3'- end to the 5'- end. Each nucleoside addition involves removal of the acid-labile dimethoxytrityl group, coupling of the activated monomer, capping of the unreacted sites, and appropriate washing and oxidation steps. This cycle is repeated to extend the oligo chain to the desired length. Deprotection is then carried out under basic conditions to cleave the oligo from the solid support and to remove base-protecting groups.

Oligo phosphorothioates can be synthesized using an automated DNA synthesizer using the β-cyanoethyl phosphoramidate approach [10, 11]. To generate phosphorothioate linkages, oxidation of the intermediate linkage obtained after each coupling can be carried out using 3H,1,2-benzodithiole-3H-one-1,1-dioxide. Alternatively, phosphorothioate oligos can be synthesized using the H-phosphonate approach [10]. In this approach, the oligo is assembled using nucleosides H-phosphonate and a suitable activator (e.g. pivaloyl chloride or adamantanecarbonyl chloride). After the required sequence, the CPG-bound oligo (bearing 5'-dimethoxytrityl [DMT] group) is then treated with concentrated ammonium hydroxide for 8–10 h at 55 °C. The crude product is then resuspended in distilled water and the absorbance at 260 nm is checked to measure the total yield.

3.2   Purification of Oligonucleotide

Purification of the oligo is a very important step since the biological activity of the oligo depends on the purity of oligo. Different protocols have been described for purification of oligos and their analogs. In our laboratory, we have optimized the protocol for purifying phosphorothioate oligos on a small scale (1–30 mgs,~ 30–1000 A260 units) as well as for large-scale synthesis (30 mg to 1 gram). Oligos purified by this protocol have been studied extensively in various cell culture systems and are non-toxic at concentrations from 1 to 10 μM.

Protocol 1: Purification of Oligonucleotide

Special materials:
– reversed phase chromatography (C18 preparative column)
– acetonitrile
– NH$_4$OAc
– acetic acid
– ethanol
– NaCl
– Dowex-50 H+ form column
– Sephadex-G 15 or -G 25
– 0.2 μ filter

- Evaporate the crude preparation after the synthesis to dryness (leave the DMT- group during synthesis).
- Resuspend the crude preparation in sterile water (1 ml) and apply it to *reversed* phase chromatography.
- For *reversed* phase purification, pack a glass column (1.5 cm × 25 cm) with preparative C18, 125 A° (Waters, Milford; 55 105 μM) in 80 % acetonitrile/water.

- Load the crude oligo on the equilibrated column (100 mM NH_4OAc) and run the column with a flow rate of 10 ml/min with a gradient of 100 % buffer A (100 mM NH_4OAc) and 50 % buffer B (80 % acetonitrile containing 20 % buffer A) for 40 min. A preparative HPLC system or any other solvent delivery system can be used.
- Collect the peak that is retained on the column. Generally, the first peak contains failure sequences and the second peak contains product.
- The product in the second peak is evaporated. DMT group from the crude preparation can be removed by either (a) treatment with 80 % acetic acid or (b) by passing the product through Dowex-50 H^+ form [11].
- The purified oligo should then be converted to Na^+ form (from NH_4^+ form) by either precipitation in ethanol/NaCl or by passing through a Dowex-50 (Na+ form) column.
- The final step in the purification of oligo, and the most critical step is desalting. This can be carried out either by size-exclusion chromatography using Sephadex-G 15 or -G 25 or by dialysis against distilled water.
- The desalted clear solution should then be filtered through a 0.2 µ filter to obtain a particle-free solution, which can then be lyophilized.
- Lyophilized oligos should be stored at –20 °C until further use.

Purification of oligo on a large scale can also be carried out by using the same steps, but require larger column and solvent delivery systems, and has been described in detail earlier [11].

3.3   Analysis of the Oligonucleotide

Purified oligos should be checked for purity, length, base composition, sequence integrity, and nature of the internucleotide linkages. In Figure 2, there is a list of analytical procedures one can use to analyze the oligo. For routine experiments, oligo can be analyzed by A260/mass ratio (for DNA content), by capillary gel electrophoresis or ion exchange HPLC (for establishing the purity of the required length of oligo), polyacrylamide gel electrophoresis (to characterize the length of oligo), melting temperature with DNA or RNA (for base sequence integrity), and ^{31}P NMR (for phosphorothioate linkage). The pH of the oligo should also be checked if higher concentrations are being used in the experiment.

Analysis of Oligonucleotides

LENGTH
- IE HPLC
- Cappillary gel electrophoresis
- Polyacrylamide gel electrophoresis
- Mass spectrometry
- Melting temperature

SEQUENCE
- Sequencing
- Mass spectrometry
- Melting temperature
- Base composition

INTERNUCLEOTIDE LINKAGES
- ^{31}P NMR

COUNTERION
- Atomic absorption
- pH

DNA CONTENT
- Extinction coefficient
- A_{260}/mass ratio

Figure 2: Methods for analyzing oligonucleotides.

4   Cellular Uptake

In order to reach the viral target, the oligo must cross the cell membrane. Despite its critical importance, oligo uptake remains poorly understood. Oligos are polyanionic (with the exception of methylphosphonate oligos) and have molecular weights in the range of 6000 to 9000 daltons (for 17- to 28-mers). Many studies have shown biological effects with antisense oligos, demonstrating that they do indeed enter cells, although they cannot passively diffuse across the cell membrane. It is believed that phosphodiester and phosphorothioate oligos enter the cell by adsorptive endocytosis and/or fluid phase endocytosis, whereas cellular uptake of methylphosphonate oligos may involve either passive diffusion or adsorptive endocytosis [12–15]. Once the oligo crosses the cell membrane, it is distributed in the cytoplasm and in the nucleus.

4.1   Cell Type Dependence

In order to study the uptake of oligos into cells, a probe of some kind is needed. Fluorescent and/or radioactive oligos have been used. The fluorescent group is usually attached covalently at the end of the oligo. Radiolabelling of the oligo can be carried out either at the 5'-end using [g-^{32}P] ATP and T4 polynucleotide kinase or internally using an ^{35}S, ^{3}H or ^{14}C-atom (see Chapter 3, Labelling). Cellular uptake studies carried out using ^{32}P-end labelled oligos cannot resolve the extent of cellular uptake and intracellular distribution because of the presence of terminal phosphatase in the cells and culture medium, which may remove the terminal phosphate. Studies of cellular uptake involving radiolabelled oligos should be carried out using internally labelled oligomers. The protocol for studying cellular uptake of oligos consists of incubating cells (about 10^5–10^6 cells/ml) with the labelled-oligo at a concentration ranging between 0.2 and 1 μM. At different time points, cells are extensively washed to remove cell surface bound oligo. The uptake is then measured either by flow cytometry for the fluorescent-labelled oligomer or by counting cell-associated radioactivity for the radio-labelled oligomer. Confocal microscopy can also be used to determine the intracellular localization of the oligo.

From the studies published to date using either fluorescent or radio-labelled oligos, it is clear that cellular uptake is sequence-dependent (certain oligo sequences have the ability to form secondary hyperstructures, e.g. oligos containing four Gs), is a saturable process, and is dependent on temperature and energy [12–16]. Uptake may vary substantially between different cell populations [16] and it might be preferable to consider cells which have an active uptake system as prime targets for oligo therapeutics. Rapidly growing cells and virally infected cells may take up oligos to a larger extent than normal cells.

4.2   Serum Concentration Dependence

The difference in cell uptake observed by different groups depends not only on labelling, cell lines and methodologies, but also on the serum concentration used. This is particularly true for phosphorothioate oligos. These oligomers bind to proteins in a non-sequence-specific manner [17]. Usually, the concentration of serum used in cell uptake studies is about 5–10 %. In one experiment, we compared the cell uptake of a fluorescent-labelled phosphorothioate oligo (25-mer) at different time points. H9 cells (10^6/ml) were incubated with labelled oligomer (0.5 μM) in RPMI medium containing various concentrations of fetal bovine serum (0 %–20 %). At 1, 4, and 24 hours, aliquots of cell culture mixtures were removed, washed and resuspended in Hank's Balanced Salt Solution (HBSS). Flow cytometric data on 5,000 viable cells were acquired. Figure 3 shows that the cell uptake is dependent on the concentration of the serum: Increasing the concentrations of serum resulted in a decrease of uptake. At very high serum concentrations the cells did not grow efficiently.

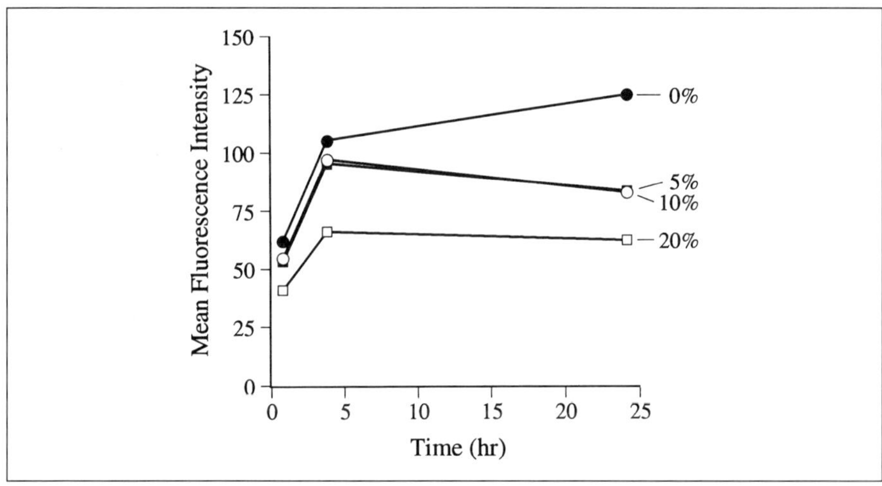

Figure 3: Effect of serum concentration on cell uptake. H9 cells were incubated with fluorescent-labelled phosphorothioate oligo at different concentrations of serum. At different time points, cell uptake was measured by flow cytometry.

5   Antiviral Activity

Antisense oligos are attractive as potential antiviral agents for several reasons: 1) the genome of viruses is much smaller than that of humans, making a choice of targeting relatively straightforward; 2) the virus has characteristics that distinguish it from its host, so that targeting the viral genome will not interfere with the expression of host sequences. Antisense oligos have been used against a wide variety of viruses. Most of the reports of antisense inhibition of viral replication have been carried out in cell culture [18–19]. These oligos inhibit viral replication apparently by different mechanisms. Well-controlled studies using antisense oligos *in vivo* are limited [60]. An important reason for this may be that the viruses studied (e.g. CMV, HIV, human hepatitis B virus) are species specific and do not productively infect a convenient small laboratory host.

5.1   Anti-HIV Activity of Oligonucleotides

HIV is the etiologic agent of the acquired immunodeficiency syndrome (AIDS). The urgent need for chemotherapy of AIDS has made HIV the major target for antisense oligos. The genome organization of HIV is more complex than that of other retroviruses: in addition to *Gag*, *Pol*, and *Env* proteins, HIV produces several other proteins, most of which are not associated with the virion but are involved in intracellular steps of the viral life cycle. Two of these proteins, *Tat* and *Rev*, are essential for viral replication.

5.1.1 Target Sequences

Several oligos have been tested for their ability to inhibit HIV replication in cell culture systems [20–36]. These oligos are targeted to various regions of the HIV genome, including the 5'-cap site, 5'-non-coding region, AUG sites, splice sites, tRNA binding sites and polyadenylation signal. For our studies we employed the *Gag* region as a target sequence. An antisense oligo phosphorothioate 25-mer (GEM 91) has been designed which is complementary to the initiation codon of the HIV-1 *Gag* gene. The rationale for selecting a *Gag* gene as a target was two-fold. First, in spite of the high mutational rate of HIVs, the *Gag* sequence is well conserved among HIV isolates [37], which is important to effectively treat HIV infected patients. Second, the *Gag* protein is a major component of the core structure of the virus. It was expected that inhibition of *Gag* protein synthesis by antisense oligos complementary to *Gag* region would profoundly affect HIV replication.

5.1.2

Protocol 2: Inhibition of HIV Replication – Short-term Assay

Special materials:
- immortalized T-cell lines
- RPMI 1640 medium
- fetal calf serum
- glutamine
- gentamicin
- virus HIV-1
- oligo
- trypan blue
- monoclonal antibody against the core protein
- fluorescein isothiocyanate-labelled goat anti-mouse IgG

- In this assay, inhibition of HIV-1 expression is carried out in immortalized T-cell lines (H9, MOLT-3). Cells are kept in culture in RPMI 1640 medium containing 10 % fetal calf serum, 2 mM glutamine, and gentamicin (250 µg/ml) in a humidified atmosphere containing 5 % CO_2/95 % air at 37 °C.
- The cells (5 × 10^5 cells/ml) are infected with 2.5–5 × 10^8 virus particles of HIV/IIIB.
- Antisense oligos are added within a few minutes of virus infection. After 4 days incubation at 37 °C, the cell supernatants are examined for the level of HIV-1 expression by measuring viral antigen p24 expression and/or syncytia formation, and/or reverse transcriptase activity.

- The cell viability is also checked by trypan blue exclusion to test for possible toxicity of oligos. The number of syncytia formed in cells is counted after triturating the cells to obtain an even distribution of syncytia in the culture. The average number of syncytia is obtained by counting several fields in duplicate cultures.
- HIV-1 virus antigen expression is measured as described earlier [22]. Briefly, cells are pelleted and spotted on toxoplasmosis slides, and fixed in methanol/acetone.
- The slides are next incubated with 10 % normal goat serum and washed. Monoclonal antibody against HIV-1 core proteins p24 or p17 is added to each well.
- The slides are then incubated with fluorescein isothiocyanate-labelled goat anti-mouse IgG and examined under a fluorescence microscope.

The earlier studies were carried out using phosphodiester oligos complementary to different regions of the HIV genome. The oligo complementary to the 5'-end region and the polyadenylation signal were the most potent, with an IC_{50} in the range of 2–10 µM [20, 21]. Since the phosphodiester oligos are unstable *in vitro* as well as *in vivo*, nuclease resistant oligos such as methylphosphonate, phosphorothioate, and phosphoramidate analogs were used. In general, modified oligos showed increased potency in inhibiting HIV-1 replication compared to unmodified oligomers. Oligos containing methlphosphonate and phosphoramidate linkages were less effective than phosphorothioate analogs. In this regard, and apart from many other properties, RNase H may be crucial in increasing the potency of phosphorothioate analogs, as the duplex formed between this analog and the RNA is substrate for RNase H. Methylphosphonate and phosphoramidate oligos do not induce RNase H. Phosphorothioate oligos are in general 10 to 100 times more potent than their phosphodiester counterparts. It was unexpectedly found, however, that sense or various random phosphorothioate oligos exhibited a similar anti-HIV activity in this assay system. In one experiment (Figure 4A), we compared the activity of different phosphorothioate oligos with the control and random oligomers. All the oligos showed efficient viral inhibition. These data strongly suggest that the protective effect of phosphorothioate oligos in this short-term assay does not require a specific sequence.

5.1.3 Inhibition of HIV Replication – Long-term Assay

To demonstrate sequence-specific inhibition of phosphorothioate oligos, long-term culture experiments are performed (see also [29]).

Protocol 3A: Inhibition of HIV Replication – Long-term Assay

Special materials:
- immortalized T-cell lines (MOLT-3)
- RPMI 1640 medium
- glutamine
- fetal bovine serum
- virus HIV-1/IIIB
- oligo
- antibiotics
- monoclonal antibody against the core protein
- fluorescein isothiocyanate-labelled goat anti-mouse IgG

- MOLT-3 cells are cultured in RPMI 1640 medium containing 13 % fetal bovine serum, 2 % glutamine, and 1 % antibiotics. The cells (5×10^5/ml) are infected with HIV-1/IIIB for 2 h, washed, and treated with phosphorothioate oligos at 1 µM[1].
- After 4 days culture supernatants are collected, the viable cells are counted by dye exclusion and split to 5×10^5 cells per ml, and the cultures are retreated with oligos.
- Virus replication is monitored in the cells by syncytia formation and p24 membrane expression (immunofluorescence as described above) and in culture supernatants by p24 ELISA.
- These procedures are repeated twice weekly, and samples are followed for up to 80 days.

The long-term antiviral activity of oligos complementary to different regions of HIV-1 was examined after 25 days (Figure 4B). At 1 µM concentration, the complementary oligos inhibited virus replication by › 90 % compared to the random or mismatched oligomers [28–30]. The inhibition of HIV-1 by phosphorothioate oligos was length-dependent. Shorter (20- or 24-mer) oligomers showed sequence-dependent inhibition when compared to the random control, although further maintenance of the infected cells in the presence of these oligomers resulted in viral breakthrough [29]. In contrast, longer oligos (28-mers) blocked HIV-1 at the same concentration.

---

[1] We used a low multiplicity of infection, which resulted in only 3 % of cells being positive for p24 membrane expression and 55 % positive for syncytia formation after 4 days, in infected cell cultures with no oligonucleotide added.

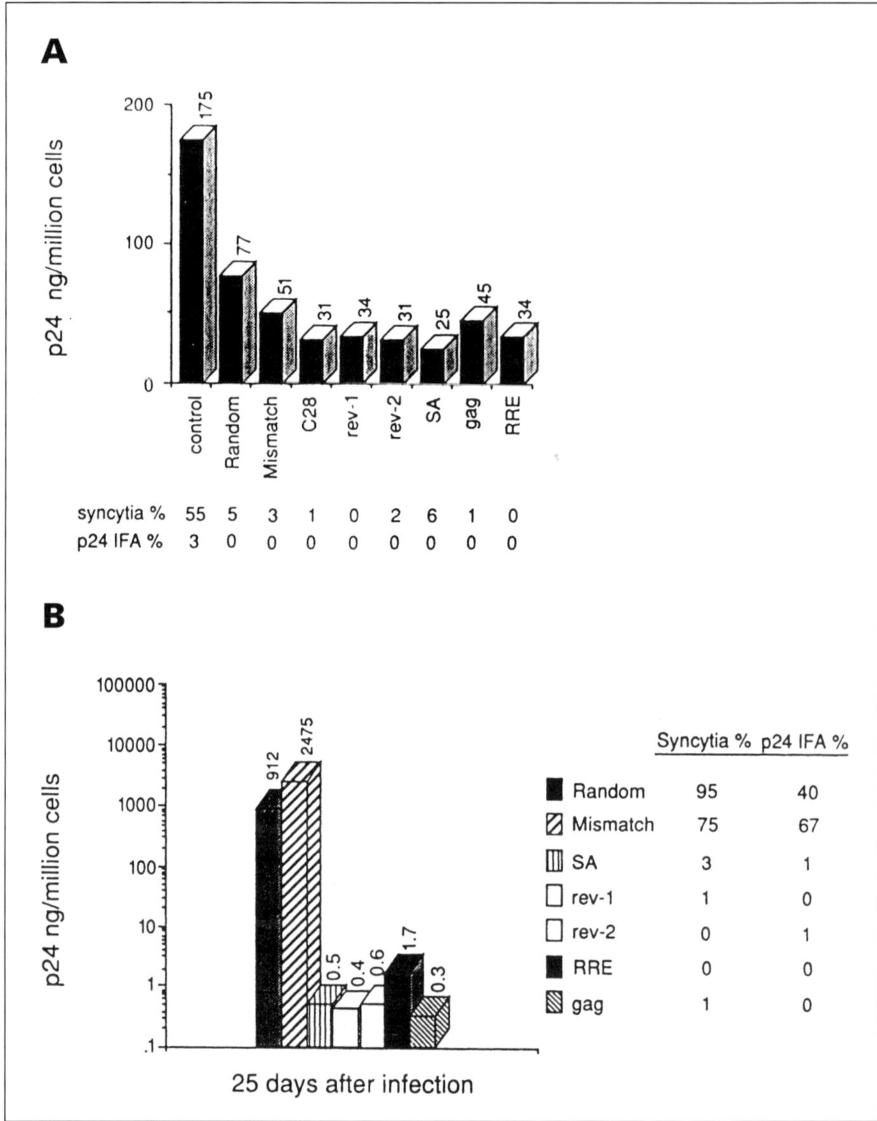

Figure 4: Antiviral activity of phosphorothioate oligos. **A:** Antiviral activity of oligomers after 4 days (short-term assay). **B:** Specific inhibition of HIV-1 replication by complementary oligos in long-term assay. (reprinted from Ref.36).

Anti-HIV activity was also dependent on oligo concentration. Phosphorothioate oligos effectively inhibited HIV-1 replication for up to 60 days at 1 µM concentration, but they failed to suppress HIV-1 replication at lower concentrations after 24 days. In one experiment (Figure 5), we tested whether a lower-dose treatment could maintain inhibition after a period of treatment with a higher concentration [29]. Cells were treated with 1 µM oligos complementary to *gag* (gag-28) or *rev* (rev 1–28; rev 2–28) re-

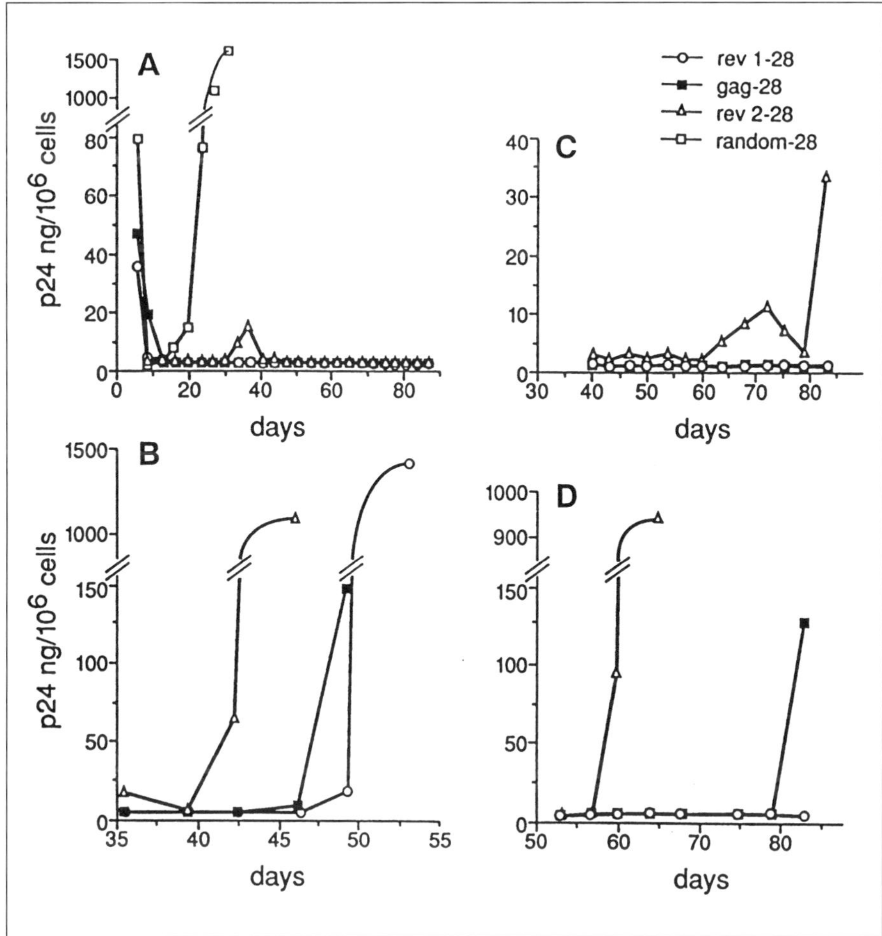

Figure 5: Long-term inhibition of HIV. **(A)** Cells infected with HIV were treated twice a week with 1 μM oligo. Every 3 days, the level of p24 was measured. **(B)** At day 35, an aliquot of cells treated with 1 μM oligo was split and maintained without antisense oligomer. **(C)** An aliquot of cells treated with 1 μM oligomer was split and incubated with 0.1 μM oligo. **(D)** An aliquot of cells treated with 0.1 μM was split at day 53 and incubated with 0.01 μM oligo. (Reprinted from Ref. 37).

gions. At day 39, an aliquot of cells were split and cultured in the presence of 0.1 μM of the same oligomers. Gag-28 and rev 1–28 oligos were able to suppress viral replication at 0.1 μM. Interestingly, the rev 2–28 oligo, which had 22 nucleotides overlapping with the rev 1–28 oligomer, was less effective. At day 53, an aliquot of cells that had been treated with 0.1 μM phosphorothioate oligos was split and maintained in the presence of 0.01 μM concentration of the same oligos. Cells treated with 0.01 μM gag-28 formed syncytia and expressed p24 antigen, in contrast to the rev-1 28 oligo-treated cells, for which there was no detectable HIV-1 replication at this time point.

5.1.4 Inhibition of HIV Replication – Chronically Infected Cells

Chronically infected cells may also be used to study the specific effect of phosphorothioate oligos [25].

Protocol 3B: Long-term Assay with Chronically Infected Cells

Special materials:
- immortalized T-cell lines (MOLT-3)
- RPMI 1640 medium
- glutamine
- fetal bovine serum
- virus HIV-1/IIIB
- oligo
- antibiotics
- monoclonal antibody against the core protein
- fluorescein isothiocyanate-labelled goat anti-mouse IgG

- Cells which are exposed to HIV-1 are kept in culture for several months in complete medium (RPMI 1640 supplemented with 15 % fetal calf serum, 4 mM L-glutamine, 50 nM 2-mercaptoethanol, and 50 units of penicillin, and 50 µg of streptomycin per ml) and do not need *de novo* infection to produce the virus[2].
- Two phosphorothioate oligos, a specific antisense oligo and a control are tested in this system. One is complementary to the HIV-1 genome (rev) and another one is used as a control (28-mer of cytidine, dC28) [25].

In these chronically infected cells, dC28 did not inhibit viral replication at the concentrations tested, whereas rev oligomer inhibited viral replication in a dose-dependent manner. The concentration that was required to inhibit 90 % replication was rather high (25 µM).

5.2   Sequence-specific and Non-sequence-specific Activity

The major attraction of the antisense approach to the development of antivirals is the rational basis with which antisense oligos can be designed. Oligos, however, have been shown to have antiviral effects that cannot be explained by Watson-Crick base pairing with the target RNA. Numerous reports of antiviral activity of antisense oligos have described that phosphorothioate oligos act by two mechanisms: non-sequence-specific inhibition and sequence-specific inhibition (Figure 6). In many studies, specificity has been inferred from the biological effects of antisense phos-

---

[2] Presumably, such chronically infected cells have an integrated HIV genome, and the cellular mechanism for viral expression would be transcription, translation, packaging and release of virions. There is no infection cycle.

phorothioate oligos compared to control oligomers. Several groups have shown that homopolymeric tracts of oligos or other sequences unrelated to the target can inhibit viral replication under certain circumstances. The mechanism of homopolymeric oligos activity is not entirely clear, but these molecules appear to inhibit the binding of HIV protein gp 120 to the receptor CD4, thereby inhibiting adsorption of the virus [38–40]. The ability of these non-specific sequences to inhibit viral replication is not limited to HIV: some reports indicate that phosphorothioate homopolymeric oligos were able to inhibit the activity *in vitro* of herpes simplex virus [41]. Although the mode of action of these non-specific oligos is not proved, some facts are consistent with a charge effect.

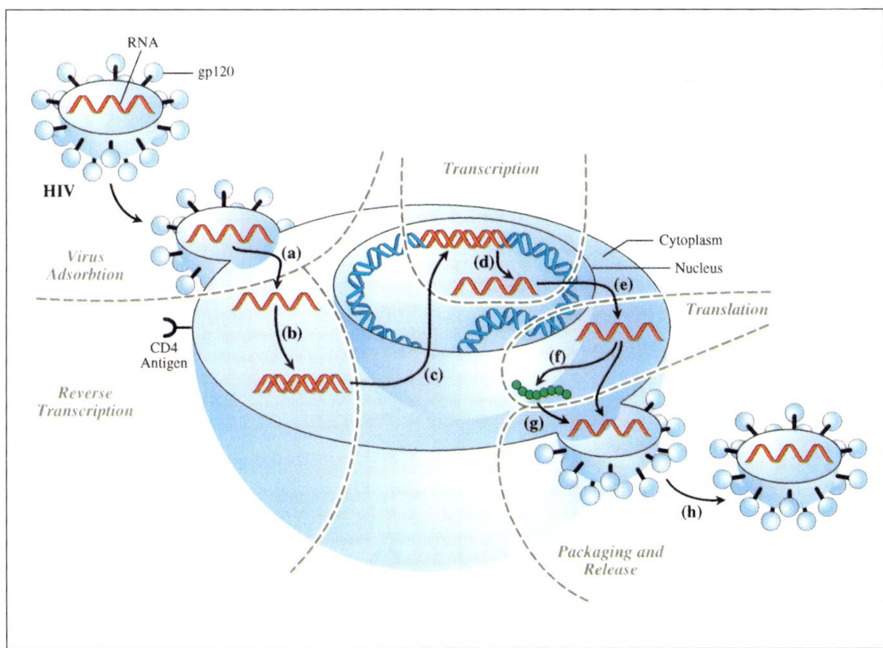

Figure 6: Sequence-specific and non-sequence-specific activities of phosphorothioate oligos. Oligos can interfere with virus adsorption and reverse transcription in a non-sequence-specific manner. But they interfere with viral mRNA processing and packaging in a sequence-specific manner.

Polyanions such as dextran sulfate or cyclodextrin sulfate inhibit viral replication in non-sequence-specific manner. Phosphorothioate oligos may also inhibit reverse transcriptase in a non-sequence-specific manner and stop the reverse transcription of the viral RNA into DNA, which has been demonstrated with purified HIV-reverse transcriptase. Sd(C)28 is the most potent inhibitor of this enzyme [42]. Phosphorothioate oligos may also act by another mechanism. For example, in one study [43], a library of phosphorothioate octanucleotides was screened for anti-HIV activity. One phosphorothioate oligo T2G4T2 was found to be the most potent inhibitor of HIV replication. This oligomer forms a parallel-stranded tetrameric guanosine-quartet structure and binds to gp120 at the V3 loop, thereby inhibiting cell-to-cell and virus-to-cell infection. In the long-term assay and chronically infected cells, phosphorothioate oligos exhibited mainly specific inhibition of HIV-1.

5.3   Defining Site of Action

Despite numerous reports of antisense inhibition of viral replication in cultured cells and in animals, the mechanism by which oligos inhibit viral replication inside the cells is still a matter of conjecture. It is difficult to prove that the antiviral activity seen with an oligo results from only an antisense mechanism. It is generally believed that when RNase H is involved in mRNA destruction, it should be possible to correlate the biological effect with levels of both targeted mRNA and protein. Analysis of the target RNA and its protein by different methods (e.g. Northern, Southern, ELISA) is important in defining the antisense mechanism. In an antisense mechanism, the biological effect should depend on the concentration of the oligo and should not change using appropriate control oligos. It is necessary therefore to test the effect of control oligos of identical base composition and length (scrambled control, reverse, or point mismatch sequence controls). A "sense" control is often inappropriate because it will ordinarily differ in base composition from the antisense oligo (see also chapter 5, Assay Systems and Controls).

In a well controlled study, phosphodiester and phosphorothioate oligos in α and β configuration were encapsulated in immunoliposomes and assessed for their inhibition of HIV-1 replication [44]. It was shown that free but not liposome-encapsulated phosphorothioate oligos can interfere with virus-mediated cell fusion in a non-sequence-specific manner. Phosphorothioate oligos also interfered with reverse transcription in a non-sequence-specific manner. However, they also interfered with viral mRNA in a sequence-specific and RNase H-dependent manner. In another study (Byrn et al. unpublished results), *in vitro* experiments were designed to examine the inhibitory effects of GEM 91 on different stages of the HIV replicative cycle. Experiments focussed on the binding of virions to the cell surface, inhibition of virus entry and reverse transcription, inhibition of HIV induction from latency, inhibition of steady-state viral mRNA levels, inhibition of virus production from chronically infected cells, and inhibition of HIV genome packaging within virions. The results showed that inhibition of virus entry/reverse transcription and reduction in steady-state viral RNA levels was sequence-dependent. The inhibition of induction of HIV from latency was dose-dependent. The inhibition of virion binding to cells and virus adsorption by uninfected cells was sequence-independent. In summary, there are stages in virus replication cycle where virus is inhibited in a sequence-specific manner and certain stages where viral replication is inhibited in a non-sequence manner. However, in a cell culture system or *in vivo*, it is difficult to define which mechanism is more effective than the other.

6   Use of Antisense Oligonucleotides as Antiviral Agents

Viruses have been used as model systems for the development of antisense oligos. Oligos have been reported to inhibit the replication of a large number of viruses in cell culture. These studies differed in the modification and length of oligo used, target gene, cells, methodologies, and various end points. Although a wide range of oligos have been used, only a few reports have contained detailed studies, some of which are summarized in Table 1. The array of antiviral effects of oligos clearly demonstrates their broad potential therapeutic effect. The efficiency of oligos varies from laboratory to laboratory for several reasons (Table 2); for example, with HIV-1, oligos are more effective in inhibiting viral replication in acutely infected cells (short-term assay) than in long-term experiments or chronically infected cells. As discussed earlier, in acutely infected cells, phosphorothioate oligos act by a specific and a non-specific mechanism, while in long-term assay they act only by a specific mechanism. We also found that primary blood mononuclear cells (PBMCs), which are the target cells for HIV-1 in the *in vivo* situation yielded more sequence-specific inhibition than other cell lines (e.g. MOLT-3, CEM, or H9) [30]. Another variable between laboratories is the viral strain and its infectivity, and the multiplicity of infection (m.o.i.) employed. A virus strain with a low infectivity will translate into apparently potent antiviral activity of oligos. Most of the reports describing the use of oligos have used low m.o.i. Third, the time of addition of the oligo varies in different studies. Incubation of cells with oligos prior or simultaneously with infection results in better inhibition of viral replication than if the oligos are added some time after infection. Finally, different modifications of oligos have been used. Oligos that are resistant to nucleases (e.g. phosphorothioate) are more potent than unstable oligos (e.g. phosphodiester). Additionally, there is a correlation between the efficacy of oligos and the ability to induce RNase H. For example, methylphosphonate, α-anomeric oligos, and phosphorothioate oligoribonu-

Table 1: Antiviral Activities of Antisense Oligonucleotides

Virus	Target	Cells	Reference
HIV	tat, rev, TAR, gag, pol, RRE, vpr	CEM, MOLT-3, H9, MT4, PBMCs, macrophages	[20–36]
Herpes Simpex Virus	Vmw65, IE 4,5,22,47; UL13, ICP4	Vero, Hela, LTK	[45–51]
Influenza	3'-end, PB1	MDCK	[51–53]
Human Papilloma	E2,E6, E7	Caski, C127, Oral cancer cells	[5, 54, 55]
Epstein Barr	DP, EBNA-1	H1, lymphoblastoid cells	[56, 57]
Hepatitis B	poly A signal, HBsAg, pre-S	HepG2, Hepatocellular carcinoma, Hepatocytes	[58–60]
Hepatitis C	5' region	cell free system	[61]
Cytomegalovirus	IE1/IE2, UL36/37	NHDF, human foreskin fibroblasts	[4, 63, 64]
Vesicular Stomatitis	N, NS, G	Mouse L, L929	[64, 65]
Rous Sarcoma	3' region	Chicken fibroblasts	[1, 2]

cleotides, which do not activate RNase H, are less efficient at inhibiting viral replication than phosphorothioate oligos.

Table 2: Factors that can influence antiviral activity of oligonucleotides.

<table>
<tr><td>1.</td><td>Cells employed
    Acutely infected
    Chronically infected
    Long-term assay</td></tr>
<tr><td>2.</td><td>Multiplicity of infection (m.o.i)</td></tr>
<tr><td>3.</td><td>Viral Strains
    Laboratory isolates
    Clinical isolates</td></tr>
<tr><td>4.</td><td>Time of oligonucleotide addition</td></tr>
<tr><td>5.</td><td>Modification of the oligonucleotide used</td></tr>
<tr><td>6.</td><td>Duration of Culture</td></tr>
<tr><td>7.</td><td>Endpoints
    Viral proteins
    Reverse Transcriptase
    Syncytia formation</td></tr>
</table>

7 Current Status of Antisense Oligonucleotides as Therapeutic Agents

Cell culture experiments, animal experiments and pharmacokinetics studies [18] seem to indicate that phosphorothioate oligos have potential as therapeutic agents. It seems likely, therefore, that antiviral therapy will be one of the first applications of antisense technology and in fact, clinical trials of oligos for cytomegaloviruses [4], and HIV [7] are underway. In the case of HIV, the oligo is injected intravenously. For treatment of cytomegalovirus, the oligo is injected intravitreally.

7.1 Pharmacokinetics of GEM 91 in Human Subjects

7.1.1

Protocol 4A: Human Subjects

- The protocols for the study was approved by the University of Alabama Institutional Review Board and the Food and Drug Administration (FDA).
- Six male homosexual or bisexual patients of age group 38–43 years old, and weight 66.7 to 102 kg participated in the study.
- All patients were HIV-1 positive without evidence of manifestation of AIDS.
- Urine was collected for a 24 hour period from individual patients to determine the urinary function including total volume, total protein, total creatinine, and creatinine clearance.

7.1.2

Protocol 4B: GEM 91 Dose Preparation

- GEM 91 (sodium salt) was checked for its purity (98 % DNA content: > 87 % 25-mer, pH 6–7) and pyrogenicity.
- ^{35}S-labelled GEM 91 was prepared by incorporating five internucleotide linkages at the 5′ terminal end. The endotoxin level was 0.94–3.9 EU per 500 µCi and the preparation was sterile in a 14-day assay.
- Actual radioactivity in dosing solution for each patient was determined immediately prior to administration, and specific activity was calculated.
- Each patient received GEM 91 by 2-hour continuous intravenous infusion at the dose of 0.1 mg/kg, including approximately 500 µCi of ^{35}S-labelled GEM 91.

7.1.3

Protocol 4C: Sample Collection

- Blood samples (20 ml for each time point) were collected in heparinized tubes from each subject, at various time points, i.e. 0 (post-infusion), 5, 30, 60 and 90 minutes, and 2, 4, 6, 8, 10, 24, 48, 72, and 96 hours after the end of 2 h infusion. Plasma was then separated by centrifugation.
- Urine was collected from each subject at intervals of 0–6, 6–12, 12–24, 24–48, 48–72, 72–96 hours after the end of infusion of GEM 91.
- To inactivate the HIV-1, these biological samples were heated at 56 °C for 2 h.
- Biological samples were stored at < –70 °C until further analysis.

In general, the safety of GEM 91 was assessed during infusion, and for 30 days after administration. Clinical observations included cardiac monitoring by telemetry, evaluation of electrocardiograms, and recording of viral signs for 4 days after infusion.

The plasma disappearance curve of GEM 91 in human patients was described by the sum of two exponentials with mean half-lives of 0.18 and 26.7 hours based on radioactivity levels. Urinary excretion represented the major pathway of elimination of oligo. On the basis of radioactivity levels, about 49 % of the administered dose was excreted within 24 hours, and about 70 % within 96 hours of oligo administration [7]. Analysis of the extracted radioactivity in plasma showed both intact and degraded forms of phosphorothioate oligo. In urine, however, all the radioactivity was associated with the degraded form of the oligo. Presently, GEM 91 is in Phase II clinical trials. HIV-1 infected patients are receiving escalating multiple doses of GEM 91 to understand its safety and pharmacokinetic profile.

8   Conclusion

Phosphorothioate oligos are the most widely studied analogs of oligos as antiviral agents. Phosphorothioate analogs show potent antiviral activity which can be attributed to both sequence-specific and non-sequence-specific activity. Pharmacology and pharmacokinetics of phosphorothioate oligos are well established.

While phosphorothioate oligos are being studied as the first generation of antisense antiviral agents, recent progress demonstrates ample opportunity to improve the properties of oligos *via* a rational medicinal chemical approach. We have designed various oligos to improve upon their pharmaceutical properties. Some of the modifications are listed below.

– Phosphorothioate oligo capped at 3'- and/or 3'- and 5' increase *in vivo* half-life [16].
– Phosphorothioate oligos containing segments of methlyphosphonate linkages at both the 3'- and 5'- end have shown improved *in vivo* stability and reduced pharmacodynamic effects (e.g. complement activation, increase in activated partial thromboplastin time, etc.) (Agrawal et al., unpublished data).
– Phosphorothioate oligos containing 3'-hairpin loop structure are more specific to their target gene. *In vivo* studies have demonstrated improved stability [51, 66].
– Phosphorothioate oligos containing segments of oligoribonucleotides and analogs have shown increased affinity and improved antisense activity. *In vivo* studies following i.v. [67] and oral [68] administration in rats showed increased *in vivo* stability of these ,hybrid' oligos.

These are some of the next generation of oligos which we have been studying, and will become the second generation of oligos [67, 68].

Acknowledgements
The authors thank K. Anthony-Pierce for excellent secretarial help.

References

1. STEPHENSON, M. L.; ZAMECNIK, P. C. Inhibition of Rous sarcoma viral RNA translation by a specific oligodeoxynucleotide. Proc. Natl. Acad. Sci. USA 75, 285 (1978).
2. ZAMECNIK, P. C.; STEPHENSON, M. L. Inhibition of Rous sarcoma virus replication and transformation by a specific oligodeoxynucleotide. Proc. Natl. Acad. Sci. USA 75, 280 (1978).
3. AGRAWAL, S.; TANG, J.-Y. GEM 91-An antisense oligonucleotide phosphorothioate as a therapeutic agent for AIDS. Antisense Res. Dev. 2, 261 (1992).
4. AZAD, R. F.; DRIVER, V. B.; TANAKA, K.; CROOKE, R. M.; ANDERSON, K. P. Antiviral activity of a phosphorothioate oligonucleotide complementary to RNA of the human cytomegalovirus major immediately-early region. Antimicrob. Agents Chemother. 37, 1945 (1993).

5. COWSERT, L. M.; FOX, M. C.; ZON, G.; MIRABELLI, C. K. *In vitro* evaluation of phosphorothioate oligonucleotides targeted to the E2mRNA of papillomavirus: Potential treatment for genital warts. Antimicrob. Agents Chemother. 37, 171 (1993).

6. BAYEVER, E.; IVERSEN, P. L.; BISHOP, M. R.; SHARP, G. J.; TEWARY, H. K.; ARNESON, M.A.; PIRRUCELLO, S. J.; RUDDON, R. W.; KESSINGER, A.; ZON, G.; ARMITAGE, J. O. Systemic administration of a phosphorothioate oligonucleotide with a sequence complementary to p53 for acute myelogenous leukemia and myelodysplastic syndrome: Initial results of a phase I trial. Antisense Res. Dev. 3, 383 (1993).

7. ZHANG, R.; YAN, J.; SHAHINIAN, H.; YAN, J.; AMIN, G.; LU, Z.; JIANG.; Z.; TEMSAMANI, J.; SAAG., M. S.; SCHECHTER, P. S.; AGRAWAL, S.; DIASIO, R. B. Pharmacokinetics of an anti-human immunodeficiency virus antisense oligodeoxynucleotide phosphorothioate (GEM 91) in HIV-infected subjects. Clinical Pharmacokinetics & Therapeutics 58, 44 (1995).

8. CROOKE, S. T. Progress in evaluation of the potential of antisense technology. Antisense Res. Dev. 4, 145 (1994).

9. ECKSTEIN, F. ed. Oligonucleotide and Analogues: A Practical Approach, IRL Press, Oxford (1991).

10. AGRAWAL, S., ed. Protocols for Oligonucleotides and Analogs: Synthesis and Properties. Humana Press: Totowa, New Jersey, 1993.

11. PADMAPRIYA, A.; TANG, J.Y.; AGRAWAL, S. Large scale synthesis, purification and analysis of oligonucleotide phosphorothioates. Antisense Res. Dev 4, 185 (1994).

12. AKHTAR, S.; JULIANO, R. L. Cellular uptake and intracellular fate of antisense oligonucleotides. Trends Cell Biol. 2, 139 (1992).

13. JAROSZEWSKI, J. W.; COHEN, J. S. Cellular uptake of antisense oligodeoxynucleotides. Adv. Drug Delivery Rev. 6, 235 (1991).

14. MILLER, P. S.; MCPARLAND, K. B.; JAYARAMAN, K.; TS'O, P. O. P. Biochemical and biological effects of nonionic nucleic acid methylphosphonates. Biochemistry 20, 1874 (1981).

15. SHOJI, Y.; AKHTAR, S.; PERIASAMY, A.; HERMAN, B.; JULIANO, R. L. Mechanism of cellular uptake of modified oligodeoxynucleotides containing methylphosphonate linkages. Nucleic Acids Res. 19, 5543 (1991).

16. TEMSAMANI, J.; KUBERT, M.; TANG, J. -Y.; PADMAPRIYA, A. A.; AGRAWAL, S. Cellular uptake of oligodeoxynucleotide phosphorothioates and their analogs. Antisense Res. Dev. 4, 35 (1994).

17. STEIN, C. A.; CHENG, Y. C. Antisense oligonucleotides as therapeutic agents – Is the bullet really magical? Science 261, 1004 (1993)

18. TEMSAMANI, J.; AGRAWAL, S. Antisense oligonucleotides as antiviral agents. In: advances in antiviral drug design, vol 2, (DeClerq, E, ed) JAI Press, Greenwich, p.1 (1996)

19. COWSERT, L. M. Antiviral activities of antisense oligonucleotides. In Antisense Research and Applications (Crooke, S. T. & Lebleu, B, eds) CRC Press, Boca Raton, 1993, 521.

20. GOODCHILD, J.; AGRAWAL, S.; CIVIERA, M. P.; SARIN, P. S.; SUN, D.; ZAMECNIK, P. C. Inhibition of human immunodeficiency virus replication by antisense oligodeoxynucleotides. Proc. Natl. Acad. Sci. USA 85, 5507 (1988).

21. ZAMECNIK, P. C.; GOODCHILD, J.; TAGUCHI, Y.; SARIN, P. S. Inhibition of replication and expression of T-cell human lymphotrophic virus (HTLV-III) in cultured cells by exogenous synthetic nucleotides, complementary to viral RNA. Proc. Natl. Acad. Sci. USA 83, 4143 (1986).

22. SARIN, P. S. , AGRAWAL, S. , CIVEIRA, M. P., GOODCHILD, J., IKEUCHI, T., ZAMECNIK. P. C. Inhibition of acquired immunodeficiency syndrome virus by oligonucleotide methylphosphonates. Proc. Natl. Acad. Sci. USA 85, 7448 (1988).

23. ZAIA, J.A.; ROSSI, J. J.; MURAKAWA, G. J.; SPALLONE, P. A.; STEPHENS, D.A.; KAPLAN, B. E.; ERITJA, R.; WALLACE, B.; CANTIN, E. M. Inhibition of human immunodeficiency virus by using an oligonucleoside methylphosphonate targeted to the tat-3 gene. J. Virol. 62, 3914 (1988).

24. MATSUKURA, M.; SHINOZUKA, K.; ZON, G.; MITSUYA, H.; REITZ, M.; COHEN, J. S.; BRODER, S. Phosphorothioate analogs of oligodeoxynucleotides: inhibitors of replication and cytopathic effects of human immunodeficiency virus. Proc. Natl. Acad. Sci. USA 84, 7706 (1987).

25. MATSUKURA, M.; ZON, G.; SHINOZUKA, K.; ROBERT-GUROFF, M.; SHIMADA, T.; STEIN, C. A.; MITSUYA, H.; WONG-STAAL, F.; COHEN, J.S.; BRODER, S. Regulation of viral expression of human immunodeficiency virus *in vitro* by an antisense phosphorothioate oligodeoxynucleotide against *rev* (art/trs) in chronically infected cells. Proc. Natl. Acad. Sci. USA 86, 4244 (1989).

26. AGRAWAL, S. Antisense oligonucleotides: A possible approach for chemotherapy of AIDS. In Prospects for antisense nucleic acid therapy of cancer and AIDS (Wickstrom, E., ed) p.143 (1991)

27. KINCHINGTON, D.; GALPIN, S.; JAROSZEWSKI, J. W.; GHOSH, K.; SUBASINGHE, C.; COHEN, J. S. A comparison of gag, pol, and rev antisense oligodeoxynucleotides as inhibitors of HIV-1. Antiviral Res. 17, 53 (1992).

28. LISZIEWICZ, J.; SUN, D.; KLOTMAN, M.; AGRAWAL, S.; ZAMECNIK, P.; GALLO, R. C. Specific inhibition of human immunodeficiency virus type I replication by antisense oligonucleotides: An *in vitro* model for treatment. Proc. Natl. Acad. Sci. USA 89, 11209 (1992).

29. LISZIEWICZ, J.; SUN, D.; METELEV, V.; ZAMECNIK, P. C.; GALLO, R. C.; AGRAWAL, S. Long-term treatment of human immunodeficiency virus-infected cells with antisense oligonucleotide phosphorothioates. Proc. Natl. Acad. Sci. USA 90, 3860 (1993).

30. LISZIEWICZ, J.; SUN, D.; WEICHOLD, F. F.; THIERRY, A. R.; LUSSO, P.; TANG, J.-Y., GALLO, R. C.; Agrawal, S. Antisense oligonucleotide phosphorothioate complementary to gag mRNA blocks HIV-1 replication in human peripheral blood cells. Proc. Natl. Acad. Sci. USA 91, 7942 (1994).

31. LI, G.; LISZIEWICZ, J.; SUN, D.; ZON, G.; DAEFLER, S.; WONG-STAAL, F.; GALLO, R. C.; KLOTMAN, M. E. Inhibition of Rev activity and human immunodeficiency virus type 1 replication by antisense oligodeoxynucleotide phosphorothioate analogs directed aginst the rev-responsive element. J. Virol. 67, 6882 (1993).

32. BALOTTA, C.; LUSSO, P.; CROWLEY, R.; GALLO, R.C.; FRANCHINI G. Antisense phosphorothioate oligonucleotides targeted to the vpr gene inhibit human immunodeficiency virus type 1 replication in primary human macrphages. J. Virol. 67, 4409 (1993).

33. MORI, K.; BOIZIAU, C.; CAZANAVE, C.; MATSUKURA, M.; SUBASINGHE, C.; COHEN, J. S.; TOULME, J. J.; STEIN, C. A. Phosphoroselenoate oligodeoxynucleotides: syn-

thesis, physico-chemical characterization, anti-sense inhibitory properties and anti-HIV activity. Nucleic Acids Res. 17, 8207 (1989).

34. SHIBAHARA, S.; MUKAI, S.; MORISAWA, H.; NAKASHIMA, H.; KOBAYASHI, S.; YAMAMOTO, N. Inhibition of human immunodeficiency virus (HIV-1) replication by synthetic oligo-RNA derivatives. Nucleic Acids Res. 17, 239 (1989).

35. MARSHALL, W. S.; BEATON, G.; STEIN, C. A.; MATSUKURA, M.; CARUTHERS, M. H. Inhibition of human immunodeficiency virus activity by phosphorodithioate oligodeoxycytidine. Proc. Natl. Acad. Sci. USA 89, 6265 (1992).

36. MORVAN, F.; PORUMB, H.; DEGOLS, G.; LEFEBVRE, I.; POMPON, A.; SPROAT, B. S.; RAYNER, B.; MALVY, C.; LEBLEU, B.; IMBACH, J.-L. Comparative evaluation of seven oligonucleotide analogues as potential antisense agents. J. Med. Chem. 36, 280 (1993).

37 LOUWAGIE, J.; MCCUTCHAN, F. E.; PEETERS, M.; BRENNAN, T. P.; SANDERS-BUELL, E.; EDDY, G. A.; VAN DER GROEN, G.; FRANSEN, K.; GERSHY-DAMET, G. M.; DELEYS, R.; BURKE, D. . Phylogenic analysis of gag genes from 70 international HIV-1 isolates provides evidence for multiple genotypes. AIDS 7, 769 (1993).

38. STEIN, C.; NECKERS, L.; NAIR, B.; MUMBAUER, S.; HOKE, G.; PAL, R. J. Phosphorothioate oligodeoxy-cytidine interferes with binding of HIV-1 gp 120 to CD4. Acquired Immune Defic. Synd. 4, 686 (1991).

39. STEIN, C. A.; CLEARY, A.; YAKUBOV, L.; LEDEMANN, S. Phosphorothioate oligodeoxynucleotides bind to the third variable loop domain (v3) of human immunodeficiency virus type 1. Antisense Res. & Dev. 3, 19 (1993).

40. LIMA, W. F.; MONIA, B. P.; ECKER, D. J.; FREIER, S. M. Implication of RNA structure on antisense oligonucleotide hybridization kinetics. Biochemistry 31, 12055 (1992).

41. GAO, W., STEIN, C. A., COHEN, J. S., DUTSHMAN, G.E., CHENG, C. Y. Effect of phosphorothioate homo-oligodeoxynucleotides on herpes simplex virus type 2-induced DNA polymerase. J. Biol. Chem 264, 11521 (1989).

42. MAJUMDAR, C.; STEIN, C. A.; COHEN, J. S.; BRODER, S.; WILSON, S. H. Stepwise mechanism of HIV reverse transcriptase: primer function of phosphorothioate oligodeoxynucleotide. Biochemistry 28, 1340 (1989).

43. WYATT, J. R.; VICKERS, T. A.; ROBERSON, J. L.; BUCKHEIT JR., R. W.; KLIMKAIT, T.; DEBAETS, E.; DAVIS, P. W.; RAYNER, B.; IMBACH, J.-L.; ECKER, D. J. Combinatorially selected guanosine-quartet structure is a potent inhibitor of human immunodeficiency virus envelope-mediated cell fusion. Proc. Natl. Acad. Sci. USA 91, 1356 (1994).

44. ZELPHATI, O., IMBACH, J. L., SIGNORET, N., ZON, G., RAYNER, B., LESERMAN, L. Antisense oligonucleotides in solution or encapsulated in immunoliposomes inhibit replication of HIV-1 by several different mechanisms. Nucleic Acids Res. 22, 4307 (1994).

45. HOKE, G. D.; DRAPER, K.; FREIER, S. M.; GONZALEZ, C.; DRIVER, V. B.; ZOUNES, M. C.; ECKER, D. J. Effects of phosphorothioate capping on antisense oligonucleotide stability, hybridization and antiviral efficacy versus herpes simplex virus infection. Nucleic Acids Res. 19, 5743 (1991).

46. SMITH, C. C.; AURELIAN, L.; REDDY, M. P.; MILLER, P. S.; TS'O, P. O. P. Antiviral effect of an oligo(nucleoside methylphosphonate) complementary to the splice junction of herpes simplex virus type 1 immediate early pre-mRNAs 4 and 5. Proc. Natl. Acad. Sci USA 83, 2787 (1986).

47. Kulka, M.; Smith, C. C.; Aurelian, L.; Fishelevich, R.; Meade, K.; Miller, P.; Ts'o, P. O. P. Site specificity of the inhibitory effects of oligo (nucleoside methylphosphonate)s complementary to the acceptor splice junction of herpes simplex virus type 1 immediate early mRNA 4. Proc. Natl. Acad. Sci. USA 86, 6868 (1989).

48. Draper, K. G.; Ceruzzi, M.; Kmetz, M. E.; Sturzenbecker, L. J. Complementary oligonucleotide sequence inhibits both Vmw65 gene expression and replication of herpes simplex virus. Antiviral Research 13, 151 (1990).

49. Jacob, A.; Duval-Valentin, G.; Ingrand, D.; Thuong, N. T.; Helene, C. Inhibition of viral growth by an α-oligonucleotide directed to the splice junction of herpes simplex virus type-1 immediate-early pre-mRNA species 22 and 47. Eur. J. Biochem. 216, 19 (1993).

50. Gao, W.-Y.; Hanes, R. N.; Vazquez-Padua, M. A.; Stein, C. A.; Cohen, J. S.; Cheng, Y. C. Inhibition of herpes simplex virus type 2 growth by phosphorothioate oligodeoxynucleotides. Antimicrob. Agents Chemother. 34, 808 (1990).

51. Agrawal, S.; Temsamani, J.; Tang, J. Y. Self-stabilized oligonucleotides as novel antisense agents. In: Delivery Strategies For Antisense Oligonucleotide Therapeutics (Akhtar, S. , ed.) CRC Press, Boca Raton, p.105 (1995).

52. Leiter, J. M. E.; Agrawal, S.; Palese, P.; Zamecnik, P. C. Inhibition of influenza virus replication by phosphorothioate oligodeoxynucleotides. Proc. Natl. Acad. Sci. USA 87, 3430 (1990).

53. Zamecnik, P. C.; Agrawal. S. Oligonucleotides hybridization inhibition of HIV and influenza viruses. Nucleic Acids Res. Symposiums 24, 127 (1991).

54. Storey, A.; Oates, D.; Banks, L.; Crawford, L; Crook, T. Antisense phosphorothioate oligonucleotides have both specific and non-specific effects on cells containing human papillomavirus type 16. Nucleic Acids Res. 19, 4109 (1991).

55. Steele, C.; Cowsert, L. M.; Shillitoe, E. J. Effect of human papillomavirus type 18-specific antisense oligonucleotides on the transformed phenotype of human carcinoma cell lines. Cancer Res. 53, 2330 (1993).

56. Yao, G.-Q.; Grill, S.; Egan, W.; Cheng, Y.-C. Potent inhibition of Epstein-Barr virus by phosphorothioate oligodeoxynucleotides without sequence specification. Antimicrob. Agents Chemother. 37, 1420 (1993).

57. Roth, G.; Curiel, T.; Lacy, J. Epstein-Barr viral nuclear antigen 1 antisense oligodeoxynucleotide inhibits proliferation of Epstein-Barr virus-immortalized B cells. Blood 84, 582 (1994).

58. Goodarzi, G.; Gross, S. C.; Tewari, A.; Watabe, K. Antisense oligodeoxyribonucleotides inhibit the expression of the gene for hepatitis B virus surface antigen. J. Gen.Virol. 71, 3021 (1990).

59. Wu, G. Y.; Wu, C. H. Specific inhibition of hepatitis B viral gene expression in vitro by targeted antisense oligonucleotides. J. Biol. Chem. 267, 12436 (1992).

60. Offensperger, W. B.; Offensperger, S.; Walter, E.; Teubner, K.; Igloi, G.; Blum, H. E.; Gerok, W. In vivo inhibition of duck hepatitis B virus replication and gene expression by phosphorothioate modified antisense oligodeoxynucleotides. EMBO J. 12, 1257 (1993).

61. Wakita, T.; Wands, J. R. Specific inhibition of hepatitis C virus expression by antisense oligodeoxynucleotides. J. Biol. Chem. 269, 14205 (1994).

62. SMITH, J.; PARI, G. Expression of human cytomegalovirus UL36 and UL37 genes is required for viral DNA replication. J. Virol. 69, 1925 (1995).

63. PARI, G. S., FIELD, K.A., SMITH, J. A. Potent antiviral activity of an antisense oligonucleotide complementary to the intron-exon boundary of human cytomegalovirus genes UL36 and UL 37. Antimicrob. Agents Chemother. 39, 1157 (1995).

64. SHEA, R. G.; MARSTERS, J. C.; BISCHOFBERGER, N. Synthesis, hybridization properties and antiviral activity of lipid-oligodeoxynucleotide conjugates. Nucleic Acids Res. 18, 3777 (1990).

65. LEMAITRE, M.; BAYARD, B.; LEBLEU, B. Specific antiviral activity of a poly(L-lysine)-conjugated oligodeoxyribonucleotide sequence complementary to vesicular stomatitis virus N protein mRNA initiation site. Proc. Natl. Acad. Sci. USA 84, 648 (1987).

66. ZHANG, R.; LU, Z.; ZHANG, X.; DIASIO, R.; LIU, T.; JIANG, Z.; AGRAWAL, S. In vivo stability and disposition of a self-stabilized oligodeoxynucleotide phosphorothioate in rats. Clinical Chemistry 41, 836 (1995).

67. ZHANG, R.; LU, Z.; ZHAO, H.; ZHANG, X.; DIASIO, R. B.; HABUS, I.; ZIANG, Z.; IYER, R. P.; YU, D.; AGRAWAL, S. In vivo stability, disposition and metabolism of a ‚hybrid' oligonucleoside phosphorothioate in rats. Biochemical Pharmacology 50, 545 (1995).

68. AGRAWAL, S.; ZHANG, X.; ZHAO, H.; LU, Z.; YAN, J.; CAI, H.; DIASIO, R. B.; HABUS, I.; JIANG, Z.; IYER, R. P.; YU, D.; ZHANG, R. Absorption, tissue distribution and In vivo stability in rats of a hybrid antisense oligonucleotide following oral administration. Biochemical Pharmacology 50, 571 (1995).

10 Animal Models in Virology

Development of Animal Models for Anti-viral Therapy

Wolf-Bernhard Offensperger, Silke Offensperger and Hubert E. Blum
Medizinische Universitätsklinik, Abt. Innere Medizin II, Freiburg,
Germany

1 **Hepatitis B Virus and the Family of Hepadnaviruses**

2 **Antisense Oligos in the Duck Hepatitis B Virus Model**

3 **Conclusions**

1   Hepatitis B Virus and the Family of Hepadnaviruses

Infection with hepatitis B virus (HBV) is endemic throughout much of the world, with an estimated 300 million persistently infected people. HBV infection is associated with a wide spectrum of clinical presentations, ranging from the healthy carrier state to acute/fulminant or chronic hepatitis and liver cirrhosis. Furthermore, chronic HBV infection clearly contributes to the development of hepatocellular carcinoma, worldwide a leading cause of death from cancer. HBV infection is ranked as the ninth most common cause of death worldwide. While HBV infection can be prevented by passive and/or active vaccination, for those chronically infected there is so far no effective therapy available. The only established therapy is interferon-alpha, with an efficacy of only 30–40 % in highly selected patients [19].

To understand the problems associated with various therapeutic strategies, the biology of HBV with its many unique features [3] as well as the pathogenesis of liver disease must be considered. Central problems are the limited immune response, the missing cytopathic effect of HBV, the presence of covalently closed, circular (ccc) HBV DNA in the nucleus, the presence of HBV in non-hepatocytes and the integration of HBV DNA into the cellular genome. The unique replicative cycle of HBV includes the cytosolic reverse transcription of a viral pregenome produced in the nucleus. The template for viral RNA synthesis is a cccDNA molecule with a supercoiled conformation found in the nucleus. This cccDNA does not self-replicate and does not undergo semiconservative replication. In order to definitely terminate HBV infection, cccDNA must be eliminated from the cell. While the synthesis of new cccDNA can be stopped by inhibitors of virion production, preexisting cccDNA is unaffected by these inhibitors. The half-life of the cccDNA forms is not known but is assumed to be very long. As a consequence, virus production often returns to pretreatment levels when drug administration is stopped. This effect was seen when nucleoside analogues were tested in vitro and in vivo in humans and animals. The clinical usefulness of this important class of antiviral compounds was further limited by often severe side-effects seen in many clinical trials.

Given the inefficiency of anti-HBV therapeutic regimens using immunomodulation or inhibition of viral DNA polymerase or reverse transcriptase newer strategies are aimed at blocking gene expression. Because the genetic material of viruses is unique and distinct from that of their hosts, the blockade of the expression of this viral genetic material is thought to be highly specific.

In hepadnaviral infections antisense oligodeoxynucleotides (oligos) were tested in vitro in hepatoma cell lines transfected with HBV and in primary duck hepatocytes infected with the duck hepatitis B virus (DHBV), and in vivo in Pekin ducks infected with DHBV. In HBV-transfected human hepatoma cells HBsAg production was significantly inhibited by an anti-sense oligo directed against the viral preS/S region [4], and viral replication was inhibited by an antisense oligo directed against the viral encapsidation signal [8]. Both the production of HBsAg and HBeAg and viral replication were inhibited by transfecting human hepatoma cells with an oligo of antisense polarity [1]. The possibility of liver-specific targeting was also tested in human hepatoma cells [20] using a soluble DNA carrier system (asialo-orosomucoid), which is targeted specifically to hepatocytes via the asialoglycoprotein receptor. The first in vivo studies using oligos in hepadnaviral infection were performed in the Pekin duck model [11]. These data are presented in this text and the conclusions drawn in respect to uptake, stability, hybridization characteristics and specificity are emphasized. These requirements must be fulfilled if an antisense oligo is to enter a clinical trial in humans.

2   Antisense Oligos in the Duck Hepatitis B Virus Model

2.1   The Duck Hepatitis B Virus Model

Duck hepatitis B virus is a member of the hepadna virus family. This family includes four viruses which have been studied in detail on the molecular level. The prototype of this family is the hepatitis B virus. Closely related viruses have been identified in woodchucks, ground squirrels and in the Pekin duck [10]. Because Pekin ducks are readily available from commercial sources and easy to house, animals infected with the duck hepatitis B virus (DHBV) represent a highly attractive system to study hepadna viruses. DHBV infection seems to be endemic in most parts of the world [13]. In France 1 % to 6 % of Pekin ducklings of different breeds were found to be DHBV infected. The host range is narrow including Pekin ducks, geese and mallards. The major route for DHBV transmission is vertical, through the eggs laid by viremic ducks. Almost 100 % of the progeny derived from infected dams are infected. Horizontal transmission is rare, if it occurs at all. Experimentally, DHBV can be efficiently transmitted by intravenous injection of the virus. Infection of animals older than 3 weeks is only rarely achieved, presumably because of immunological defense mechanisms. Experimental infection during the first 5 days after hatching leads to chronic infection in almost all birds. In infected animals, DHBV is found in two morphologically distinct forms: as spheres approximately 40 nm in diameter which represent complete virions and as pleomorphic particles which represent empty viral particles. The virion consists of an electron-dense internal core structure (nucleocapsid) and an envelope. The nucleocapsid contains DHB core antigen, the viral DNA and a DNA polymerase/reverse transcriptase. The DHBV genome is a small circular DNA molecule of about 3.0 kbp in length. The DNA is only partially double-stranded. On the DNA minus strand 3 overlapping open reading frames are present, which encode the viral nucleocapsid protein, the envelope proteins and the DNA polymerase. Replication of hepadna viruses is strikingly different from all other animal DNA viruses and involves the reverse transcription of an RNA intermediate by a virus-encoded reverse transcriptase [14]. This replication strategy, involving RNA-directed DNA synthesis, is central to the life cycle of the RNA-containing retroviruses.

Various treatments have been used in an attempt to alter the natural course of chronic hepatitis B virus infection in human patients but none has been proven to be efficacious. DHBV infected animals are a useful animal system because chronically infected ducks are readily available, because many aspects of the life cycle of DHBV, mainly the replication pathway, are very similar to those of HBV and because antiviral effects can be easily monitored. Antiviral agents can be tested *in vitro* in primary duck hepatocytes isolated from DHBV infected duck liver or *in vivo* in DHBV infected Pekin ducks. In primary duck hepatocytes infected by DHBV many nucleoside analogues were tested [2, 15, 21], and *in vivo*, adenine arabinoside [6], 2',3'-dideoxycytidine [7], 3'-azido-3'-deoxythymidine [5], the carbocyclic analog of 2'-deoxyguanosine [9], ganciclovir [18] and famciclovir [16] were tested.

2.2   Inhibition of Duck Hepatitis B Virus by Antisense Oligos [11]

As a first step, 9 different antisense oligos directed against different regions of the DHBV genome were tested *in vitro* in DHBV infected primary duck hepatocytes. These antisense oligos were synthesized as phosphorothioate analogs and between 16 to 18 nucleotides long. The final concentration of the antisense oligo in the culture medium was 1.5 µM. After 10 days in culture the cells were harvested, the DNA was isolated and analysed by Southern blot hybridization. A very efficient inhibition of viral replication was seen with antisense oligo #2 (AS2), directed against the start of the pre-S-region (nucleotides 795–812 of the DHBV genome). To demonstrate the antisense specificity of the inhibition of DHBV replication *in vitro* the respective sense oligo was synthesized. The parallel analysis of AS2 and the respective sense oligo demonstrated the specificity of the inhibition by AS2.

In the next step, DHBV infected Pekin ducklings were treated with AS2 over a time period of 10 days by daily intravenous injection. Then liver DNA was analysed for the presence of DHBV DNA replicative intermediates by Southern blot analysis. As shown in Figure 1, the *in vivo* administration of AS2 resulted in a dose-dependent inhibition of viral replication with a nearly complete elimination of viral DNA from liver cells at a daily dose of 20 mg per gram body weight, applied for a total of 10 days. To test the effect of AS2 on viral gene expression *in vivo*, the production of viral surface antigen (DHBsAg) in serum and of viral core antigen (DHBcAg) in liver was analysed by Western blot using polyclonal antibodies. Figure 2 demonstrates the effective *in vivo* inhibition of viral gene expression with disap-

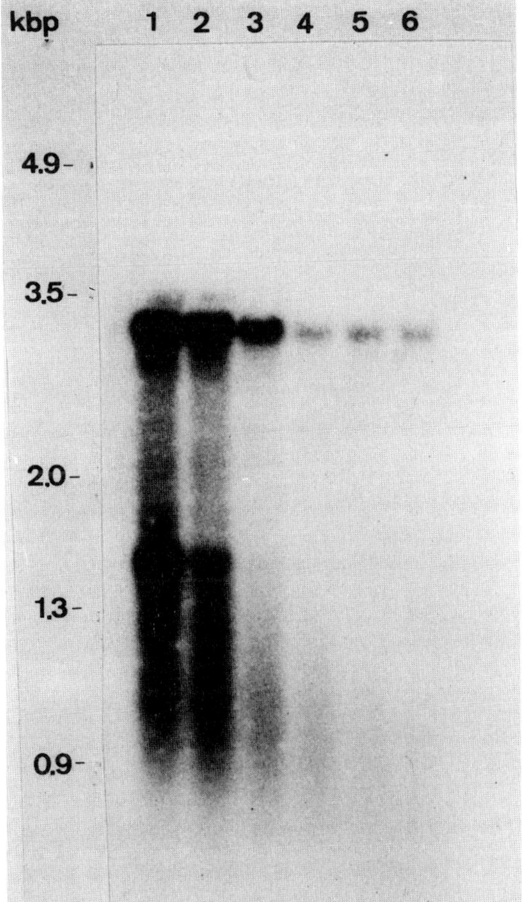

Figure 1: Effect of antisense oligo AS2 on DHBV replication *in vivo* in DHBV infected ducks. Lane 1, control duck; lanes 2–6, ducks treated for 10 days with daily intravenous injection of AS2 at a concentration of 5 mg (lane 2), 10 mg (lane 3) and 20 mg per gram body weight (lanes 4–6). Southern blot analysis was performed with 20 mg total liver DNA per lane.

pearance of viral pre-S and S-antigens from serum and viral pre-C and C-antigens from liver. To detect possible side-effects of this *in vivo* therapy with oligos several clinico-chemical parameters including alanine aminotransferase, aspartate aminotransferase, gamma-glutamyl transpeptidase, cholinesterase, total protein and albumin were measured in serum without showing differences between the control ducks and the ducks treated with AS2.

Several aspects of this *in vivo* experiment should be discussed: Certain conditions which are necessary for antisense oligos to be active have been fulfilled: (a) AS2 was taken up efficiently by the target cell. The exact mechanism by which antisense oligos enter the cells is not known. A receptor-mediated uptake is postulated. (b) AS2 was stable. This is mainly due to the

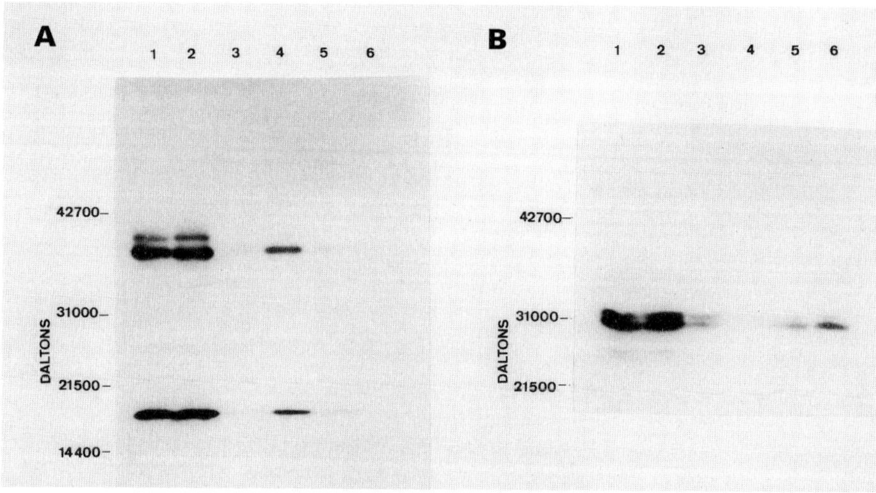

Figure 2: A: Western blot analysis of sera obtained from DHBV-infected ducks treated with AS 2 *in vivo* using a polyclonal antibody against native DHBsAg. The experimental design corresponds to figure 1. Lane 1, control duck; lanes 2–6, ducks treated with AS2 at a concentration of 5 mg (lane 2), 10 mg (lane 3) and 20 mg per gram body weight (lanes 4–6). **B:** Western blot analysis of liver extracts obtained from DHBV-infected ducks treated with AS2 *in vivo* using a polyclonal antibody against DHBcAg. The numbers on top of the lanes correspond to those in panel **A**.

modification of these antisense oligos as phosphorothioates. (c) AS2 possessed favorable hybridization conditions. These depend on the oligo, the target RNA, the intracellular environment and the intracellular localization of the oligo and the target RNA. In the special case of hepadna virus infection the dual effect of the antisense strategy, acting at the post-transcriptional level on viral gene expression and viral replication, can be explained by the replicative pathway including reverse transcription of a RNA intermediate. (d) Treatment with AS2 was specific and without side-effects. Although modified oligos were found to have some non-sequence-specific inhibitory effects, probably mediated by independent mechanisms (e.g. inhibition of viral DNA polymerase/reverse transcriptase), control oligos (sense, random) did not show comparable inhibitory effects.

The following questions are currently being addressed in our laboratory in the *in vivo* DHBV system: Since this short-time treatment did not completely eliminate the so-called cccDHBV DNA in the nucleus, a long-time treatment with AS2 over a time period of 8 weeks will be done. Should it be possible to eliminate this cccDNA completely, no reactivation of viral replication after withdrawal of the therapy should occur. Further, to deliver oligos specifically to liver cells, work is in progress to encapsidate the antisense oligos into liposomes or into liposomes conjugated with an asialoglycoprotein. Asialoglycoprotein can be targeted specifically to liver cells via a cell surface associated asialoglycoprotein receptor unique to hepatocytes.

2.3   Practical Approach

2.3.1 The Ducks

One day old ducklings were obtained from a German duck farm. Two days after hatching blood was taken from the foot veins and analysed for the presence of DHBV DNA by dot blot analysis. On the same day the ducklings were infected by intravenous injection of 100 µl of DHBV DNA positive serum. This DHBV DNA positive serum contains about 10^9 virions per ml as determined by dot blot analysis comparing the bound radioactive material of the serum dot and of cloned DNA. Under these circumstances (infection during the first 5 days after infection) almost 100 % of the ducklings become chronic DHBV carriers. If the dot blot of the serum 2 days after hatching shows congenital DHBV infection of a duck, this duck will be excluded from the study. The percentage of congenitally infected ducks is small (below 10 %).

2.3.2 In vitro Studies

Primary duck hepatocytes were isolated from 10 day old DHBV infected Pekin ducks according to Tuttleman et al. [17] with only minor modifications.

Protocol 1:** In vitro **Analysis of Duck Hepatocytes

Special materials/equipment:
- Ducklings, obtainable from duck farms
- Sodium pentobarbital (Nembutal)
- Solutions and media as described in the protocol

- Anesthetize the animal by intraperitoneal injection of 0.8 ml sodium pentobarbital (Nembutal).
- Perform *in situ* perfusion of the liver via the portal vein with 200 ml Swimms 77 medium buffered with 20 mM HEPES (pH 7.4) and containing 0.5 mM EGTA, followed by 200 ml Williams medium E buffered with 20 mM HEPES (pH 7.4) and containing 0.5 mg collagenase type 1/ml and 2.5 mM CaCl2.
- Keep all solutions at 39 °C to maintain the perfusate at 37 °C.
- Prepare a single cell suspension in Williams medium E buffered with 20 mM HEPES (pH 7.4) by gently passing the perfused liver through a fine wire-gauze mesh.
- Centrifuge the cell suspension at 50 g for 2 min and wash three times with the above medium.
- Seed the hepatocytes at a density of 2 10^5 cells/cm^2 on Primaria tissue culture dishes using Williams medium E supplemented with 20 mM HEPES (pH 7.4), 5 mM glutamine, 0.066 µM insulin, 10 mM dexamethasone, 100 mg/ml penicillin, 100 mg/ml streptomycin and 1.5 % dimethyl sulfoxide.

> • Maintain the cultures at 37 °C and 5 % CO2 in a humidified incubator with daily medium changes for the duration of the experiments, usually 10 days.
> • Determine the cell viability by trypan blue exclusion.

2.3.3 *In vivo* **Studies**

At the age of 14 days, daily intravenous injection of the antisense oligos was started and continued for 10 days, when the ducks were sacrificed or the therapy was withdrawn. The vein of choice is the readily accessible foot vein. Because of the rapid growth of the ducks during the first weeks the weight was determined daily to calculate the exact dosage of the drug to be tested.

Dosing scheme for AS2 at a concentration of 20 µg/g for a time period of ten days:

Age of the duck	Weight in g	AS 2 in µg
14 days	200	4000
15 days	220	4400
16 days	240	4800
17 days	260	5200
18 days	280	5600
19 days	300	6000
20 days	330	6600
21 days	360	7200
22 days	390	7800
23 days	420	8400

2.3.4 *Assays for Virus Detection*

The following assays to detect DHBV parameters can be done:

- DHBV DNA in liver (Southern blot analysis, dot blot analysis, PCR, in situ hybridization)
- DHBV RNA in liver (Northern blot analysis, dot blot analysis, in situ hybridization)
- DHBV proteins (DHBsAg, DHBcAg) in liver (Western blot analysis, immunohistochemistry)
- DHBV DNA in serum (Southern blot analysis, dot blot analysis, PCR)
- DHBV proteins (DHBsAg, DHBcAg) in serum (Western blot analysis)
- DHBV DNA, RNA and proteins in extrahepatic organs (spleen, pancreas, kidney)

In our hands the most useful parameters are the detection of viral DNA in the liver by Southern blot analysis, and the detection of viral DNA in serum by dot blot analysis and the detection of DHBsAg in serum by Western blot analysis using a polyclonal antibody. These techniques follow standard procedures [12].

Protocol 2: Southern blot analysis of DHBV DNA from liver

Special materials/equipment:
- Solution A: 1 % SDS, 20 mM Tris-HCl (pH 7.4), 10 mM EDTA, 150 mM NaCl
- Further solutions as described in the protocol
- Proteinase K
- Phenol: chloroform:isoamyl alcohol (25:24:1)
- Chloroform: isoamyl alcohol (24:1)
- 0.3 M sodium acetate, pH 5.2
- ^{32}P-labelled, full-length DHBV DNA probe.

- Lyse the tissue with solution A (1 ml per 100 mg tissue).
- Digest with proteinase K (0.4 mg/ml) at 37 °C for 12 h.
- Extract with phenol:chloroform:isoamyl alcohol and chloroform:isoamyl alcohol.
- Precipitate nucleic acids with 2 vol of 100 % ethanol in the presence of 0.3 M sodium acetate, pH 5.2.
- Lyophilize.
- Dissolve nucleic acids in 50–200l 10 mM Tris-HCl (pH 7.4), 1 mM EDTA.
- Hybridize with ^{32}P-labeled DNA probe according to standard laboratory protocols.

Protocol 3: Western blot analysis of DHBsAg and DHBcAg

Special materials/equipment:
three different polyclonal antibodies were used
- An antibody against native DHBsAg
- An antibody against DHB-preSAg
- An antibody against DHBcAg, all raised in rabbits.

- Denature serum or liver samples.
- Separate by SDS-PAGE (12 %) and transfer to nitrocellulose.
- Block and incubate the membrane with the primary antibody overnight.
- Use the enhanced chemiluminiscence detection method and carry out the sequential steps according to the instructions of the manufacturer (Amersham Buchler, Braunschweig, Germany).

Using anti-DHBsAg, a major band with a molecular weight of 36 kDa (preS/S protein), some minor bands of different molecular weights due to the different degree of glycosylation and a major band of molecular weight 18 kDa (S protein) can be detected. Using anti-DHBcAg, preC/C proteins with molecular weights of 30 and 33 kDa can be detected.

3   Conclusions

The data demonstrate that DHBV replication and gene expression in liver cells can be blocked by antisense oligos *in vivo*. The action is antisense-specific, i.e. the corresponding sequence of sense polarity does not significantly affect viral replication or gene expression. In addition, the effect of the antisense oligos is sequence-dependent with the strongest inhibition observed for an oligo binding to a DHBV mRNA region 5' to the preS-gene. The amount of oligo required to obtain these effects *in vivo* corresponds to the requirements for inhibition *in vitro* and suggests comparable uptake, metabolism and degradation of the antisense oligos *in vivo*. Toxic side-effects of the therapy with antisense oligos could not be seen in respect to growth, behaviour of the ducks and clinico-chemical parameters. With respect to the translation of our antisense oligo findings into clinical practice, several issues should be addressed: The effect of antisense oligos on viral pathogenesis should be tested *in vivo* in woodchucks infected by the woodchuck hepatitis virus. Chronic infection with this hepadnavirus leads to liver inflammation and in almost 100 % of these animals to hepatocellular carcinoma. Furthermore, long-time treatment of DHBV infected ducks with antisense oligos should show if elimination of the viral cccDNA species by this strategy is possible. To reduce the amount of oligos needed and to deliver oligos specifically to liver cells, delivery systems such as liposomes, lipopolyamines, receptor-mediated endocytosis or viral vectors should be tested.

Acknowledgements

This work was supported by a grant of the Deutsche Forschungsgemeinschaft (SFB 154, TP A5). The expert technical assistance of Birgit Hockenjos and Petra Kary is gratefully acknowledged.

References

1. BLUM, H. E., E. GALUN, F. von WEIZSÄCKER, J. R. WANDS: Inhibition of hepatitis B virus by antisense oligodeoxynucleotides. Lancet 337 (1991) 1230.
2. CIVITICO, G., Y. WANG, C. LUSCOMBE, N. BISHOP, G. TACHEDJIAN, I. GUST, S. Locarnini: Antiviral strategies in chronic hepatitis B virus infection: II. Inhibition of duck hepatitis B virus *in vitro* using conventional antiviral agents and super-coiled-DNA active compounds. J. Med. Virol. 31 (1990) 90.
3. GERLICH, W.: STRUCTURE and MOLECULAR VIROLOGY; IN ZUCKERMAN, A. J., THOMAS, H. C. (eds): Viral Hepatitis. Edinburgh, Churchill Livingstone (1993) 83.
4. GOODARZI, G., S. C. GROSS, A. TEWARI, A. WATABE: Antisense oligodeoxyribonucleotides inhibit the expression of the gene for hepatitis B virus surface antigen. J. Gen. Virol. 71 (1990) 3021.
5. HARITANI, H., T. UCHIDA, Y. OKUDA, T. SHIKATA: Effect of 3'-Azido-3'-Deoxythymidine on replication of duck hepatitis B virus *in vivo* and *in vitro*. J. Med. Virol. 29 (1989) 244.

6. HIROTA, K., A. SHERKER, M. OMATA, O. YOKOSUKA, K. OKUDA: Effects of adenine arabinoside on serum and intrahepatic replicative forms of duck hepatitis B virus in chronic infection. Hepatology 7 (1987) 24.

7. KASSIANIDES, C., J. H. HOOFNAGLE, R. MILLER, E. DOO, H. FORD, S. BRODER, H. MITSUYA: Inhibition of duck hepatitis B virus replication by 2',3'-dideoxycytidine. Gastroenterology 97 (1989) 1275.

8. KORBA, B., F. WELLS, K. JONES, R. ENGLE, A. BUCKLER-WHITE, J. Gerin: Inhibition of hepatitis B virus replication in vitro by antisense oligonucleotides. Antiviral Res. 23, Suppl. 1 (1994) 78.

9. MASON, W., J. CULLEN, J. SAPUTELLI, T.-T. WU, C. LIU, W. LONDON, E. LUSTBADER, P. SCHAFFER, A. O'CONNELL, I. FOUREL, C. ALDRICH, A. JILBERT: Characterization of the antiviral effects of 2' carbodeoxyguanosine in ducks chronically infected with duck hepatitis B virus. Hepatology 19 (1994) 398.

10. MASON, W., G. SEAL, J. SUMMERS: Virus of Pekin ducks with structural and biological relatedness to human hepatitis B virus. J. Virol. 36 (1980) 829.

11. OFFENSPERGER, W.-B., S. OFFENSPERGER, E. WALTER, K. TEUBNER, G. IGLOI, H. E. BLUM, W. GEROK: *In vivo* inhibition of duck hepatitis B virus replication and gene expression by phosphorothioate modified antisense oligodeoxynucleotides. EMBO J. 12 (1993) 1257.

12. OFFENSPERGER, W.-B., E. WALTER, S. OFFENSPERGER, C. ZESCHNIGK, H. E. BLUM, W. GEROK: Duck hepatitis B virus: DNA polymerase and reverse transcriptase activities of replicative complexes isolated from liver and their inhibition *in vitro*. Virology 164 (1988) 48.

13. SPRENGEL, R., H. WILL: DUCK HEPATITIS B VIRUS. In: DARAI, G. (ed.) Virus diseases in laboratory and captive animals. Nijhoff Publishers, Boston (1988) 363–386.

14. SUMMERS, J., W. MASON: Replication of the genome of a hepatitis B-like virus by reverse transcription of an RNA intermediate. Cell 29 (1982) 403.

15. SUZUKI, S. , B. LEE, W. LUO, D. TOVELL, M. ROBINS, D. TYRRELL: Inhibition of duck hepatitis B virus replication by 2',3'-dideoxynucleosides. Biochem. Biophys. Res. Commun. 156 (1988) 1144.

16. TSIQUAYE, K., M. SLOMKA, M. MAUNG: Oral famciclovir against duck hepatitis B virus replication in hepatic and nonhepatic tissues of ducklings infected in ovo. J. Med. Virol. 42 (1994) 306.

17. TUTTLEMAN, J. S. , J. PUGH, J. SUMMERS: *In vitro* experimental infection of primary duck hepatocyte cultures with duck hepatitis B virus. J. Virol. 58 (1986) 17.

18. WANG, Y., S. BOWDEN, T. SHAW, G. CIVITICO, Y. CHAN, M. QIAO, S. LOCARNINI: Inhibition of duck hepatitis B virus replication *in vivo* by the nucleoside analogue ganciclovir. Antiviral Chem. Chemother. 2 (1991) 107.

19. WONG, D. K. H., A. M. CHEUNG, K. O'ROURKE, C. D. NAYLOR, A. S. DETSKY, J. HEATHCOTE: Effect of alpha-interferon treatment in patients with hepatitis B e antigen-positive chronic hepatitis B. Ann. Intern. Med. 119 (1993) 312.

20. WU, G. Y., C. H. WU: Specific inhibition of hepatitis B viral gene expression in vitro by targeted antisense oligonucleotides. J. Biol. Chem. 267 (1992) 12436.

21. YOKOTA, T., K. KONNO, E. CHONAN, S. MOCHIZUKI, K. KOJIMA, S. SHIGETA, E. DE CLERCQ: Comparative activities of several nucleoside analogs against duck hepatitis B virus *in vitro*. Antimicrobial Agents and Chemotherapy 34 (1990) 1326.

11 Parasites

Inhibition of Replication of Drug-resistant P. falciparum *in vitro* by Specific Antisense Phosphorothioate Oligodeoxynucleotides

Paul Zamecnik[1], Eliezer Rapaport[1], Valeri Metelev[1] and
Robert H. Barker Jr.[1,2]
[1]Worcester Foundation for Biomedical Research, Shrewsbury, USA
[2]Hybridon Inc., Worcester, USA

1   Introduction

A number of years have now passed since introduction of the concept of inhibiting genomic messages in living organisms by means of synthetic exogenous oligonucleotides (oligos) [1, 2]. Viruses and parasites are promising targets, since their genomes are smaller than those of humans, and they are dependent on their hosts for supplying proteins and metabolites necessary for replication and expression. Their own genomes code for particular and specific protein structures, thus offering a special opportunity for therapeutic exploitation. A number of investigators, [3] including us [4, 5, 6] have described such antisense oligo inhibitions.

In the present chapter we describe an antisense approach toward inhibition of replication of the malaria parasite *P. falciparum*. Malaria is endemic throughout most tropical regions and is estimated to afflict more than 200 million people annually. The disease is caused by infection with one or more species of *Plasmodium*. Three species (*P. vivax, P. ovale,* and *P. malariae*) produce relatively mild symptoms consisting of spiking periodic fever, anemia, and some jaundice. In contrast, infection with *P. falciparum* can lead to coma and death via occlusion of cerebral capillaries unless chemotherapy is initiated, and is responsible for about 800,000 deaths per year among African children under 5 years.

Recent surveys indicate a growing resistance of the malarial parasite to chloroquine, pyrimethamine, and mefloquine, major therapeutic agents, until recently used effectively against malaria. Immunization against the malarial parasite has thus far been ineffective, and prospects for future success are at the moment not bright. The antisense technique offers an entirely new approach for the treatment of malaria since the agents may be directed against specific genes responsible for replication.

In 1976 Trager and Jensen [7] developed a method for studying replication of the erythrocytic stages of the malarial parasite *P. falciparum in vitro*, by using fresh human blood, diluting it with modified tissue culture medium, adding an aliquot of infected human blood, and incubating at 37 °C in a CO_2-air mixture. Under these laboratory conditions, the malarial parasite is able to achieve parasitemias of ten to fifteen percent. By repeated transfers of infected cells to fresh units of diluted human blood, the erythrocytic cycle can be maintained indefinitely. In this way, a convenient system has become feasible for assay of chemotherapeutic agents, under non-hazardous conditions.

The clinical life cycle of *P. falciparum*, a major pathogen for human malaria, involves the bite of a mosquito, development of the sexual forms of the parasite in its gut, injection into a human host, passage through a cycle in the human liver, and then invasion of human erythrocytes. At this last-mentioned point replication of the parasite may now be studied in this *in vitro* system, short-circuiting the difficult stage which involves the mosquito.

It has thus become possible for biochemists working in a microbiological hood to look for a unique feature of the *P. falciparum* metabolism to which to direct an antisense oligo bullet.

2   Rationale

The malarial parasite is unable to use preformed thymidylic acid (TMP) for synthesis of DNA, and must convert deoxyuridylic acid to TMP, which is then phosphorylated to TTP, the monomer precursor employed by DNA polymerase in replication. In contrast, humans utilize both a *de novo* synthetic pathway leading to TTP, and a salvage pathway in which thymidylic acid can be phosphorylated to TTP. Dihydrofolate reductase – thymidylate synthase is therefore a more critical enzyme in the pathway leading to DNA synthesis and replication for *P. falciparum* than is the case for humans. Decades ago, Hitchings and colleagues [8] found that antifolate analogs inhibit the donation of an active methyl group from dihydrofolate to deoxyuridylate, an activity catalyzed by thymidylate synthase. Unfortunately, a single or a maximum of two point mutations in the DHFR-TS gene renders the parasite resistant to drugs such as pyrimethamine which target malarial DHFR-TS. Recently, resistance to chloroquine and mefloquine has also become a major problem in treatment and control of malaria.

We have targeted the DHFR-TS enzyme complex by synthesizing antisense oligos complementary to a portion of the *P. falciparum* genome [9]. In keeping with previous experience, we have used phosphorothioate-modified oligomers (S-ODNs) to minimize oligo degradation while still retaining the ability to activate RNase H. As controls, we have employed mismatched oligomers, sense strand oligomers, and oligomers designed for targets other than malaria.

Although investigators agree (cf. [4, 10]) that oligos are able to enter virtually all eukaryotic cells, a striking exception is the adult human erythrocyte [11, 12]. The malarial parasite penetrates this cell by indenting and invading the erythrocyte membrane [13] (See Figures 1 and 2). The parasite also introduces proteins into the erythrocyte membrane and cytoskeleton which modify its permeability [14]. Whatever the route, oligos are able to enter the parasitized human red cell, a circumstance fortunate for antisense oligo chemotherapy.

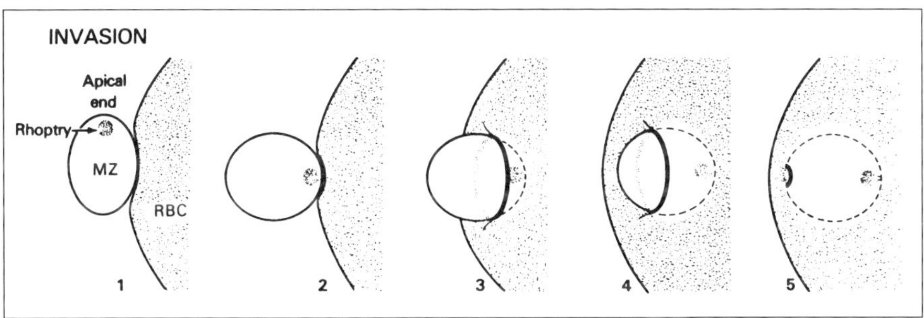

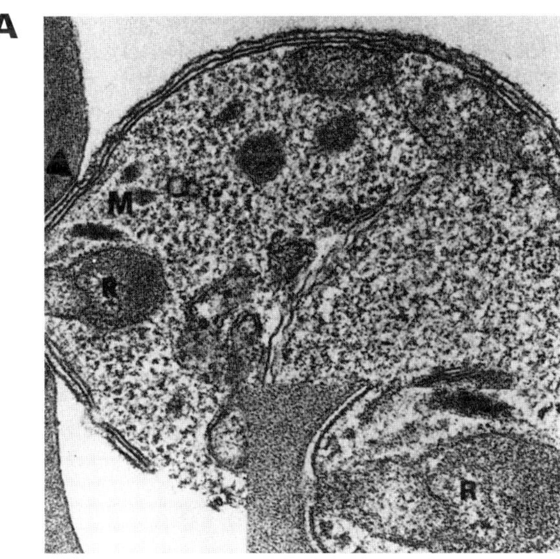

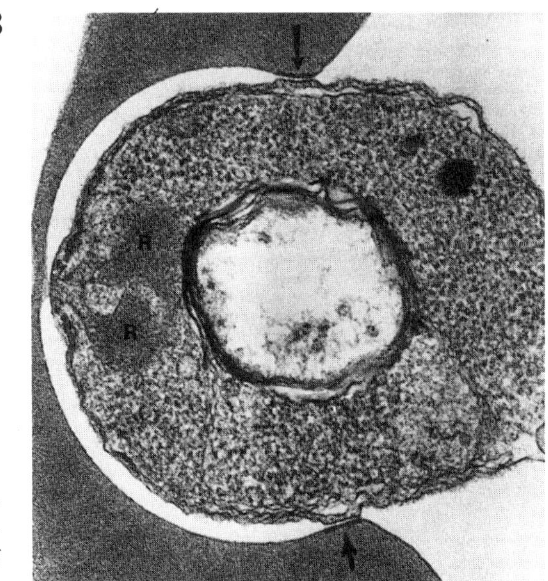

Figures 1 and 2 A, B Malarial parasites penetrating human erythrocyte cell wall (with permission of Dr. Louis H. Miller) [13].

3 Experimental Procedure

3.1

Protocol 1: Oligonucleotide Uptake

Special materials:
- RPMI 1640 medium (GIBCO Laboratories, Grand Island, NY)
- Sodium bicarbonate
- TES (N-tris[hydroxymethyl]-methyl-2-aminoethane-sulfonic acid)-sodium salt
- Pyruvate
- Dextrose
- Glutamine
- Hypoxanthine
- Gentamycin
- Type A-positive red blood cells
- Type A-positive human plasma

- Parasites (1–10 % parasitemia) are maintained in 75 cm^2 tissue culture flasks, in an atmosphere of 2 % O_2/8 % CO_2/90 % N_2 in a 5–8 % (hematocrit) washed human erythrocytic suspension, and in medium composed of RPMI 1640 supplemented with 1 g sodium bicarbonate, 3 mg of TES sodium salt, 2 mg of glucose, 110 mg of sodium pyruvate, 300 mg of glutamine, 5 mg of hypoxanthine, and 25 mg of gentamycin per ml, plus 10 % (vol/vol) human plasma, at 37 °C. Fresh human plasma and erythrocytes, type A+, for propagation of the parasites, may be obtained from the Red Cross or clinical sources. For experiments, parasites are diluted to 0.4 % parasitemia (5 % hematocrit), and are cultured in 48-well microtiter plates at a total volume of 1 ml/well.
- Unmodified oligos are labelled with ^{32}P 2.4 × 10^7 cpm/mg) and incubated overnight with adult human erythrocytes in a modified Trager-Jensen system [7], (see also chapter 3, Labelling)
- A thin blood smear of the erythrocytes is then prepared: approximately 5 ml of blood is placed on a microscope slide, then a second slide is positioned over the drop at an angle of about 30 degrees, and is drawn back until it just begins to contact the drop of blood. Capillary action then draws the blood along the junction between the two slides. The thin smear is formed by pushing the top slide briskly away from the blood drop, resulting in an even monolayer of erythrocytes. Slides are fixed and stained, (using the Diff-Quick™ staining procedure (Baxter Healthcare Corp, McGraw Park, IL) according to the manufacturer's instructions. After staining, slides are dipped in photoemulsion, and exposed to film at –70 °C for several days.
- Slides are developed and fixed according to standard techiques. Silver grains are studied by light microscopy.

3.2

Protocol 2: Assessment of Parasite Replication

Special materials: see Protocol 1

- To study the inhibitory effect, erythrocytes are grown in a modified Trager-Jensen system [7] as described above.
- Cells are incubated with a specific antisense, control or no oligo at different concentrations (1.0, 0.5, 0.1, and 0.05 µM final concentrations).
- Oligos are aliquotted into wells of the microtiter plate, then parasites are added (0.4 % parasitemia, 5 % hematocrit in modified RPMI medium, as described), and are cultured for 48 hours.
- A blood smear (see above) is stained with Diff-Quik Fix and Stain Set (Baxter Scientific Products, McGraw Park, IL). This is similar to Wright's Giemsa, and produces dark purple-stained parasites against a pink erythrocyte cytoplasm.
- Slides are then viewed under oil immersion, and the percentage of infected cells is determined by counting the stained replicative forms.

4   Results

Silver grains were consistently seen in infected erythrocytes overlying the stained parasites, but not in uninfected cells. When oligos were covalently labelled with fluorescein [10] and similarly incubated with parasitized erythrocytes, the fluorescein label lit up the infected cells under ultraviolet light, but not the uninfected ones. Interestingly, the fluorescence did not occur throughout the red cell, but illuminated only a smaller area surrounding the parasite, demarcated by a circular membrane. This observation is illustrated in Figure 3.

Our earliest studies showed good inhibition of parasite replication, demonstrated by far fewer erythrocytes showing stained replicative forms in the S-ODN treated cultures versus the control or untreated cultures. At an S-ODN concentration of 1 µM, however, there was also considerable nonspecific inhibition by mismatched S-ODNs [12]. A similar observation was subsequently reported by Clark et al. [15]. In more recent experiments [16] we have done numerous dose-response curves, and have found a significantly greater inhibition with the sequence dependent S-ODNs than with the mismatched or apparently random S-ODNs, at concentrations below 1 µM. Such a comparison is shown in Figure 4. The investigators [15] who reported the non-sequence-specific inhibition of *P. falciparum* by S-ODNs, did not report carrying out a dose-response curve, and stopped at a concentration of 1 µM. Thus, their general conclusion that inhibition of P. falciparum replication by S-ODNs is non-sequence-specific is incompatible with the present results. At a concentration around 0.01 µM, there is still statistically significant (~15–20 percent) inhibition from the DHFR-TS

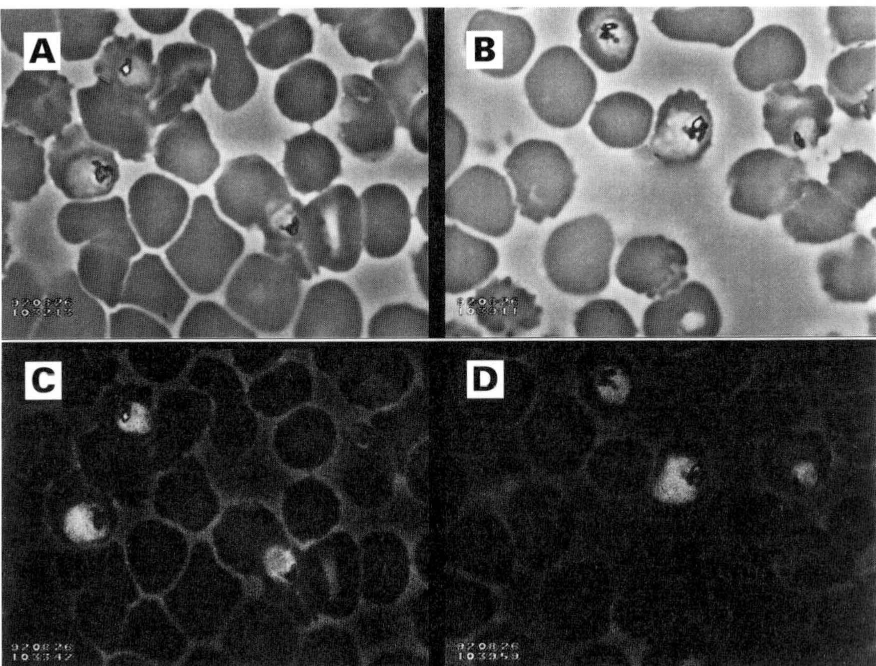

Figure 3. Human erythrocytes were incubated with fluorescently labeled ODNs, stained as described and photographed microscopically under ultraviolet light. Uninfected erythrocytes showed no fluorescence, indicating no cell entry of ODN; infected cells showed fluorescence overlying the parasites, plus within a small circular area surrounding the parasites. **A, B:** Illumination by both visible and 495 nM light. **C, D:** illumination just by 495 nM light.

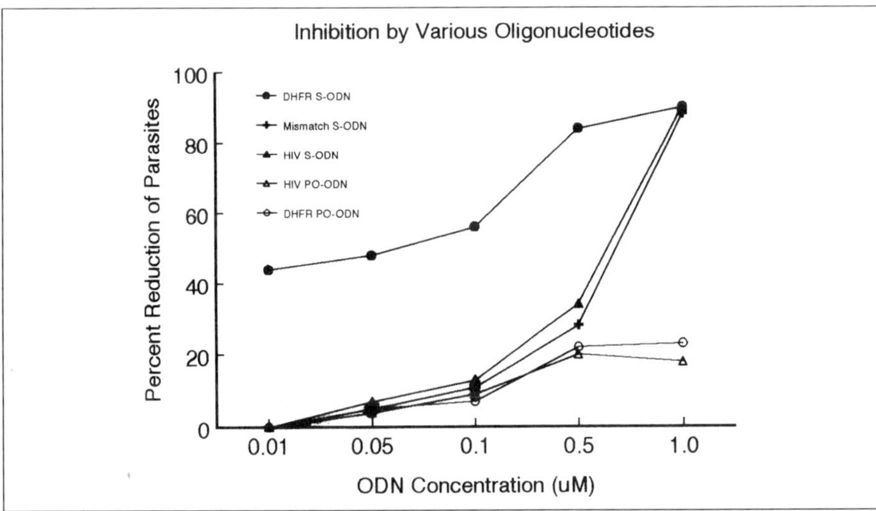

Figure 4. Comparison of inhibition of drug-resistant and drug-sensitive malarial parasite replication measured by visual microscopy, as a function of concentration of added S-ODNs. Note that at 0.01–0.05 μM concentration the specific dihydrofolate reductase – thymidylate synthase (DHFR-TS) S-ODN is more inhibitory than the "random" and mismatch S-ODNs. Both of the unmodified ODNs are only feebly inhibitory.

Table 1: Inhibition of Drug-Resistant Malaria Using ODNs

ODN Conc (uM)	ODN 105				ODN HIV				ODN "mismatch"				ODN "sense"			
	W2		W2mef		W2		W2mef		W2		W2mef		W2		W2mef	
	%	sd	%	sd	%	sd	%	sd	%	sd	%	sd	%	sd	%	sd
0.5	76	15	82	11	25	11	23	12	42	13	36	12	37	18	33	15
0.1	57	7	60	9	11	9	13	11	21	9	23	11	5	5	7	7
0.05	53	9	55	9	4	4	6	5	12	12	9	8	8	8	6	7
0.01	41	13	44	8	5	5	5	5	3	5	6	6	8	10	7	10
0.005	23	10	25	11	4	7	5	6	4	4	5	5	5	5	4	5

ODN 105:	Targets DHFR of *P. falciparum*. (TCT-TAA-AAA-TAA-TTT-CTT-CGT-AGT-TAA)
ODN HIV:	Targets *gag* region of HIV. (TCT-TCC-TCT-CTC-TAC-CCA-CGC-TCT-C)
ODN "mismatch":	Derived from 105, but containing 8 mismatsches. (TCA-TAT-AAT-TAT-TTA-CTA-CGT-TGT-AAA)
ODN "sense":	Complementary to ODN 105; sense strand. (TTA-ACT-ACG-AAG-AAA-TTA-TTT-TTA-AGA)
W2:	W2 strain of *P. falciparum*. Resistant to chloroquine, sensitive to mefloquine.)
W2mef:	W2mef strain of *P. falciparum*. This strain was derived from W2, and is resistant to both chloroquine and mefloquine.
%:	Percent reduction in parasite growth compared with controls; average of 5 experiments.
sd:	Standard deviation.

targeted antisense S-ODN, and none from the non-sequence-specific S-ODN. The IC_{50} for this specific inhibition is 20–50 × nM (cf. Figure 4).

Two other genomic targets for *P. falciparum*, dihydropteroate synthetase [17] an enzyme in the folate biosynthetic pathway, and ribonucleoside reductase [18,19] show similar degrees of inhibitory potency, no greater than for DHFR-TS but nearly as good. In particular, we have compared the efficacy of S-ODNs in drug-resistant and drug-sensitive strains of parasites. The sensitivity to inhibition is the same for both (cf. Table 1).

5   Conclusions

These *in vitro* studies suggest that the use of antisense oligos may offer a new approach to the chemotherapy of drug-resistant malaria. A next step is to carry such studies into an animal malaria model, for which the chick system, *P. gallinaceum*, appears to be a possible choice. It is also pertinent to test third generation oligo derivatives – hybrids, chimerics, and self-stabilized [20], both *in vitro* and *in vivo*. These newer oligo modifications are as effective or more so in HIV than are the all-phosphorothioate oligos. In animal systems, they minimize that part of toxicity related to the more electronegative charge of the phosphorothioate, and its greater tendency to associate with electropositively charged proteins within cells.

Acknowledgements

The authors are indebted to Dr. Yu-li Wang for expert assistance with fluorescence microscopy.

References

1. ZAMECNIK, P. C., and STEPHENSON, M. L. Inhibition of Rous sarcoma virus replication and transformation by a specific oligodeoxynucleotide. Proc. Natl. Acad. Sci. USA 75 (1978) 280–284.

2. STEPHENSON, M. L., and ZAMECNIK, P. C. Inhibition of Rous sarcoma viral RNA translation by a special oligodeoxynucleotide. Proc. Natl. Acad. Sci. USA 75 (1978) 285–288.

3. WICKSTROM, E. Ed. Prospects for antisense Nuclcic Acid Therapy of Cancer and AIDS. Wiley-Liss, New York (1991).

4. ZAMECNIK, P. C., GOODCHILD, J., TAGUCHI, Y., and SARIN, P. S. Inhibition of replication and expression of human T-cell lymphotropic virus type III in cultured cells by exogenous synthetic oligonucleotides complementary to viral RNA. Proc. Natl. Acad. Sci. USA 83 (1986) 4143–4146.

5. LISZIEWICZ, J., SUN, D., METELEV, V., ZAMECNIK, P., GALLO, R. C. and AGRAWAL, S. Long-term treatment of human immunodeficiency virus – infected cells with antisense oligonucleotide phosphorothioates. Proc. Natl. Acad. Sci. USA 90 (1993) 3860–3864.

6. LEITER, J. M. E., AGRAWAL, S. , PALESE, P. and ZAMECNIK, P. C. Inhibition of influenza virus replication by phosphorothioate oligodeoxynucleotides. Proc. Natl. Acad. Sci. USA (1990) 3430–3434.

7. TRAGER, W., and JENSEN, J .B. Human malaria parasites in continuous culture. Science, 193 (1976) 673–675.

8. HITCHINGS, G. H., ELION, G.B., VANDERWERFF, H., and FALCO, E. A. Pyrimidine derivatives as antagonists of Pteroylglutamic acid. J. Biol. Chem. 174 (1948) 765–766.

9. BZIK, D.J., L. I., W.-B., HORII, T., and INSELBURG, J. Molecular cloning and sequence analysis of the Plasmodium falciparum dihydrofolate reductase-thymidylate synthase gene. Proc. Natl. Acad. Sci. USA 84 (1987) 8360–8364.

10. AGRAWAL, S., SARIN, P. S. ZAMECNIK, M. and ZAMECNIK, P. C. Cellular uptake and Anti-HIV activity of oligonucleotides and their analogs in gene regulation: In R.P. Erickson and J.G. Izant (ed.) Biology of antisense RNA and DNA. Raven Press, Ltd. New York (1992) 273–283.

11. GOODCHILD, J., KIM, B., and ZAMECNIK, P. C. The clearance and degradation of oligodeoxynucleotides following intravenous infection into rabbits. Antisense Res. Dev. 1 (1991) 153–160.

12. RAPAPORT, E., MISIURA, K., AGRAWAL, S. , and ZAMECNIK, P. C. Antimalarial activities of oligodeoxynucleotide phosphorothioates in chloroquine-resistant Plasmodium falciparum. Proc. Natl. Acad. Sci. USA 89 (1992) 8577–8580.

13. HADLEY, T. J., KLOTZ, F. W., and MILLER, L. H. Invasion of erythrocytes by malaria parasites: a cellular and molecular overview. Ann. Rev. Microbiol. 40 (1986) 451–477.

14. UPSTON, J. M., and GERO, A .M. Parasite-induced permeation of nucleosides in

Plasmodium falciparum malaria. Biochimica et Biophysica Acta 1236 (1995) 249–258.

15. CLARK, D. L., CHRISEY, L. A., CAMPBELL, J. R., and DAVIDSON, E. A. Non-sequence-specific antimalarial activity of oligodeoxynucleotides. Mol. Biochem. Parasitol. 63 (1994) 129–134.

16. BARKER, R. H., Jr., METELEV, V. RAPAPORT, E., and ZAMECNIK, P. Inhibition of *Plasmodium falciparum* malaria using antisense oligodeoxynucleotides. Proc. Natl. Acad. Sci. USA 93 (1996) 514–518.

17. ZHANG, Y., and MESHNICK, S. R. Inhibition of Plasmodium falciparum dihydropteroate synthetase and growth *in vitro* by sulfa drugs. Antimocrib. Agents Chemother. 35 (1991) 267–271.

18. RUBION, H., SALEM, J. S., L. I., L.-S., YANG, F-D., MAMA, S. WANG, Z.-M., FISHER, A., HAMANN, C. S. and COOPERMAN, B. S. Cloning, sequence determination, and regulation of the ribonucleotide reductase subunits from Plasmodium falciparum: A target for antimalarial therapy. Proc. Natl. Acad. Sci. USA 90 (1993) 9280–9284.

19. CHAKRABARTI, D., SCHUSTER, S. S., and CHAKRABARTI, R. Cloning and characterization of subunit genes of ribonucleotide reductase, a cell-cycle-regulated enzyme, from Plasmodium falciparum. Proc. Natl. Acad. Sci. USA 90 (1993) 12020–12024.

20. AGRAWAL, S., TEMSAMANI, J., and TANG-, J-.Y. Self-stabilized oligonucleotides as novel antisense agents. In Akhtar, S. , ed. Delivery Strategies: Antisense Oligonucleotide Therapeutics. CRC Press, Boca Raton, FL (1995) 105–121.

21. LILLEY, D. M. J. Kinking of DNA and RNA by base bulges. Proc. Natl. Acad. Sci. USA 92 (1995) 7140–7142.

12 Oncology

Developing Antisense Therapeutics for Tumor Treatment

Karl-Hermann Schlingensiepen and Reimar Schlingensiepen
Max-Planck-Institut für Biophysikalische Chemie, Göttingen, Germany

1   Introduction

1.1   Molecular Basis of Antisense Therapy

Development of antisense oligonucleotides (oligos) as therapeutic agents is a novel concept for rationally designing drugs for the treatment of patients with oncological diseases. The development of antisense therapeutics is particularly attractive since it targets the molecular roots of disease without altering the genome by gene transfer.

In recent years, the molecular basis of tumor development has been increasingly well understood. In cells which acquire a malignant phenotype, growth control mechanisms are dysfunctional. The tight balance between stimuli that promote cell proliferation and those which inhibit excessive cell growth is altered.

Molecular alterations in the genes which control cell growth form the root of tumor development. The imbalance between positive and negative growth control genes can either be initiated by tumor viruses as in cervical carcinoma, or, more commonly, by mutations in the cell's own genome. Mutations can either inactivate tumor suppressor genes (also called anti-oncogenes) or they may activate proto-oncogenes which then become oncogenes.

Usually, cells acquire their malignant phenotype in several steps. A single mutation leads to a growth advantage which facilitates the acquisition of further mutations in proto-oncogenes. In addition, the regulation of genes that control cell anchorage, extracellular matrix deposition, angiogenesis and suppression of immune surveillance are altered. This process has been particularly well characterized for colon carcinoma from adenoma formation to the development of metastasizing carcinoma [15, 19, 29]. The deciphering of the molecular basis for tumor development has provided us with new tools for diagnosis, more accurate prognosis and, most important, with the basis for rationally designing novel drugs.

1.2   Antisense Oligonucleotide Properties Compared to Other Established or Experimental Therapeutics

While the basis for conventional chemotherapy is the non-specific inhibition of cell proliferation (suppressing growth of both normal and malignant cells), antisense therapeutics are being designed to selectively target genes that form the molecular basis of the deregulated growth of tumor cells, of tumor progression and for the escape of tumor cells from immune surveillance.

Antisense oligos differ from two other new therapeutic approaches, neutralizing antibodies and gene therapy (gene delivery by viral or other vectors), in that they are orders of magnitude smaller, with a molecular weight of around 5,000 Dalton (5 kD). Since tumors are not just lumps of malignant cells, but consist to a varying degree of extracellular matrix, stroma and blood vessels, the size of therapeutic molecules is of importance for penetrating the capillary walls as well as the interstitial space to reach the tumor cells.

It was recently shown that oligos injected into tumor-bearing mice easily penetrate tumor compartments. Diffusion was unimpeded by the extracellular matrix, both in tumors with low and high stromal content. One hour after injection, oligos could be detected in the tumor stroma as well as in the cytoplasm of tumor cells. After 24 h, oligos were still detected in the cytoplasm of tumor cells following a single 70 nmol bolus injection [51]. Further, the charged nature of the oligos did not interfere with their distribution.

Considering the difficulty to deliver genes in conventional gene therapy to tumors or even to a high proportion of cells in other tissues like *e.g.* liver, the easy and even penetration of oligos into the tumor cells of various different tumors *in vivo* in the above study are of great importance.

The size of the oligos (approx. 5 kD) may have an additional advantage: In contrast to blood vessels in normal tissue, tumor vessels are pathologically leaky for molecules of a size between approximately 4 kD and 10 kD [33, 39, 75]. Thus, oligos may preferentially penetrate tumor blood vessels over normal ones. This notion is substantiated by the fact that oligo labelling was as intense over tumor tissue, as it was over the the normal liver [51], the preferential organ for oligo uptake.

Furthermore, the cellular uptake of oligos into the cytoplasm under *in vivo* conditions is a great advantage over antibodies. While neutralizing antibodies have to be targeted to extracellular proteins or the extracellular domains of membrane proteins, antisense therapy can potentially target any protein, including cytoplasmic enzymes and also transcription factors, the molecular "switches" which regulate gene expression.

1.3 Strategies for Antisense Oligonucleotide Tumor Treatment

Four strategies for antisense oligo treatment of tumors are described in some detail below.
These are:
- Suppression of oncogenes
- Induction of tumor cell differentiation
- Inhibition of angiogenesis
- Reversal of immunosuppression by tumor-secreted molecules

Since activation of oncogenes is a key molecular mechanism for tumor development, inhibition of oncogene expression is the most obvious goal for rationally designed antisense drugs for tumor treatment (cf. [14, 47, 52]). Secondly, selective inhibition of gene expression opens up the possibility for another strategy: to induce differentiation of tumor cells. A few transcription factors play a central role in regulating gene expression to allow either cell proliferation or terminal differentiation. These molecular "switches" of gene expression thus form important targets for antisense mediated differentiation therapy.

Thirdly, factors that induce the ingrowth of blood vessels into solid tumors, *i.e.* neoangiogenesis may be suppressed by antisense oligos. Solid tumors depend on the ingrowth of blood vessels to supply them with oxygen, nutrients, etc. and to remove metabolites. Without blood vessel formation, tumors cannot grow beyond the size of approximately 1 mm in diameter [22, 23]. Inhibiting secretion of factors inducing endothelial cell proliferation and blood vessel formation is thus another strategy to develop treatment for solid tumors.

A fourth strategy, which is particularly attractive, is the reversal of immunosuppression: Many tumors secrete molecules which shield them from destruction by the immune system [31, 32, 68]. While the immune system can often recognize tumor cells and their mutant proteins, secretion of immunosuppressive factors, in particular TGF-β1 and TGF-β2, by the tumor cells renders immune surveillance ineffective [68]. Tumor cell overexpression of TGF-β suppresses the cytotoxic function of tumor infiltrating lymphocytes (TIL) as well as their proliferation. Reversal of the immunosuppressive effects with antisense oligos can restore lymphocyte proliferation and even more importantly, the ability of a patient's cytotoxic lymphocytes to kill autologous tumor cells [31].

This chapter will concentrate on recent developments, briefly review some exemplary studies and then give in-depth examples with experimental protocols[1] on the four different strategies outlined above, including several *in vivo* experiments in immunocompromised mice transplanted with human tumors.

2   Suppression of Oncogenes

Oncogenes may be viral genes expressed in infected cells. Alternatively, cellular growth regulating genes (proto-oncogenes) may be mutated to become oncogenes. This activation by mutation may occur through different mechanisms.

Genomic amplification of proto-oncogenes, for example, of c-*erb*B-2 [63, 64] is one important mechanism. Alternatively, gene translocation to a strong promoter region *e.g.* the t(14;18) bcl-2/immunoglobulin fusion gene in B-cell lymphoma, can occur. Both, amplification and this form of translocation, then lead to overexpression, resulting in abnormal growth regulation.

---

[1] All experiments desribed in the protocols of this chapter were performed with fully phosphorothioated oligodeoxynucleotides (S-ODN), purified to pharmacological grade purity (Biognostik, Germany)

Point mutations act differently and usually lead to constitutive activation of the encoded growth regulating proteins. Thus, the rat homologue of the human c-*erb*B-2 gene called *neu* is constitutively activated by a point mutation causing a single amino acid change in the membrane spanning domain of the encoded receptor protein[2]. The most common example of oncogene formation through point mutation are *ras* point mutations. In one famous example mutations may even convert a tumor suppressor gene, *p53*, into an oncogene (reviewed in [40]).

Yet another form of mutation is the formation of fusion genes, where gene translocation leads to the generation of fusion proteins like bcr-abl [66].

Mutated gene sequences may provide target regions for antisense oligos since theoretically they would confer specificity by binding to the mutated sequence more stringently than to the wild type sequence expressed in normal cells of the same patient.

The disadvantage of antisense oligos targeted to a mutated region, however, be it a point mutation site or the breakpoint region of a fusion gene, is that the antisense sequence cannot be chosen according to criteria for optimal effectiveness and selectivity [11, 12] (see also chapter 1, Introduction). Therefore, many more successful studies inhibiting overexpressed oncogenes compared to mutated genes have been performed. The specificity of this approach is much higher than anticipated since tumor cells often depend on high expression levels of the respective overexpressed gene, while normal cells do not [7, 8, 10, 13] (see also chapter 13, Hematology).

2.1 Oncogenes with Point Mutations

Point mutations in the *ras* gene family are amongst the most common activating mutations. They are found in a wide variety of tumors. Antisense oligos complementary to the mRNA region containing a Ha-*ras* point mutation were found to inhibit cell growth in T24 bladder carcinoma cells, expressing mutated Ha-*ras* but not in cells with wild type Ha-*ras* gene expression [58]. In this study, 9mer oligos were linked to the intercalating agent 5´-acridine and/or to a hydrophobic 3´dodecanol tail. The results, however, have to be interpreted with some caution, since a data bank search shows several hundred completely complementary binding sites for these 9mer sequences in a wide variety of different mRNA molecules. Thus, enhancement of oligo binding by intercalating agents will reduce specificity, leading to suppression of many different genes by oligomer sequences as short as 9mers.

---

[2] This mutation does not occur in humans, since the respective c-*erb*B-2 gene segment has a different sequence, in which a point mutation cannot cause the same amino acid substitution.

2.2   Fusion Genes

The bcr-abl fusion protein is frequently found in chronic myeloid leukemia. (CML). CML cells can be transplanted into immunodeficient mice, *e.g.* severe combined immuno-deficiency (Scid) mice. This transplantation of human leukemic cells leads to a disease process in Scid mice resembling that seen in leukemic patients [17, 54, 62]. Treatment of these mice with anti-bcr-abl antisense oligo resulted in increased survival compared to both untreated animals as well as to mice treated with sense or mismatch oligos [62].

2.3   Overexpressed Oncogenes: c-*erb*B-2

2.3.1   *The Role of c-erbB-2 in Tumor Development and Progression*

In a series of seminal studies, the c-*erb*B-2 gene (also called *HER*-2 or *neu*) was found to be amplified and/or overexpressed in about 30 %–45 % of human mammary carcinomas [24, 63, 64, 67, 72].

Amplification, as well as overexpression without amplification, was strongly correlated with a poor clinical prognosis: i.e. lymph node metastasis, shorter times to relapse and decreased patient survival time. The rate of c-*erb*B-2 overexpression correlates with disease progression. It increases from 30–45 % in primary mammary carcinomas, to more than 50 % in lymph node metastasis and about 67 % in bone micrometastasis, suggesting a major role in tumor development, progression and metastasis [77, 63, 24, 72, 44, 49].

Furthermore, it was discovered that c-*erb*B-2 overexpression is sufficient to cause normal fibroblasts to become tumorigenic [30] and surprisingly, c-*erb*B-2 is 100-fold more transforming than the related EGF-receptor [56].

After the initial finding of a key role of c-*erb*B-2 overexpression in mammary carcinoma, it was soon discovered, that c-*erb*B-2 was also overexpressed in about 50 % of pancreas carcinomas [71, 73], as well as in ovarian carcinomas [64, 5], gastric carcinomas [74, 41], non-small-cell lung cancer [35], oral squamous cell carcinomas [28], endometrial tumors [6, 27] and the 2nd most common tumors in males, prostatic carcinomas, [69, 46]. It was shown in a variety of these tumors that overexpression of the c-*erb*B-2 gene was the strongest molecular predictor for malignant progression, development of tumor metastasis and reduced patient survival times.

The p185/ErbB-2 protein encoded by the c-*erb*B-2 gene is a transmembrane receptor with a single membrane spanning domain and an intracellular tyrosine kinase domain. In SK-Br-3 human mammary carcinoma cells, which overexpress c-*erb*B-2, treatment with anti-c-*erb*B-2 oligos was used to determine the effects on ErbB-2/p185 protein expression by Western blot analysis, measurement of protein activity (tyrosine kinase activity) and with cell proliferation assays.

Furthermore, anti-c-*erb*B-2 oligos also showed marked effects *in vivo* in mice transplanted with human tumor cells.

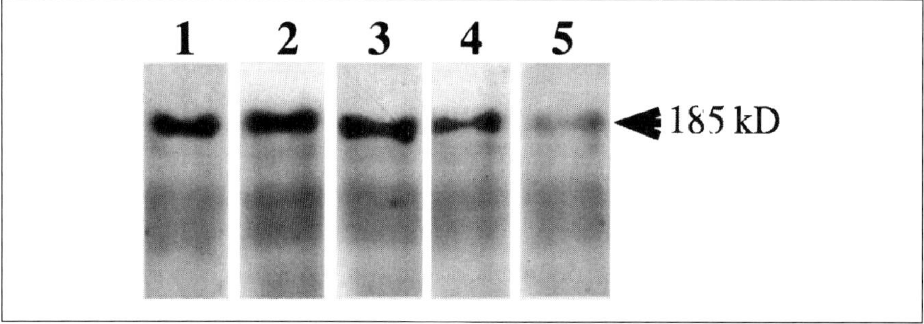

Figure 1: Western blot detecting ErbB-2 protein in SK-Br 3 carcinoma cells. Cells were treated with no oligo (lane 1), 2 µM randomized control oligo (lane 2 = 24 h; lane 3 = 48 h) or 2 µM anti-c-*erb*B-2 oligo (lane 4 = 24 h; lane 5 – 48 h). Cells were harvested and lysates separated by gel electrophoresis according to standard techniques. The c-ErbB-2 protein (p185) was detected with an anti-c-ErbB-2 specific antibody. Antisense treatment leads to a reduction in ErbB-2 protein expression at 24 h and an even more pronounced inhibition at 48 h. In cells treated with control oligo, ErbB-2 protein expression is not reduced compared to untreated control cells.

2.3.2

Protocol 1: Western Blot Analysis of p185/ErbB-2 Protein Inhibition

Special materials:
- Anti-c-*erb*B-2 oligo (Biognostik, Göttingen)
- SK-Br-3 mammary carcinoma cells (American Type Culture collection, ATCC)
- RPMI 1640 medium (Gibco)
- Fetal calf serum (FCS)
- penicillin, streptomycin (Gibco)
- 250 ml cell culture flasks (Nunc)
- 96-well flat bottom microtiter plates (Nunc)
- Anti-ErbB-2 antibody (Oncogene Science)
- Standard Western Blotting equipment

- Culture SK-Br-3 mammary carcinoma cells in RPMI 1640 medium supplemented with 100 U/ml penicillin, 100 µg/ml streptomycin and 10 % FCS.
- Prepare one 250 ml cell culture flask for each condition and time point. Plate 10^6 cells per flask in RPMI 1640 medium/100 U/ml penicillin/100 µg/ml streptomycin/10 % FCS and incubate with 2 µM antisense oligo, 2 µM control oligo or no oligo (untreated control cells).
- Harvest the cells by scraping off with a sterile cell scraper. Do not trypsinize the cells since this would lead to degradation of the extracellular domain of the ErbB-2 receptor protein.
- Centrifuge the cells, incubate with lysis buffer and prepare for gel electrophoresis.

- Perform Western blotting and immunodetection according to standard techniques. For important considerations regarding Western blotting in antisense experiments refer to chapter 6, Cell Culture Protocols.

Figure 1 shows the results of a Western blotting experiment with strong reduction in p185/ErbB-2 expression after 48 h of incubation with c-*erb*B-2 antisense oligo.

2.3.3

Protocol 2: Functional Assay of p185/ErbB-2: Cellular Tyrosine Kinase Activity

Special materials:
- "Buffered rinsing solution":
  137 mM NaCl; 5.4 mM KCL, 0.3 mM sodium phosphate; 0.4 mM potassium phosphate, 1 mM $CaCl_2$; 1 mg/ml D-glucose, 20 mM HEPES, pH 7.4
- "Tyrosine kinase substrate solution":
  Supplement above "buffered rinsing solution" with 50 µg/ml digitonin; 10 mM NaF; 10 mM $MgCl_2$; 1mM EGTA; 20 µM Na-vanadate, 25 µg/ml protein-kinase-A inhibitor IP-20, 25 µM -glycerophosphate, 2µCi -^{32}P-ATP (3000 Ci/mmol, Amersham) and 100 µM synthetic tyrosine kinase peptide (RRLIEDAEYAARG) as substrate.

- Culture SK-Br-3 mammary carcinoma cells as above.
- Plate 2500 cells per well into 96-well flat bottom microtiter plates in 100 µl of RPMI 1640 medium supplemented with 100 U/ml penicillin, 100 µg/ml streptomycin and 5 % FCS. To determine dose dependence of the antisense effect incubate the cells with 3 different concentrations of antisense oligo (we used 0.5 µM, 1 µM and 2 µM concentration). In control conditions incubate the cells with 2 µM control oligo and also prepare untreated control cells, plated into the same microtiter plate, as a reference condition (6 wells per condition).
- At different time points (we used 24 h and 48 h) perform *in situ* tyrosine kinase assay: Remove culture medium, rinse the wells with 100 µl of "buffered rinsing solution" (see materials above) and incubate the cells in 100 µl of the same solution for 10 min.
- Incubate the cells in 40 µl of "Tyrosine kinase substrate solution" (see materials above).
- Stop the reaction after 15 min with 10 µl tricchloracetic acid (25 %).
- Spot the supernatant onto Whatman P-81 phosphocellulose paper, wash with 10 % phosphoric acid and determine the amount of incorporated radioactivity in a liquid scintillation counter.

Figure 2 shows a dose- and time-dependent reduction in tyrosine kinase activity after treatment with different concentrations of antisense oligo for 24 h or 48 h.

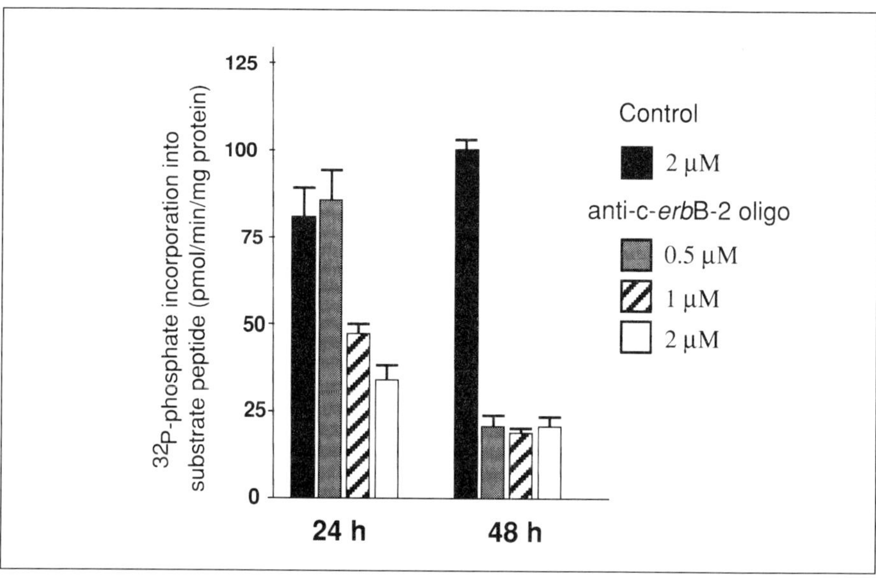

Figure 2: Tyrosine kinase activity. Tyrosine kinase activity in SK-Br-3 mammary carcinoma cells treated either with 2 µM control oligo (black bars), or increasing concentrations of anti-c-*erb*B-2 oligo: 0.5 µM (gray bars), 1 µM (hatched bars) or 2 µM (white bars). A dose-dependent effect is seen after 24 h and the inhibition of tyrosine kinase activity is even stronger after 48 h.

2.3.4 Cell Proliferation

Protocol 3: Thymidine Incorporation

Special materials:
- see Protocol 1
- ³H-thymidine (Amersham)
- Liquid scintillation counter
- 96-well flat bottom microtiter plates

- Culture SK-Br-3 mammary carcinoma cells as above.
- Plate 2500 cells per well into 96-well microtiter plates in 100 µl of RPMI 1640 medium/100 U/ml penicillin/100 µg/ml streptomycin/5 % FCS with 2 µM final concentration of oligos. Cells incubated with antisense oligos and control oligos as well as untreated control cells can be plated into the same microtiter plate (4–6 wells per condition; see chapter 6, Cell Culture Protocols).
- Prepare one plate for each time point *e.g.* 24 h, 48 h, 120 h.
- Incubate the cells with 0.15 µCi ³H-thymidine/well for 6 h before harvesting.
- Lyse the cells by storing the microtiter plates at –20 °C overnight. If cells are frozen at different time points following treatment with antisense

Figure 3 shows pronounced and long-lasting inhibition of thymidine incorporation after incubation with anti-c-*erb*B-2 oligo in SK-Br-3 mammary carcinoma cells. As a further control for the specificity of effects a different antisense oligo, targeting the growth inhibitory anti-oncogene *p53*, was used. Since inhibition of an oncogene and an anti-oncogene would be expected to have opposite effects, this provides a mean for a "positive" control, the most stringent control in antisense experiments, which should lead to opposite effects by two different antisense oligos. Indeed, the anti-*p53* oligo leads to a marked increase in thymidine incorporation by up to ten-fold.

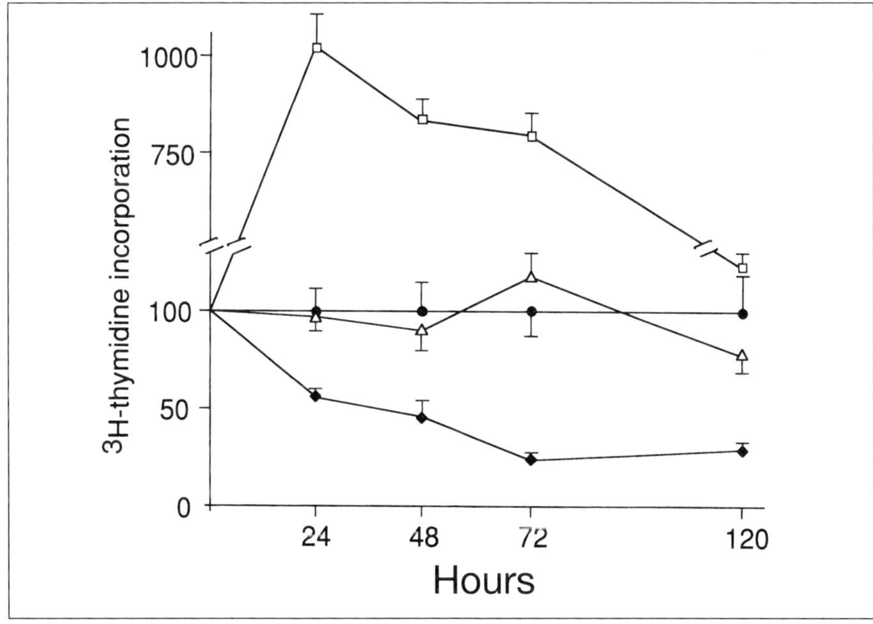

Figure 3: Thymidine incorporation in SK-Br-3 mammary carcinoma cells. Cells were incubated either with 2 µM anti-c-*erb*B-2 oligo (diamonds), 2 µM control oligo (triangles), no oligo (filled circles) of 2 µM anti-*p53* oligo (squares). Values are given as a percentage of control values in the untreated cells. Opposite effects are observed after inhibition a ErbB-2 expression with anti-c-*erb*B-2 oligo, leading to a prolonged reduction in thymidine incorporation compared to the application of anti-*p53* oligo, leading to an up to ten-fold increase in thymidine incorporation, consistent with the anti-proliferative action of the p53 protein. The different effects following inhibition of the c-*erb*B-2 oncogene versus suppression of the *p53* anti-oncogene demonstrates the specificity of the technique.

Protocol 4: Cell Counting

Special materials:

- see Protocol 1
- Trypan blue (Sigma)
- Neubauer counting chamber or automatic cell counter
- 96-well flat bottom microtiter plates (Nunc)

- Plate 2500 cells/well into 96-well plates in 100 µl of RPMI 1640 medium/100 U/ml penicillin/100 µg/ml streptomycin/5 % FCS. Add oligos at 2 µM final concentration. Cells incubated with antisense oligos and control oligos as well as untreated control cells can be plated into the same microtiter plate.
- Prepare one microtiter plate for each time point.
- Harvest the cells by trypsinization and mix with an equal volume of trypan blue.
- Determine cell numbers either by using a Neubauer counting chamber and trypan blue dye exclusion to determine the cell viability or by using an automatic cell counter.

Figure 4 shows a proliferation arrest of SK-Br-3 cells treated with antisense oligo four days after the beginning of treatment.

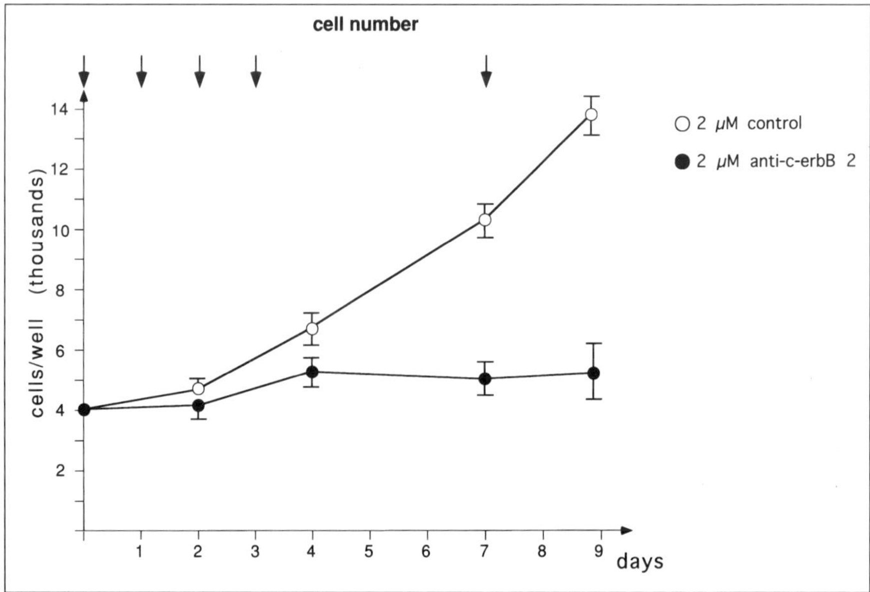

Figure 4: Cell growth arrest of SK-Br-3 carcinoma cells treated with anti-c-*erb*B-2 oligo. Cells were either treated with 2 µM anti-c-*erb*B-2 oligo (filled circles) or 2 µM control oligo. Because of the duration of the experiment, oligos were added repeatedly (indicated by the arrows). Cells treated with anti-c-*erb*B-2 oligo show a growth arrest following day 4, while control cells continue to proliferate.

2.3.5 *In vivo* *Treatment of Human Tumor-bearing SCID Mice with* *anti-c-erbB-2 Oligonucleotide*

For animal studies we chose human pancreas carcinoma as a model system. Pancreas carcinoma is the 6th most common malignancy in western countries and a tumor with increasing incidence. Approximately 85 % of cases are inoperable at the time of diagnosis and the mean survival rate of these patients is only 6 months.

As noted above, human pancreas carcinomas show c-*erb*B-2 overexpression in about 50 % of the cases.

Because of the extremely poor prognosis of patients with pancreas carcinoma (the 5-year survival rate is < 5 %) clinical studies should reveal effects rapidly. Furthermore, since there is practically no treatment for this disease, development of novel therapeutics is mandatory.

Protocol 5:** In vivo **Treatment of Tumor Bearing SCID Mice with anti-c-erbB-2 Oligonucleotide

- Grow low passage pancreas carcinoma cells in RPMI 1640 medium (Gibco) to subconfluency (70–80 % of density at confluence) with 100 U/ml penicillin, 100 µg/ml streptomycin, 2 mM Glutamine, 1 mM sodium pyruvate and 10 % FCS.
- Harvest cells by trypsinization.
- Stop trypsinization with FCS, centrifuge the cells and resuspend in 5 ml of serum-free RPMI 1640 medium per culture flask.
- Count the cells and aliquot 2×10^6–10^7 cells per animal into microfuge tubes.
- Centrifuge the cells and remove the supernatant.
- Add 350 µl of RPMI 1640 medium to the cells fill into 1 ml insulin syringes (*e.g.* Norm Ject, Henke-Sass Wolf GmbH, Tuttlingen) using a 27 gauge needle.
- Inject cells subcutaneously between the scapulae into mice using a 21 gauge needle.
- Let subcutaneous tumors grow to a size of 500–2000 mm³. Start treatment with antisense oligos by injecting 50 nmols of antisense oligo subcutaneously at a site removed from the tumor. Inject the animals once daily for 7 days. Measure tumor in three diameters according to [45] twice weekly.

In pilot experiments we found daily subcutaneous injection of 50 nmols of antisense oligo for seven days to be more effective than continuous application of 50 nmol per day with an Alzet osmotic minipump (Pump 2002, Alza Corp., Palo Alto, CA, USA) over 14 days. This may be due to markedly increased relative uptake into liver and kidney at low oligo doses, compared to high oligo doses, suggesting a saturable uptake mechanism by these two organs (A. Rifai, W. Brysch and K.-H. Schlingensiepen, unpublished observations).

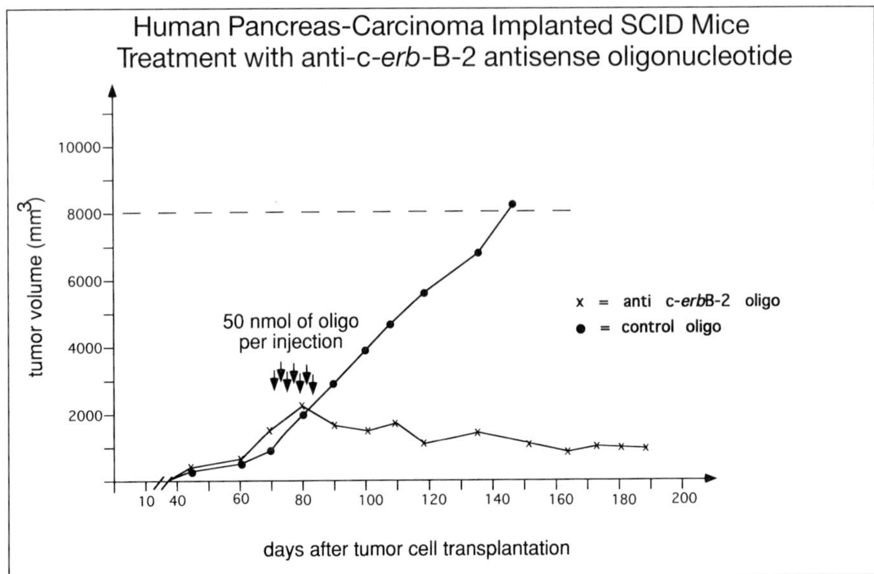

Figure 5: *In vivo* treatment of SCID mice bearing human pancreas carcinomas with anti-c-*erb*B-2 oligo. SCID mice were transplanted with human pancreas carcinoma cells. Animals were treated with seven subcutaneous (s.c.) injections (see arrows, 50 nmol per injection) of either anti-c-*erb*B-2 oligo (crosses) or randomized control oligo (filled circles). To minimize animal suffering, animals had to be killed when the tumors reached a volume of 8,000 mm³ (indicated by the broken line). Tumor size decreased after anti-c-*erb*B-2 oligo treatment, while the tumor grew rapidly after injection of control oligo.

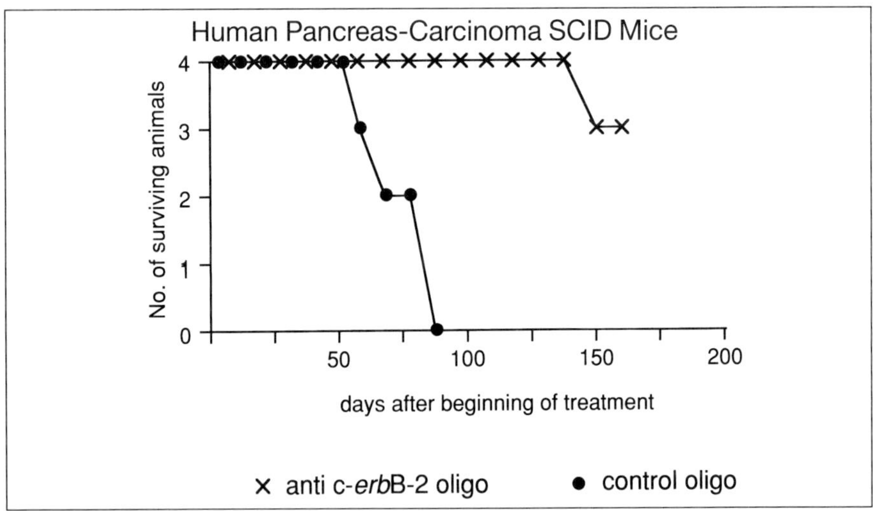

Figure 6: Survival of tumor-bearing SCID mice. SCID mice were transplanted with human pancreas carcinoma cells. Groups of four animals were treated either with seven s.c. injections (50 nmol per injection) of anti-c-*erb*B-2 oligo (crosses) or with randomized control oligo (filled circles). All control animals died or had to be killed due to tumor size by day 85 after beginning of treatment. In contrast, 160 days after initiation of treatment, three animal were still surviving, while one animal had died of an infection.

Thus, subcutaneous bolus injections lead to a strongly reduced percentage of oligo uptake by liver and kidney, allowing for higher concentrations to reach the tumor.

Figure 5 shows the comparison of tumor growth after antisense oligo treatment of a tumor-bearing mouse compared to control oligo treatment.

Figure 6 shows survival times of groups of four tumor-bearing mice after antisense treatment, compared to control oligo treatment.

3   Inhibition of Tumor Secreted Angiogenic Factors

Solid tumors depend on the ingrowth of blood vessels to grow more than 1 mm in diameter. Neoangiogenesis, the formation of new blood vessels supplying solid tumors is initiated by the secretion of angiogeneic factors like basic fibroblast growth factor (bFGF), transforming growth factor beta (TGF-β), platelet derived growth factors (PDGF) or vascular endothelial growth factor (VEGF). Several of these factors cooperate during the processes involved in induction of endothelial cells to migrate or proliferate and in the initiation of vascular growth with formation of differentiated capillaries [50, 65]. Selective inhibition of one or a combination of these angiogenic growth factors may offer a strategy to inhibit solid tumor growth.

In the experiment described below we used anti-bFGF antisense oligos, which had previously been shown to selectively suppress bFGF expression and autocrine growth inhibition of tumor cells by bFGF secretion [4, 21, 26].

SCID mice transplanted with human medulloblastoma cells were treated *in vivo* with antisense oligos targeted to bFGF mRNA. Constant low dose delivery of antisense oligo was achieved with an osmotic minipump. Later experiments with other antisense oligos have shown subcutaneous bolus injection of oligos to be more effective than constant low dose delivery by minipump, most likely due to a higher percentage of oligo distribution into the tumor compared to liver and kidney (see above, sections 1.2 and 2.3.5).

Protocol 6: *In vivo **Inhibition of Tumor Secreted Angiogenic Factors***

Special materials:
- Anti-bFGF S-ODN (Biognostik, Göttingen)
- HTB-186 medulloblastoma cells (American Type Culture Collection, ATCC)
- Eagle's minimal essential medium (MEM) supplemented with Earl's salts and 1 mM sodium pyruvate (Gibco)
- SCID mice
- Alzet osmotic minipump mod. 2002 (Alza Corp., Palo Alto, CA)

- Keep HTB-186 medulloblastoma cells according to ATCC instructions. Grow the cells to subconfluency (70–80 % of density at confluence) as above, see also chapter 6, Cell Culture Protocols, protocol 11.
- Prepare and aliquot 10^7 cells per animal and inject the tumor cell suspension subcutaneously between the scapulae of SCID-mice using a 21 gauge needle.
- Let subcutaneous tumors grow to a size of 500–3000 mm³.
- Load the osmotic minipump under sterile conditions according to the manufacturer's guidelines avoiding air bubble formation. We loaded pumps with 875 nmol of oligo dissolved in 250 µl of sterile buffering solution (while the pump delivers 200 µl over 14 days, in practice 250 µl are required for loading). This should lead to a constant delivery of 50 nmol per day for 14 days.
- Incubate the pump in sterile physiologic saline solution overnight in order to prime it (this ensures constant delivery rates throughout the experiment).
- Under deep anesthesia make a skin incision over the neck region and (by blunt dissection with a hemostat clamp) create a subcutaneous pouch along the body axis, large enough to accommodate the 2002 pump.
- Insert the pump and close the incision with two surgical clips or by suturing.
- Measure the tumor at regular intervals in three diameters as above (Protocol 5).

Figure 7 shows the results of treatment with anti-bFGF oligo delivered for 14 days via minipump compared to treatment with randomized control oligo. Constant delivery of anti bFGF oligo leads to a temporary halt in tumor growth followed again by a period of rapid growth. This is in contrast to the sustained long-lasting growth suppression in transplanted mammary carcinomas treated subcutaneously with anti c-*erb*B-2 oligo for 7 days (see Figure 5).

Tumor regrowth in the anti-bFGF oligo experiment may be due to several reasons. Firstly, as described above, the mode of delivery appears not to be optimal. Secondly, suppression of a single angiogenic factor may not be sufficient to induce a long-term inhibition of solid tumor growth. Possibly repression of a combination of different angiogenic factors would be required to achieve more continued success. This conclusion is further supported by results in another human tumor (see below, Figure 10). Thirdly, further experiments will be required to determine in detail whether the temporary reduction in tumor growth is due to anti-angiogenic effects or may be related to other properties of bFGF, *e.g.* autocrine growth stimulation of tumor cells [4].

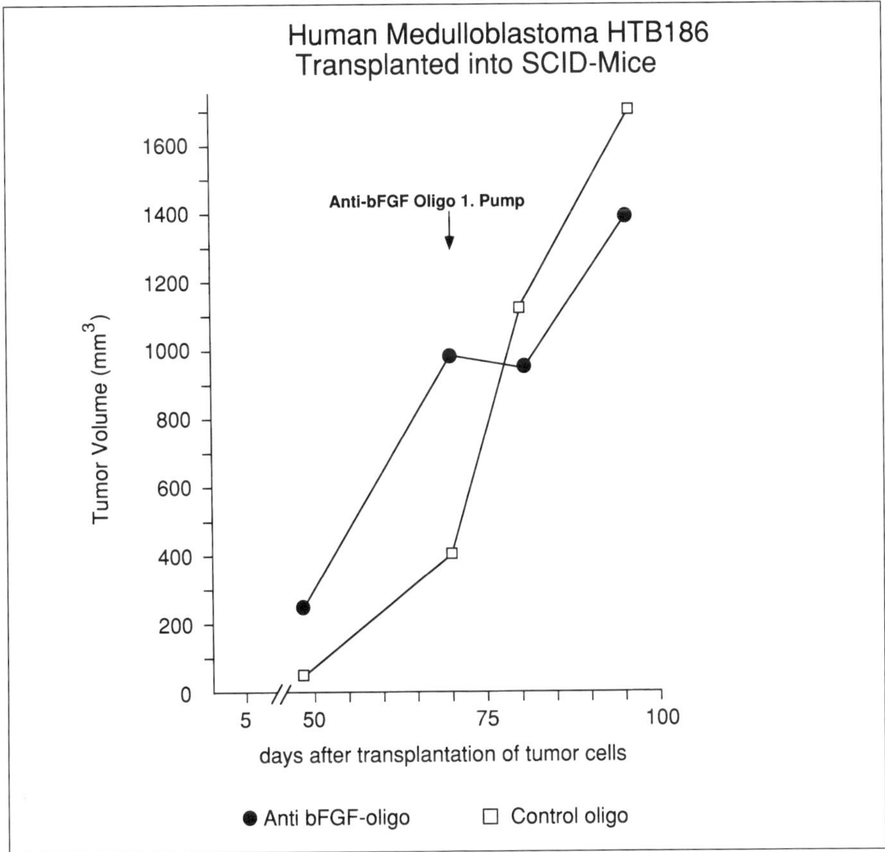

Figure 7: Treatment of medulloblastoma bearing SCID mice with anti-bFGF oligo: Continuous infusion. Animals were implanted with an osmotic minipump (see arrow), delivering oligos at a constant rate of 50 nmol per day for 14 days. Treatment with anti bFGF oligo led to a temporary arrest of tumor growth, followed by a rapid continuation of tumor growth at the end of the treatment. This is in contrast to the experiment described in Figure 5 which showed a persistent reduction in pancreas carcinoma growth after anti-c-*erb*B-2 oligo treatment.

Anti-angiogenesis remains an important area for antisense research beyond tumor therapy *e.g.* for the treatment of diabetic retinopathies or pathologic blood vessel formation in polyarthritis.

4   Induction of Tumor Cell Differentiation

The goal of differentiation therapy is to promote terminal cell differentiation in tumor cells. Differentiation and malignant proliferation are usually inversely correlated; the more dedifferentiated or "primitive" tumor cells are, the more malignant they behave.

Induction of terminal differentiation of tumor cells is a concept that is already clinically used to treat some forms of human leukemia. For example, treatment with retinoic acid can lead to long-lasting remission. The differentiation of tumor cells, just like the differentiation of normal cells from their stem cell precursors, requires a switch in gene expression. Genes which are required for proliferation have to be downregulated, while genes required for the differentiation process have to be switched on. Activation of the new gene expression pattern does not occur in a single step. First, regulatory genes encoding transcription factors have to be induced. These then switch on the expression of genes required for growth arrest and terminal differentiation.

The expression of only a few transcription factors is a key to changing cellular programs *e.g.* from proliferation to differentiation. By binding to numerous regulatory DNA regions they orchestrate the expression of a large number of downstream target genes. Consequently, they represent key molecules for switching cells from proliferation to terminal differentiation. Results shown below demonstrate that there is a tightly regulated balance between transcription factors promoting proliferation and those inducing differentiation (see schematic drawing in Figure 8).

The antisense technique opens up the possibility of altering the balance between the various transcription factors that regulate gene expression. By changing the expression of these molecular "switch-genes", cells may be induced to leave the proliferative cycle and start differentiation.

Since no specific agonists or antagonists of most transcription factors are known (with the exception of hormone receptor transcription factors), and since they cannot be inactivated by antibodies, antisense oligos represent the only molecules capable of modulating expression of most transcription factors.

4.1 Differentiation of PC-12 Tumor Cells Treated with anti c-*jun* or anti *jun*B Oligonucleotide

Experimentally we used the antisense strategy to modulate tumor cell differentiation by altering transcription factor expression in the phaeochromocytoma cell line PC-12. These cells differentiate, albeit to a low degree, upon treatment with nerve growth factor (NGF) into neuronal like cells. Treatment with antisense oligos designed to inhibit transcription factors of the Jun family could markedly alter the differentiation reaction of the tumor cells by NGF [59].

Protocol 7: Modulation of Tumor Cell Differentiation with Antisense Oligonucleotides

Special materials:
– Anti-c-*jun* S-ODN, anti-*jun*B S-ODN (Biognostik, Göttingen)
– PC-12 cells
– RPMI 1640 medium (Gibco)
– Fetal calf serum

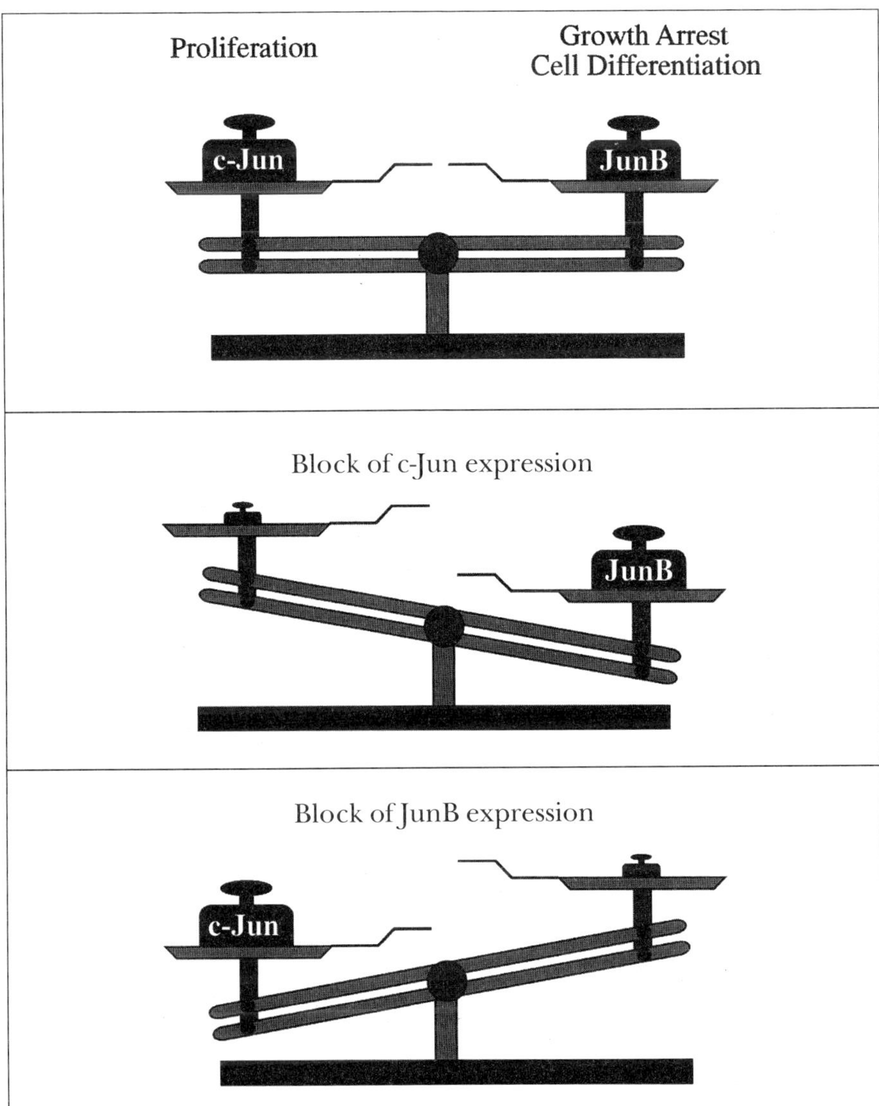

Figure 8: Antagonistic function of anti-c-*jun* and anti-*jun*B oligos on cell proliferation, growth arrest and cell differentiation. The two related transcription factors c-Jun and JunB are key regulators of cell proliferation and the differentiation process. The two proteins appear to be antagonists. The balance between the two transcription factors is an important regulatory mechanism to switch cellular programs from cell proliferation to growth arrest and differentiation.

- poly-L-lysine (Sigma) for coating of culture plates to allow neurite outgrowth
- Nerve growth factor (NGF) 2.5S (Sigma)
- 96-well flat-bottom microtiter plates

- Grow PC-12 cells in RPMI 1640 supplemented with 100 U/ml penicillin, 100 µg/ml streptomycin and 10 % FCS.
- Coat 96-well flat-bottom microtiter plates with poly-L-lysine to allow for better adherence of cells and outgrowth of neurites: Incubate the wells with a 0.1 % solution of poly-L-lysine overnight. Remove poly-L-lysine, wash the wells twice with sterile water and once with culture medium.
- Plate 2500 cells/well in RPMI 1640 medium with 5 % FCS into the poly-L-lysine coated wells and add phosphorothioate oligos at 2 µM concentration
- 2–24 h after plating of the cells add 10 ng/ml of NGF.
- Depending on the PC-12 cell clone used, neurite outgrowth takes different time spans, from a few days to more than a week, and occurs to a different extent.
- Observe and photograph the cells daily under the microscope.

Figure 9 shows the opposite effects of inhibiting expression of the two transcription factors c-Jun and JunB. In control cells, incubated with randomized control cells, few cells started to morphologically differentiate and extend neurites. Inhibition of JunB expression almost completely prevented morphological differentiation of NGF treated PC-12 cells. In contrast, inhibition of c-Jun expression strongly enhanced cell differentiation. The number of cells extending neuritic processes was strongly increased and neurite outgrowth was markedly enhanced (Figure 9).

Induction of tumor cell differentiation has also been described after treatment of neuroblastoma cells and-HL60 leukemic cells with antisense oligos against c-*myc* [3, 38].

4.2   Treatment of SCID Mice Transplanted with Human Tumor Cells with anti c-*jun* Oligonucleotide

In an *in vivo* experiment we further used antisense oligo against c-*jun* to experimentally treat SCID mice transplanted with human colon carcinoma cells.

Protocol 8: Treatment of Tumor Bearing Mice with anti c-jun Oligonucleotide

Special materials:
- see protocol 5
- Anti-c-*jun* S-ODN (Biognostik, Göttingen)
- Colo-357 colon carcinoma cells

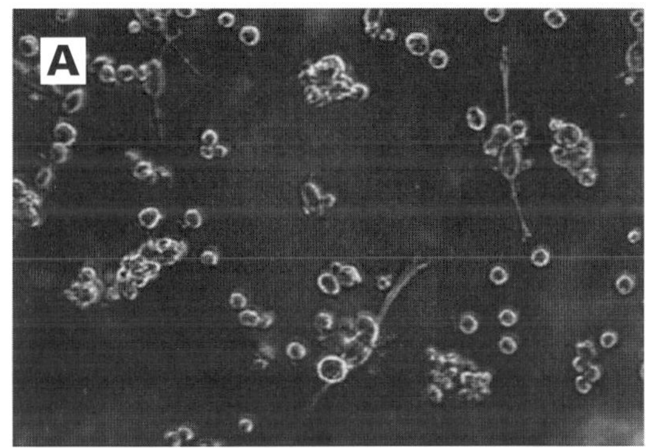

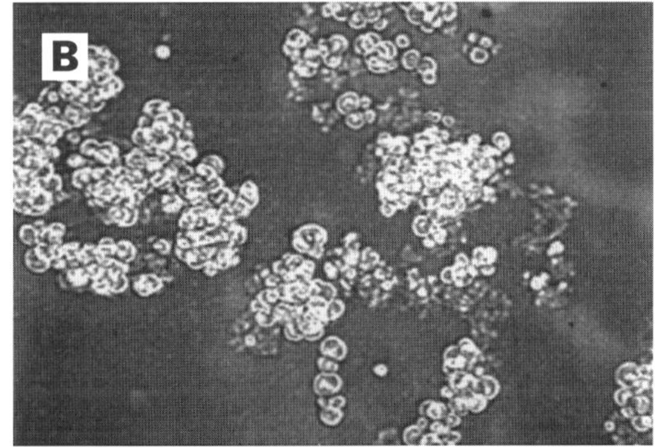

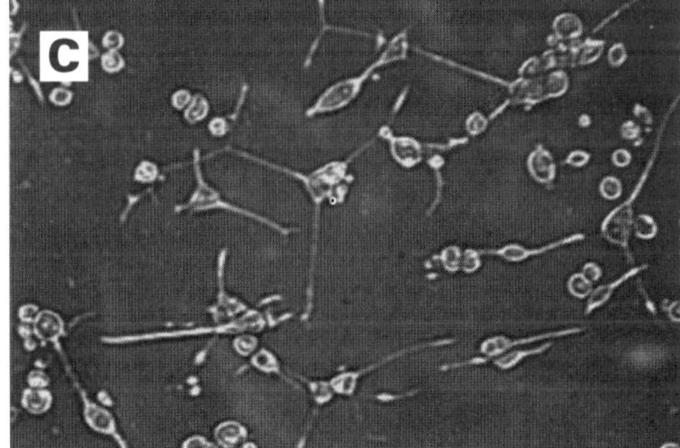

Figure 9: PC-12 tumor cell differentiation. NGF treated PC-12 tumor cells incubated with **A:** 2 µM randomized control oligo; **B:** 2 µM anti-*jun*B oligo; **C:** 2 µM anti-c-*jun* oligo. Treatment with anti-*jun*B oligo completely prevents tumor cell differentiation and the cells proliferate rapidly. In contrast, treatment with anti-c-*jun* oligo led to more pronounced morphological differentiation and the number of differentiating tumor cells is increased.

- Grow Colo-357 tumor cells in RPMI 1640 medium supplemented with penicillin, streptomycin, glutamine and sodium pyruvate to subconfluency (70–80 % of density at confluence) as described above in protocol 5, see also chapter 6, Cell Culture Protocols, protocol 11.
- Prepare and aliquot 10^7 cells per animal and inject the tumor cell suspension subcutaneously between the scapulae of mice using a 21 gauge needle.
- Let the subcutaneous tumors grow to a size of 500–3000 mm³. Start treatment with antisense oligos by injecting 50 nmols of antisense oligo subcutaneously at a site removed from the tumor. Inject the animals once daily for 7 days. Measure the tumor in three diameters twice weekly.

Results from a pilot experiment are shown in Figure 10. Survival of an animal treated with anti-c-jun oligo is compared with anti-bFGF oligo treatment and a saline injected control mouse. The oligos used have the same length and CG content.

Anti-c-*jun* treatment of animals transplanted with this aggressive human tumor led to markedly increased survival compared to anti-bFGF oligo injection and saline control treated animals.

In future experiments tumors will be examined for markers of cell differentiation after *in vivo* treatment with anti-c-*jun* oligo.

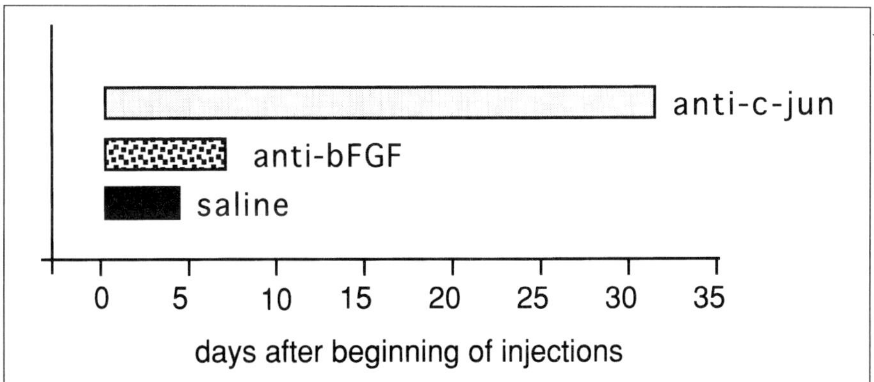

Figure 10: Survival of colon carcinoma bearing SCID mice treated with anti-bFGF oligo or anti-c-*jun* oligo. One mouse per condition was injected with 7 injections of either physiologic saline solution, or 50 nmol per injection of anti-bFGF oligo, or 50 nmol per injection of anti-c-*jun* oligo.

5   Reversal of Immunosuppression by Tumor Secreted Factors

5.1   Secretion of TGF-β by Tumors Leads to Escape from Immune Surveillance

Secretion of immunosuppressive factors by tumor cells impairs the function of tumor infiltrating lymphocytes. Thus, lysis of tumor cells by cytotoxic T-cells is suppressed, despite the presence of mutated antigens in tumor cells. A well-characterized example of tumor mediated immunosuppressive effects is malignant glioma.

Immunological deficiencies in glioma patients include reduced antibody production, cutaneous anergy and diminished numbers of circulating T-lymphocytes. T-cell activation is impaired and there are quantitative and qualitative defects in T-lymphocyte subsets [9, 57]. Secretion of members of the transforming growth factor β (TGF-β) family by the tumor cells is a major mechanism involved in inducing the immunosuppressive effects [18, 34].

Immunosuppressive effects, mediated by TGF- include the following: Inhibition of T-cell proliferation, inhibition of natural killer (NK) cell activation, reduced IL-1 and IL-2 receptor expression and depression of T-cell mediated tumor cytotoxicity [37, 43, 48, 53, 60].

Clinically, TGF-β expression correlates with the development of deep invasion, tumor progression, metastasis, immunosuppression of tumor infiltrating lymphocytes and markedly decreased patient survival in a variety of tumors, including melanoma [55], breast carcinoma [1, 2, 16, 42, 70], mesothelioma [20], pancreas carcinoma [25] and colonic carcinoma [29]. Furthermore, plasma levels of TGF-β also correlate with tumor aggressiveness and metastasis [36, 61].

The importance of TGF-β secretion by tumor cells for escaping immune surveillance was impressively demonstrated by the fact that highly immunogenic tumors escape immune surveillance if transfected with TGF-β [68].

We constructed different antisense oligos for the selective inhibition of TGF-β1 and TGF-β2 expression [31, 32]. Reversal of the immunosuppressive effects of tumor secreted TGF-β molecules was tested in a co-culture assay of tumor cells and autologous lymphocytes [31].

5.2

Protocol 9: Reversal of Immunosuppression in Tumor-lymphocyte Co-cultures

Special materials:
- Low passage TGF-β2 overexpressing glioblastoma cells
- Peripheral blood mononuclear cells (PBMCs) from the same patient (see chapter 7, Immunology, protocol 1)
- Ficoll-Hypaque centrifugation system (Pharmacia, Uppsala, Sweden)
- RPMI 1640/Glutamax and DMEM culture media (Gibco)
- Pooled human AB serum (Flow Laboratories, McLean, VA)

- IL-1α (R & D Systems, Inc. Minneapolis, MN)
- IL-2 (Biotest AG, Frankfurt, Germany)
- 48-well and 96-well flat-bottom-microtiter plates (Nunc)
- Anti-TGF-β2 polyclonal antibody
- ^{3}H-thymidine (Amersham)

- Establish tumor cell cultures cf. [31].
- Prepare PBMCs from heparinized venous blood, taken from the glioblastoma patient at the time of surgery, using Ficoll-Hypaque centrifugation according to the manufacturer's guidelines (see also chapter 7, Immunology, protocol 1) and store the cells in liquid nitrogen.
- Grow the PBMCs in RPMI 1640/Glutamax with 10 % pooled human AB serum.
- Plate 10^6 cells per well into 48-well flat-bottom microtiter plates and prepare lymphokine activated killer (LAK) cells by preactivation with 10U/ml IL-1 and 100 U/ml IL-2.
- Add antisense or control oligo to the tumor cells at 1 µM concentration 12 hours before both determination of lymphocyte proliferation and assay of cytotoxic effects against tumor cells.

Lymphocyte proliferation in mixed lymphocyte/tumor cell (MLTC) co-culture:

- Plate 1.5×10^4 lethally irradiated^3 tumor cells into 96-well flat-bottom microtiter plates in RPMI 1640/Glutamax with 10 % pooled human AB serum supplemented with 10U/ml IL-1 & 100 U/ml IL-2.
- Add 2.5×10^4 preactivated LAK cells. Determine LAK cell proliferation by adding 1 µCi ^{3}H-thymidine/well for 6 h before harvesting and liquid scintillation counting.

Cytotoxicity in mixed lymphocyte/tumor cell (MLTC) co-culture assay:

- Plate 1.5×10^4 tumor (target) cells into 96-well flat-bottom microtiter plates in RPMI 1640/Glutamax with 10 % pooled human AB serum without addition of lymphokines.
- Add antisense oligos 12 later and 1.5×10^5 irradiated LAK cells (30 Gy) 24 h after tumor cell plating.
- Determine the tumor cell cytotoxicity 3 days after adding lymphocytes to the culture by trypan blue staining and ^{3}H-thymidine incorporation.

In our experiments [31], lymphocyte proliferation in the presence of tumor cells was 3-fold increased after treatment with anti-TGF-β2 oligo com-

---

^{3} Irradiation of one cell type in co-culture assays prevents DNA replication in this cell type, *e. g.* the tumor cells. Thus, it allows to only determine the thymidine incoporation rate of the other cell type, *e. g.* the lymphocytes, while the irradiated cell type does not contribute to thymidine uptake.

pared to both treatment with control oligo and treatment without oligo. From the therapeutic perspective even more important was a marked increase in cytotoxic effect of lymphocytes on the tumor cells. The results, together with a schematic drawing of the experiment, are shown in Figure 11.

The ability to selectively reverse tumor-mediated immunosuppression may represent a further important step forward also for therapeutic application of tumor vaccines. The development of cancer vaccines has made considerable progress recently (cf. abstracts of the conference on Cancer Vaccines, Cambridge Healthtech Institute, 1995)[4]. However, initial success with destruction of tumor by the immune system is overcome with time through overexpression of TGF-β by the tumor cells. TGF-β thus reverses the therapeutic effects of vaccination (K. Melief, oral communication) *in vivo*.

In a recent review, the reversal of tumor mediated immunosuppression with anti-TGF-β oligos was considered to be of outstanding interest. The review concludes that "Since TGF-β 2 has been shown to be a potent immunosuppressive factor secreted by human malignant gliomas, the ability to inhibit TGF-β 2 expression by antisense oligos may have far reaching clinical implication in the development of molecular immunotherapy" [76].

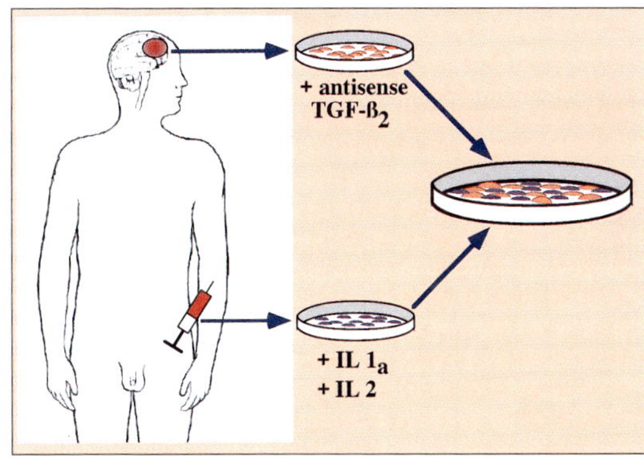

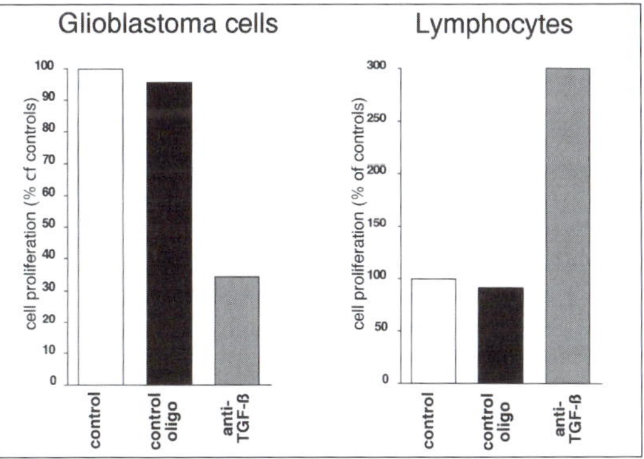

Figure 11: Reversal of immunosuppressive effects of tumor secreted TGF-β2 with anti-TGF-β2 oligo. Coculture assays of glioma cells with lymphocytes from the same patient. Pretreatment of tumor cells with anti TGF-β2 oligo reverses the suppression of lymphocyte proliferation by tumor secreted TGF-β. Furthermore, the lymphocytes also regain their cytotoxic function and lyse the tumor cells, leading to a reduction in the number of surviving glioma cells.

---

[4] Cambridge Healthtech Institute, 100 Winter Street, Suite 3700, Waltham, MA 02154, USA

6   Conclusions

Selective inhibition of gene expression with antisense oligos opens the way for causal treatment of tumors. We have described four different strategies to develop antisense agents as drugs for tumor treatment.

1. Inhibition of oncogenes, which play a pivotal role in tumor growth, can lead to a strong reduction in tumor cell proliferation in cell culture and *in vivo*.
2. Suppression of secreted factors involved in inducing tumor vascularization may halt solid tumor growth at least temporarily. Further experiments with combinations of antisense oligos against different angiogenic factors will be needed to further elucidate the therapeutic relevance of this strategy.
3. Differentiation of tumor cells may be induced by modulating expression of molecular "switches" like members of the Jun transcription factor family.
4. Finally, a particularly attractive form of therapy is the reversal of tumor mediated immunosuppression. Reactivation of the cytotoxic activity of autologous lymphocytes against tumor cells by TGF-β directed antisense oligos may yield therapeutic effects even after a short course of treatment.

Highly selective antisense oligos may thus become important therapeutic agents in the next few years.

References

1. ARTEAGA, C. L., T. C. DUGGER, A. R. WINNIER, and J. T. FORBES: Evidence for a positive role of transforming growth factor-beta in human breast cancer cell tumorigenesis. J. Cell Biochem. 17G(187–93) (1993).
2. ARTEAGA, C. L., S. D. HURD, A. R. WINNIER, M. D. JOHNSON, B. M. FENDLY, and J. T. FORBES: Anti-transforming growth factor (TGF)-beta antibodies inhibit breast cancer cell tumorigenicity and increase mouse spleen natural killer cell activity. Implications for a possible role of tumor cell/host TGF-beta interactions in human breast cancer progression. J. Clin. Invest. 92(6) (1993) 2569.
3. BACON, T. A. and E. WICKSTROM: Daily addition of an anti-c-myc DNA oligomer induces granulocytic differentiation of human promyelocytic leukemia HL-60 cells in both serum-containing and serum-free media. Oncogene Res. 6(1) (1991) 21.
4. BEHL, C., J. WINKLER, U. BOGDAHN, J. MEIXENSBERGER, K. H. SCHLINGENSIEPEN, and W. BRYSCH: Autocrine growth regulation in neuroectodermal tumors as detected with oligodeoxynucleotide antisense molecules. Neurosurgery 33(4) (1993) 679.
5. BERCHUCK, A., A. KAMEL, R. WHITAKER, B. KERNS, G. OLT, R. KINNEY, J. T. SOPER, *et al.*: Overexpression of HER-2/neu is associated with poor survival in advanced epithelial ovarian cancer. Cancer Res. 50(13) (1990) 4087.

6. Berchuck, A., G. Rodriguez, R. B. Kinney, J. T. Soper, R. K. Dodge, P. D. Clarke, and R. J. Bast: Overexpression of HER-2/neu in endometrial cancer is associated with advanced stage disease. Am. J. Obstet. Gynecol. 164 (1991) 15.

7. Bertram, J., M. Killian, W. Brysch, K. H. Schlingensiepen, and M. Kneba: Reduction of erbB2 gene product in mamma carcinoma cell lines by erbB2 mRNA-specific and tyrosine kinase consensus phosphorothioate antisense oligonucleotides. Biochemical & Biophysical Research Communications 200(1) (1994) 661.

8. Bertram, J., K. Palfner, M. Killian, W. Brysch, K. H. Schlingensiepen, W. Hiddemann, and M. Kneba: Reversal of multiple drug resistance in vitro by phosphorothioate oligonucleotides and ribozymes. Anti Cancer Drugs 6(1) (1995) 124.

9. Brooks, W. H., M. G. Netsky, D. E. Normansell, and et. al.: Depressed cell-mediated immunity in patients with primary intracranial tumors. Characterization of a human immunosuppressive factor. J. Exp. Med. 136 (1972) 1631.

10. Brysch, W., E. Magal, J. C. Louis, M. Kunst, I. Klinger, R. Schlingensiepen, and K. H. Schlingensiepen: Inhibition of p185c-erbB-2 proto-oncogene expression by antisense oligodeoxynucleotides down-regulates p185-associated tyrosine-kinase activity and strongly inhibits mammary tumor-cell proliferation. Cancer Gene Therapy 1(2) (1994) 99.

11. Brysch, W., A. Rifai, W. Tischmeyer, and K. H. Schlingensiepen: Rationale drug design, pharmacokinetics, intracerebral application and organ uptake of phosphorothioate oligodeoxynucleotides. In: S. Agrawal, ed. Antisense Therapeutics. Humana Press Totowa, NJ (in press).

12. Brysch, W. and K. H. Schlingensiepen: Design and application of antisense oligonucleotides in cell culture, in vivo, and as therapeutic agents. [Review] Cellular & Molecular Neurobiology 14(5) (1994) 557.

13. Calabretta, B., R. B. Sims, M. Valtieri, D. Caracciolo, C. Szczylik, D. Venturelli, M. Ratajczak, *et al.*: Normal and leukemic hematopoietic cells manifest differential sensitivity to inhibitory effects of c-myb antisense oligodeoxynucleotides: an in vitro study relevant to bone marrow purging. Proc. Natl. Acad. Sci. USA 88(6) (1991) 2351.

14. Carter, G. and N. R. Lemoine: Antisense technology for cancer therapy: does it make sense? Br. J. Cancer 67(5) (1993) 869.

15. Cho, K. R. and B. Vogelstein: Genetic alterations in the adenoma-carcinoma sequence. Cancer (1992).

16. Dalal, B. I., P. A. Keown, and A. H. Greenberg: Immunocytochemical localization of secreted transforming growth factor-beta 1 to the advancing edges of primary tumors and to lymph node metastases of human mammary carcinoma. Am. J. Pathol. 143 (1993) 381.

17. De, L. C., R. D. Clutterbuck, R. L. Powles, T. Min, J. C. Titley, and J. L. Millar: Human Philadelphia chromosome-positive chronic myeloid leukemia: a potential model for antisense therapy. Experimental Hematology 21(6) (1993) 826.

18. de Martin, R., B. Haendler, R. Hofer-Warbinek, H. Gaugitsch, M. Wrann, H. Schlusener, J. M. Seifert, *et al.*: Complementary DNA for human glioblastoma-derived T cell suppressor factor, a novel member of the transforming growth factor-beta gene family. Embo. J. 6(12) (1987) 3673.

19. FEARON, E. R. and B. VOGELSTEIN: A genetic model for colorectal tumorigenesis. Cell 61(5) (1990) 759.

20. FITZPATRICK, D. R., H. BIELEFELDT-OHMANN, R. P. HIMBECK, A. G. JARNICKI, A. L. MARZO, and B. W. ROBINSON: Transforming growth factor-beta: antisense RNA-mediated inhibition affects anchorage-independent growth, tumorigenicity and tumor-infiltrating T-cells in malignant mesothelioma. Growth Factors 11 (1994) 29.

21. FLOTT, R. B., W. GERDES, P. A. PECHAN, W. BRYSCH, K. H. SCHLINGENSIEPEN, and W. SEIFERT: bFGF induces its own gene expression in astrocytic and hippocampal cell cultures. Neuroreport 3(12) (1992) 1077.

22. FOLKMAN, J.: The role of angiogenesis in tumor growth. Semin. Cancer Biol. 3(2) (1992) 65.

23. FOLKMAN, J. and D. INGBER: Inhibition of angiogenesis. Semin. Cancer Biol. 3(2) (1992) 89.

24. FONTAINE, J., M. TESSERAUX, V. KLEIN, G. BASTERT, and N. BLIN: Gene amplification and expression of the neu (c-erbB-2) sequence in human mammary carcinoma. Oncology 45(5) (1988) 360.

25. FRIESS, H., Y. YAMANAKA, M. BUCHLER, M. EBERT, H. G. BEGER, L. I. GOLD, and M. KORC: Enhanced expression of transforming growth factor beta isoforms in pancreatic cancer correlates with decreased survival. Gastroenterology 105(6) (1993) 1846.

26. GERDES, W., W. BRYSCH, K. H. SCHLINGENSIEPEN, and W. SEIFERT: Antisense bFGF oligodeoxynucleotides inhibit DNA synthesis of rat astrocytes. Neuroreport 3(1) (1992) 43.

27. HETZEL, D. J., T. O. WILSON, G. L. KEENEY, P. C. ROCHE, S. S. CHA, and K. C. PODRATZ: HER-2/neu expression: a major prognostic factor in endometrial cancer. Gynecol. Oncol. 47(2) (1992) 179.

28. HOU, L., D. SHI, S. M. TU, H. Z. ZHANG, M. C. HUNG, and D. LING: Oral cancer progression and c-erbB-2/neu proto-oncogene expression. Cancer Lett. 65(3) (1992) 215.

29. HSU, S. , F. HUANG, M. HAFEZ, S. WINAWER, and E. FRIEDMAN: Colon carcinoma cells switch their response to transforming growth factor beta 1 with tumor progression. Cell Growth Differ. 5(3) (1994) 267.

30. HUDZIAK, R. M., J. SCHLESSINGER, and A. ULLRICH: Increased expression of the putative growth factor receptor p185HER2 causes transformation and tumorigenesis of NIH 3T3 cells. Proc. Natl. Acad. Sci. USA 84(20) (1987) 7159.

31. JACHIMCZAK, P., U. BOGDAHN, J. SCHNEIDER, C. BEHL, J. MEIXENSBERGER, R. APFEL, R. DORRIES, et al.: The effect of transforming growth factor-beta 2-specific phosphorothioate-anti-sense oligodeoxynucleotides in reversing cellular immunosuppression in malignant glioma. J. Neurosurg. 78(6) (1993) 944.

32. JACHIMCZAK, P., K. FABEL-SCHULTE, B. HESSDÖRFER, W. BRYSCH, K. H. SCHLINGENSIEPEN, A. BLESCH, and U. BOGDAHN: Transforming growth factor-beta-mediated regulation of human peripheral blood mononuclear cell proliferation as detected with phosphorothioate antisense oligodeoxynucleotides. Cellular Immunology 165(1) (1995) 125.

33. JAIN, R. K.: Delivery of novel therapeutic agents in tumors: physiological barriers and strategies. J. Natl. Cancer Inst. 81(8) (1989) 570.

34. JENNINGS, M. T., R. J. MACIUNAS, R. CARVER, C. C. BASCOM, P. JUNEAU, K.

Misulis, and H. L. Moses: TGF beta 1 and TGF beta 2 are potential growth regulators for low-grade and malignant gliomas in vitro: evidence in support of an autocrine hypothesis. Int. J. Cancer 49(1) (1991) 129.

35. Kern, J. A., D. A. Schwartz, J. E. Nordberg, D. B. Weiner, M. I. Greene, L. Torney, and R. A. Robinson: p185neu expression in human lung adenocarcinomas predicts shortened survival. Cancer Res. 50(16) (1990) 5184.

36. Kong, F. M., M. S. Anscher, T. Murase, B. D. Abbott, J. D. Iglehart, and R. L. Jirtle: Elevated plasma transforming growth factor-beta 1 levels in breast cancer patients decrease after surgical removal of the tumor. Ann. Surg. 222 (1995) 155.

37. Kuppner, M. C., M. F. Hamou, Y. Sawamura, S. Bodmer, and T. N. de: Inhibition of lymphocyte function by glioblastoma-derived transforming growth factor beta 2. J. Neurosurg. 71(2) (1989) 211.

38. Larcher, J. C., M. Basseville, J. L. Vayssiere, L. L. Cordeau, B. Croizat, and F. Gros: Growth inhibition of N1E-115 mouse neuroblastoma cells by c-myc or N-myc antisense oligodeoxynucleotides causes limited differentiation but is not coupled to neurite formation. Biochem Biophys Res Commun 185(3) (1992) 915.

39. Less, J. R., T. C. Skalak, E. M. Sevick, and R. K. Jain: Microvascular network architecture in a mammary carcinoma. Exs 61(74) (1992) 74.

40. Levine, A. J., J. Momand, and C. A. Finlay: The p53 tumour suppressor gene. Nature 351(6326) (1991) 453.

41. Lin, J. T., M. Wu, C. T. Shun, W. J. Lee, J. C. Sheu, and T. H. Wang: Occurrence of microsatellite instability in gastric carcinoma is associated with enhanced expression of erbB-2 oncoprotein. Cancer Res. 55(7) (1995) 1428.

42. MacCallum, J., J. M. Bartlett, A. M. Thompson, J. C. Keen, J. M. Dixon, and W. R. Miller: Expression of transforming growth factor beta mRNA isoforms in human breast cancer Br. J. Cancer 69 (1994) 1006.

43. Maxwell, M., T. Galanopoulos, G. J. Neville, and H. N. Antoniades: Effect of the expression of transforming growth factor-beta 2 in primary human glioblastomas on immunosuppression and loss of immune surveillance. J. Neurosurg. 76(5) (1992) 799.

44. May, E., H. Mouriesse, L. F. May, J. F. Qian, P. May, and J. C. Delarue: Human breast cancer: identification of populations with a high risk of early relapse in relation to both oestrogen receptor status and c-erbB-2 overexpression. Br. J. Cancer 62(3) (1990) 430.

45. Morrison, S. D.: In vivo estimation of size of experimental tumors. J. Natl. Cancer Inst. 71(2) (1983) 407.

46. Myers, R. B., S. Srivastava, D. K. Oelschlager, and W. E. Grizzle: Expression of p160erbB-3 and p185erbB-2 in prostatic intraepithelial neoplasia and prostatic adenocarcinoma. J. Natl. Cancer Inst. 86(15) (1994) 1140.

47. Neckers, L., L. Whitesell, A. Rosolen, and D. A. Geselowitz: Antisense inhibition of oncogene expression. Crit. Rev. Oncog. 3(1–2) (1992) 175.

48. Palladino, M. A., R. E. Morris, H. F. Starnes, and A. D. Levinson: The transforming growth factor-betas. A new family of immunoregulatory molecules. Ann. NY Acad. Sci. 593(181) (1990) 181.

49. Pantel, K., G. Schlimok, S. Braun, D. Kutter, F. Lindemann, G. Schaller, I. Funke, et al.: Differential expression of proliferation-associated molecules in

individual micrometastatic carcinoma cells. J. Natl. Cancer Inst. 85(17) (1993) 1419.

50. PERTOVAARA, L., A. KAIPAINEN, T. MUSTONEN, A. ORPANA, N. FERRARA, O. SAKSELA, and K. ALITALO: Vascular endothelial growth factor is induced in response to transforming growth factor-beta in fibroblastic and epithelial cells. J. Biol. Chem. 269(9) (1994) 6271.

51. PLENAT, F., N. KLEIN-MONHOVEN, B. MARIE, J.-M. VIGNAUD, and A. DUPREZ: Cell and tissue distribution of sythetic oligonucleotides in healthy and tumor-bearing nude mice. Am. J. Pathol. 147 (1995) 124.

52. PRINS, J., V. E. DE, and N. H. MULDER: Antisense of oligonucleotides and the inhibition of oncogene expression. Clin. Oncol. R. Coll. Radiol. 5(4) (1993) 245.

53. RANGES, G. E., I. S. FIGARI, T. ESPEVIK, and M. J. PALLADINO: Inhibition of cytotoxic T cell development by transforming growth factor beta and reversal by recombinant tumor necrosis factor alpha. J. Exp. Med. 166(4) (1987) 991.

54. RATAJCZAK, M. Z., J. A. KANT, S. M. LUGER, N. HIJIYA, J. ZHANG, G. ZON, and A. M. GEWIRTZ: In vivo treatment of human leukemia in a scid mouse model with c-myb antisense oligodeoxynucleotides. Proc. Natl. Acad. Sci. USA 89(24) (1992) 11823.

55. REED, J. A., N. S. McNUTT, V. G. PRIETO, and A. P. ALBINO: Expression of transforming growth factor-beta 2 in malignant melanoma correlates with the depth of tumor invasion. Implications for tumor progression. Am. J. Pathol. 145(1) (1994) 97.

56. ROMANO, A., W. T. WONG, M. SANTORO, P. J. WIRTH, S. S. THORGEIRSSON, and P. P. DI FIORE: The high transforming potency of erbB-2 and ret is associated with phosphorylation of paxillin and a 23 kDa protein. Oncogene 9(10) (1994) 2923.

57. ROSZMAN, T., L. ELLIOTT, and W. BROOKS: Modulation of T-cell function by gliomas. Immunol. Today 12 (1991) 370.

58. SAISON-BEHMOARAS, T., B. TOCQUE, I. REY, M. CHASSIGNOL, N. T. THUONG, and C. HELENE: Short modified antisense oligonucleotides directed against Ha-ras point mutation induce selective cleavage of the mRNA and inhibit T24 cells proliferation. Embo. J. 10(5) (1991) 1111.

59. SCHLINGENSIEPEN, K. H., R. SCHLINGENSIEPEN, M. KUNST, I. KLINGER, W. GERDES, W. SEIFERT, and W. BRYSCH: Opposite functions of jun-B and c-jun in growth regulation and neuronal differentiation. Developmental Genetics 14(4) (1993) 305.

60. SCHWYZER, M. and A. FONTANA: Partial purification and biochemical characterization of a T cell suppressor factor produced by human glioblastoma cells. J. Immunol. 134(2) (1985) 1003.

61. SHIRAI, Y., S. KAWATA, S. TAMURA, N. ITO, H. TSUSHIMA, K. TAKAISHI, S. KISO, et al.: Plasma transforming growth factor-beta 1 in patients with hepatocellular carcinoma. Comparison with chronic liver diseases. Cancer 73(9) (1994) 2275.

62. SKORSKI, T., S. M. NIEBOROWSKA, N. C. NICOLAIDES, C. SZCZYLIK, P. IVERSEN, R. V. IOZZO, G. ZON, et al.: Suppression of Philadelphia1 leukemia cell growth in mice by BCR-ABL antisense oligodeoxynucleotide. Proc. Natl. Acad. Sci. USA 91(10) (1994) 4504.

63. SLAMON, D. J., G. M. CLARK, S. G. WONG, W. J. LEVIN, A. ULLRICH, and W. L.

McGuire: Human breast cancer: correlation of relapse and survival with amplification of the HER-2/neu oncogene. Science 235(4785) (1987) 177.

64. Slamon, D. J., W. Godolphin, L. A. Jones, J. A. Holt, S. G. Wong, D. E. Keith, W. J. Levin, *et al.*: Studies of the HER-2/neu proto-oncogene in human breast and ovarian cancer. Science 244(4905) (1989) 707.

65. Sunderkotter, C., K. Steinbrink, M. Goebeler, R. Bhardwaj, and C. Sorg: Macrophages and angiogenesis. J. Leukoc. Biol. 55(3) (1994) 410.

66. Szczylik, C., T. Skorski, N. C. Nicolaides, L. Manzella, L. Malaguarnera, D. Venturelli, A. M. Gewirtz, *et al.*: Selective inhibition of leukemia cell proliferation by BCR-ABL antisense oligodeoxynucleotides. Science 253(5019) (1991) 562.

67. Tommasi, S. , C. Giannella, A. Paradiso, A. Barletta, A. Mangia, G. Simone, A. T. Primavera, *et al.*: HER-2/neu gene in primary and local metastatic axillary lymph nodes in human breast tumors. Int. J. Biol. Markers 7(2) (1992) 107.

68. Torre-Amione, G., R. D. Beauchamp, H. Koeppen, B. H. Park, H. Schreiber, H. L. Moses, and D. A. Rowley: A highly immunogenic tumor transfected with a murine transforming growth factor type beta 1 cDNA escapes immune surveillance. Proc. Natl. Acad. Sci. USA 87(4) (1990) 1486.

69. Veltri, R. W., A. W. Partin, J. Epstein, G. M. Marley, C. M. Miller, D. S. Singer, K. P. Patton, *et al.*: Quantitative nuclear morphometry, Markovian texture descriptors, and DNA content captured on a CAS-200 Image analysis system, combined with PCNA and HER-2/neu immunohistochemistry for prediction of prostate cancer progression. J. Cell Biochem. Suppl. 19 (1994) 249.

70. Walker, R. A., S. J. Dearing, and B. Gallacher: Relationship of transforming growth factor beta 1 to extracellular matrix and stromal infiltrates in invasive breast carcinoma. Br. J. Cancer 69(6) (1994) 1160.

71. Williams, T. M., D. B. Weiner, M. I. Greene, and H. J. Maguire: Expression of c-erbB-2 in human pancreatic adenocarcinomas. Pathobiology 59(1) (1991) 46.

72. Yamada, Y., M. Yoshimoto, Y. Murayama, M. Ebuchi, S. Mori, T. Yamamoto, H. Sugano, *et al.*: Association of elevated expression of the c-erbB-2 protein with spread of breast cancer. Jpn. J. Cancer Res. 80(12) (1989) 1192.

73. Yamanaka, Y., H. Friess, M. S. Kobrin, M. Buchler, J. Kunz, H. G. Beger, and M. Korc: Overexpression of HER2/neu oncogene in human pancreatic carcinoma. Hum. Pathol. 24 (1993) 1127.

74. Yonemura, Y., I. Ninomiya, S. Ohoyama, H. Kimura, A. Yamaguchi, S. Fushida, T. Kosaka, *et al.*: Expression of c-erbB-2 oncoprotein in gastric carcinoma. Immunoreactivity for c-erbB-2 protein is an independent indicator of poor short-term prognosis in patients with gastric carcinoma. Cancer 67(11) (1991) 2914.

75. Yuan, F., M. Leunig, D. A. Berk, and R. K. Jain: Microvascular permeability of albumin, vascular surface area, and vascular volume measured in human adenocarcinoma LS174T using dorsal chamber in SCID mice. Microvasc. Res. 45(3) (1993) 269.

76. Yung, W. K. A.: New approaches in brain tumor therapy using gene transfer and antisense oligonucleotides Current Opinion in Oncology 6 (1994) 235.

77. Zhou, D., H. Battifora, J. Yokota, T. Yamamoto, and M. J. Cline: Association of multiple copies of the c-erbB-2 oncogene with spread of breast cancer. Cancer Res. 47(22) (1987) 6123.

13 Hematology

Perturbing Hematopoietic Cell Gene Expression with Oligodeoxynucleotides – Research and Clincial Applications

Alan M. Gewirtz
University of Pennsylvania, Dep. of Pathology, Philadelphia, USA

1   Introduction

Antisense oligodeoxynucleotides (oligos) are used for loss-of-function studies in basic research and eventually they may be developed as therapeutic agents. In this chapter I describe the development of antisense oligos, from their use as investigative reagents in cell culture to their use as drugs in clinical trials, as it has been done in my laboratory. We have been using antisense oligos since 1988 to understand the role of *c-myb* and other protooncogenes in normal and malignant human hematopoiesis. These investigations were initially designed to elucidate the role of Myb protein in regulating hematopoietic cell development. Because the results obtained from these studies had obvious clinical relevance, more translationally oriented studies were undertaken. These have now culminated in Phase I clinical trials which are presently ongoing at the Hospital of the University of Pennsylvania. Below, I summarize the development of the *c-myb* targeted antisense oligo, and provide general laboratory protocols for carrying out the work described. In addition, I will also provide considerations for setting up a clinical trial.

1.1   The Antisense Strategy in Hematology

The last ten years have been a "golden age" in the field of hematology. During this time many of the extracellular growth factors which regulate hematopoietic cell development have been molecularly cloned, expressed as active proteins, and have made their way into the clinic. In addition, the intracellular events which are triggered when hematopoietic growth factors interact with their receptors has also begun to be defined. Nevertheless, most of the molecular machinery which regulates blood cell development remains hidden in the black box which lies below the cell surface. This machinery, in normal blood cells in particular, has been largely inaccessible because of the difficulty in applying modern molecular analytic techniques to the small numbers of cells which are ordinarily available for such investigations.

Two main strategies have been applied to understanding gene function in hematopoietic cells. One approach involves infecting cells with a vector engineered to express the gene of interest at high levels. If the cell's phenotype/behavior changes in the infected cell, but not a sham transfected cell, one can tentatively attribute the change in phenotype to the newly expressed gene. Function may therefore be inferred. This approach, while potentially informative, has a number of drawbacks. First, it is not certain that the changes observed are directly related to the gene's function as any effect observed may be an indirect one. Second, the experiment is not really "physiologic" since overexpression of the gene may exaggerate its normal function or impart new functions by leading to an excess of the encoded protein. Finally, this approach is often limited to leukemic cell lines since normal cells are often much more problematic to work with in terms of both infection and expression of the vector. An alternative experimental approach is to either physically "knock-out" the gene of interest or to interfere with its function at the mRNA level with antisense oligos. This approach has the theoretical advantage of being more predictive of function under physiologic conditions since any resulting phenotype is presumably the direct result of the loss of function of the targeted gene. In addition, these "loss of function" experiments may be more easily carried out in normal cells. For these reasons, this approach has become an increasingly popular strategy for exploring gene function in cells of virtually any type.

1.2   Selection of the Appropriate Target for the "Appropriate Disease"

For the past several years, we have been engaged in trying to develop a therapeutically effective strategy of disrupting gene function with antisense oligos. This pursuit has focused on finding an appropriate disease in which to apply this approach, a suitable gene target, and development of "scale-up" methods so that techniques developed in the laboratory could be applied in the clinic. As detailed in the preliminary results section, human leukemias appeared to be particularly amenable to this therapeutic strategy. Their growth characteristics are compatible with adequate oligo uptake, and a great deal is known about their cell and molecular biology. The latter in particular facilitates choice of a gene target.

Of the genes that we have targeted for disruption using the antisense oligo strategy [6, 11, 25, 26] one that has been of particular scientific interest to both of our laboratories, and one where therapeutically motivated disruptions are now in clinical trial, is the c-*myb* gene [14]. C-*myb* is the normal cellular homologue of v-*myb*, the transforming oncogene of the avian myeloblastosis virus (AMV) and avian leukemia virus E26. It is a member of a family composed of at least two other highly homologous genes designated A-myb and B-myb [22]. Located on chromosome 6q in humans, c-myb's predominant transcript encodes a ~75 kDa nuclear binding protein (Myb) [14] which recognizes the core consensus sequence 5'-PyAAC(G/Py)G-3' [3]. Myb consists of three primary functional regions [28] (Figure 1).

At the NH$_2$ terminus is the DNA binding domain. This region consists of three imperfect tandem repeats (R1, R2, R3) each consisting of 51–52

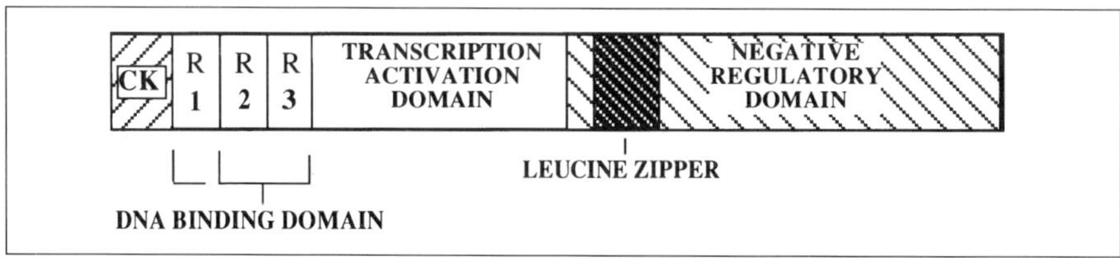

Figure 1: Functional map of the c-myb protein. See text for details

amino acids. Three perfectly conserved tryptophan residues are found in each repeat. Together they form a cluster in the hydrophobic core of the protein which maintains the DNA binding helix-turn-helix structure. The mid-portion of the protein contains an acidic transcriptional activating domain. The DNA binding portion of the protein is required for these transcriptional effects to be observed. The protein also contains a negative regulatory domain which has been localized to the carboxy terminus. Interestingly, the carboxy terminus is deleted in v-myb and this has been thought to contribute to v-myb's transforming ability. Recently reported experiments have been among those to confirm this hypothesis and have further demonstrated that amino terminal deletions give rise to a protein with even more potent transforming ability [9]. Deletions of both the amino and carboxy termini create a protein with the greatest transforming ability, and one which induces the formation of hematopoietic cells that are more primitive than those produced by amino terminal deletions alone [9]. These data suggest that the simultaneous loss of Myb's ability to bind DNA and interact with as yet unidentified proteins are potent transforming stimuli. Nevertheless, this simple hypothesis is complicated by the observation that overexpression of the c-terminal portion of c-*myb* can also be oncogenic [23] wherease overexpression of the whole protein is not [9]. At the least, one may conclude that the sequestration of certain potential Myb binding proteins may also be an oncogenic event.

Recently, a putative leucine zipper structure was described within the amino terminal portion of Myb's carboxy terminal domain [13]. Leucine zippers, such as those found in the transcription factors Jun, Fos, and Myc are thought to facilitate the protein-protein interactions which permit heterodimerization of DNA binding proteins. Such dimerization is thought to play a key role in regulating the transcriptional activity of these factors. A Myb dimerizing binding partner has yet to be identified but Myb-Myb homodimerization, which likely occurs through its leucine zipper, does lead to loss of DNA binding and transactivation ability [21]. Accordingly, one could reasonably postulate that Myb driven transactivation and/or transformation might be regulated by the binding of additional protein partners in the leucine zipper domain [13]. Alternatively, loss of the ability of Myb to dimerize with a putative regulatory partner might also contribute, directly or indirectly, to cellular transformation and leukemogenesis. Point mutations in the Myb negative regulatory domain might be one mechanism for bringing about such a loss [13]. Finally, interaction (not physical dimerization) with other nuclear binding proteins such as the CCAAT enhancer binding protein (C/BEP) [4], and the related myeloid nuclear factor NF-M [17] may also regulate Myb's transactivation or repressor functions.

The above discussion suggests that c-myb might play a role in leukemogenesis. Additional, albeit indirect, evidence also supports this contention. For example, *c-myb* amplification in AML and overexpression in 6q$^-$ syndrome has been reported [2]. The mechanism whereby overexpressed Myb might be leukemogenic is uncertain but points out the important difference of working with primary cells as opposed to cell lines. As noted above, it has been reported that overexpressing Myb is not by itself leukemogenic [9] but this work was carried out in cell lines which may give results that are valid only for the lines tested. As was also noted above, one could reasonably postulate that Myb driven transformation might be regulated by the binding of additional protein partners in the leucine zipper domain [13]. Recent evidence demonstrating that Myb interacts with other nuclear binding proteins, and that Myb's carboxy-terminus may interact with a cellular inhibitor of transcription supports this hypothesis [4, 23]. Other potential mechanisms might relate to Myb's ability to regulate hematopoietic cell proliferation [10], perhaps by its effects on important cell cycle genes including c-myc [7], PCNA [1], and cdc2 [15]. Finally, Myb also plays a role in regulating hematopoietic cell differentiation [31]. It functions as a transcription factor for several cellular genes including the neutrophil granule protein mim-1 [20], CD4 [19], IGF-1 [30], and CD34 [18] and possibly other growth factors [14], including *c-kit* [26]. The latter is of particular interest since it has been shown that when hematopoietic cells are deprived of *c-kit* ligand (Steel Factor) they undergo apoptosis [32]. Our preliminary data suggest that malignant hematopoietic cells may be more susceptible to this phenomenon then normal cells. Accordingly, *c-myb* is clearly an important hematopoietic cell gene which may, directly or indirectly, contribute to the pathogenesis or maintainence of human leukemias. For this reason it is a rational target for therapeutically motivated disruption strategies. Nevertheless, elucidating Myb's targets or cooperating proteins, may be even more rational, especially if leukemic specific partners can be identified.

2   Targeting the *c-myb* Gene

2.1   Role of c-myb Encoded Protein in Normal Human Hematopoiesis

Attempts to exploit the c-*myb* gene as a therapeutic target for antisense oligos began as an outgrowth of studies which were seeking to define the role of Myb protein in regulating normal human hematopoiesis. During the course of these studies it was determined that exposing normal bone marrow mononuclear cells (MNC) to c-*myb* antisense oligomers resulted in a decrease in cloning efficiency and progenitor cell proliferation. The effect was lineage indifferent since c-*myb* antisense DNA inhibited granulocyte-macrophage colony forming units (CFU-GM), CFU-E (erythroid), and CFU-Meg (megakaryocyte). In contrast, c-*myb* sense oligomers had no consistent effect on hematopoietic colony formation when compared to growth in control cultures (p = 0.778, p = 0.796, p = 0.375 for CFU-GM,

CFU-E, and CFU-Meg respectively). Inhibition of colony formation was also dose related and sequence specific. If progenitor cells were exposed to 0.3, 1.4, 7, and 14 µM of c-*myb* oligos the sense oligomers did not significantly effect colony formation while the antisense oligomers at the above concentrations led to the formation of 21 ± 3, 17 ± 1, 10 ± 2, and 4 ± 2 megakaryocyte colonies, respectively. Antisense oligos at concentrations of 1.4, 7, and 14 µM led to the formation of 775 ± 127, 515 ± 9, and 211 ± 25 myeloid colonies. The requirement for Myb during normal hematopoiesis has also been confirmed in gene targeting experiments carried out using the technique of homologous recombination. In other investigations, it was also determined that hematopoietic progenitor cells appeared to require Myb protein during specific stages of development, in particular when they were actively cycling.

2.2   Myb Protein is also Required for Leukemic Hematopoiesis

Since the c-*myb* antisense oligomers inhibited normal cell growth, we were also interested in determining their effect on leukemic cell growth. Can one reasonably postulate that aberrant c-*myb* expression or Myb function might play a role in carcinogenesis? Amplification of c-*myb* in acute myelogenous leukemia (AML) and c-*myb* overexpression in 6q⁻ syndrome has been reported suggesting that altered c-*myb* expression may indeed play a role in leukemogenesis. The exact nature of this role remains to be defined. One potential mechanism is through Myb's ability to regulate hematopoietic cell proliferation. Indeed, regulating progression through the G_1/S gate of the cell cycle seems to be a key function of c-Myb. The mechanism whereby c-*myb* regulates cell proliferation is uncertain but might well relate to its role in regulating a number of important cell cycle genes including c-*myc*, PCNA (proliferating cell nuclear antigen), and cdc2. Myb may also play a role in regulating differentiation. Evidence to support this hypothesis includes the observations that as primitive hematopoietic cells mature c-*myb* expression declines, and that constitutive expression of c-*myb* usually, but not always, inhibits the ability of hematopoietic cells to undergo differentiation. More directly, c-Myb has been shown to function as a transcription factor for several cellular genes including the neutrophil granule protein mim-1, CD4, IGF-1, and CD34, and possibly other growth factors or growth factor receptors such as c-Kit.

To address the question of Myb's role in leukemic hematopoiesis, we employed a variety of leukemic cell lines, including those of myeloid and lymphoid origin . In addition, we also employed primary patient material. We first detemined the effect of *myb* sense and antisense oligomers on the growth of HL-60, K562, KG-1, AND KG-1a cell lines. The antisense oligomer inhibited the proliferation of each leukemia cell line, although the effect was most pronounced on HL-60 cells. Specificity of this inhibition was demonstrated by the fact that the sense oligomer had no effect on cell proliferation, nor did "antisense" sequences with 2 or 4 nucleotide mismatches. To determine whether the treatment with c-*myb* antisense modified cell cycle distribution of HL-60 cells, we measured the DNA content in exponentially growing cells exposed to either sense or antisense *myb* oligomers. Control cells, and cells treated with c-*myb* sense oligomers had twice the DNA content of HL-60 cells exposed to the antisense oligomers. The majority of these cells appeared to reside either in the G1 compartment, or were blocked at the G1/S boundary. To examine the effect of the c-*myb* oligomers on lymphoid cell growth we employed a lymphoid leukemia cell line, CCRF-CEM. As we noted in the case of normal lymphocytes, the CCRF-CEM cells were extremely sensitive to the anti-proliferative effects of the c-*myb* antisense DNA. When exposed to the sense oligomers, we found negligible effects on CEM cell growth in short-term suspension cultures. In contrast, exposure to c-*myb* antisense DNA resulted in a daily decline in cell numbers. Compared to untreated controls, antisense DNA inhibited growth by ~ 2 logs. Growth reduction was not a cytostatic effect since cell viability was reduced by only ~ 70 % after exposure to the antisense oligomers, and CEM cell growth did not recover when cells were left in culture for an additional nine days. The c-*myb* antisense oligomers also appeared effective in inhibiting cell growth of primary patient material when derived from patients with both acute and chronic myeloid leukemias. Colony inhibition was ~ 58–95 %, and as determined by elimination of *bcr-abl* expressing CML clones, apparently quite efficient.

2.3 Normal and Leukemic Progenitor Cells Rely Differentially on Myb Function

In order to be useful as a therapeutic target, leukemic cells would have to be more dependent on Myb protein than their normal counterparts. To examine this critical issue we incubated phagocyte and T cell depleted normal human marrow mononuclear cells (MNC), human T lymphocyte leukemia cell line blasts (CCRF-CEM), or 1:1 mixtures of these cells with sense or anti-sense oligos to codons 2–7 of human c-myb mRNA. Oligomers were added to liquid suspension cultures at Time 0 and at Time +18 h. Control cultures were untreated. In controls, or in cultures to which "high" doses of sense oligomers were added, CCRF-CEM proliferated rapidly, whereas MNC numbers and viability decreased < 10 %. In contrast, when CCRF-CEM were incubated for four days in c-myb antisense DNA, cultures contained $4.7 \pm 0.8 \times 10^4$ cells/ml (mean ± SD; n=4) compared to $285 \pm 17 \times 10^4$/ml in controls. At the effective antisense dose, MNC were largely unaffected. After four days in culture, remaining cells were transferred to methylcellulose supplemented with recombinant hematopoietic growth factors. Myeloid colonies/clusters were enumerated at day ten of culture inception. Depending on cell number plated, control MNC formed from 31 ± 4 to 274 ± 18 colonies. In dishes containing equivalent numbers of untreated or sense oligomers exposed CCRF-CEM, colonies were too numerous to count. When MNC were mixed 1:1 with CCRF-CEM in antisense oligomer concentrations of ≤ 5 µg/ml, only leukemic colonies could be identified by morphologic, histochemical and immunochemical analysis. However, when antisense oligomer exposure was intensified, normal myeloid colonies could now be found in the culture while leukemic colonies could no longer be identified with certainty using the same analytic methods. Finally, at antisense DNA doses used in the above studies, AML blasts from eighteen of twenty-three patients exhibited a ~75 % decrease in colony and cluster formation compared to untreated or sense oligomer treated controls. When 1:1 mixing experiments were carried out with primary AML blasts and normal MNC, we were again able to preferentially eliminate AML blast colony formation while normal myeloid colonies continued to form.

2.4   Use of c-*myb* Oligomers as Bone Marrow Purging Agents

The above experiments suggested that leukemic cell growth could be preferentially inhibited after exposure to c-*myb* antisense oligos. In contemplating a clinical use for our findings, application in the area of bone marrow transplantation seemed compelling. In this application, exposure conditions are entirely under the control of the investigator. In addition, the patient's exposure to the antisense DNA is minimal. This circumstance would also make approval by regulatory agencies less difficult. We therefore determined if the antisense oligomers could be utilized as *ex vivo* bone marrow purging agents. To test this hypothesis, normal MNC were mixed (1:1) with primary AML or CML blast cells and then exposed to the oligomers using a slightly modified protocol designed to test the feasibility of a more intensive antisense exposure. With this in mind, an additional oligomer dose (20 µg/ml) was given just prior to plating the cells in methylcellulose. In control growth factor-stimulated cultures leukemic cells formed 25.5 ± 3.5 (mean $\pm$ SD) colonies and 157 ± 8.5 clusters (per 2×10^5 cells plated). Exposure to c-*myb* sense oligomers did not significantly alter these numbers (19.5 ± 0.7 colonies and 140.5 ± 7.8 clusters; $p > 0.1$). In contrast, equivalent concentrations of antisense oligomers totally inhibited colony and cluster formation by the leukemic blasts. Colony formation was also inhibited in the plates containing normal MNCs, but only by ~50 % in comparison to untreated control plates (control colony formation, 296 ± 40 per 2×10^5 cells plated; treated colony formation, 149 ± 15.5 per 2×10^5 cells).

2.5   Evaluating the Efficiency of an Antisense Purge

To assess the potential effectiveness of an antisense purge in CML patients, we carried out co-culture studies with cells obtained from patients in blast crisis and in chronic phase of their disease. RNA was extracted from cells cloned in methylcellulose cultures after exposure to the highest c-*myb* antisense oligomer dose. The RNA was then reverse transcribed and resulting cDNA amplified. For each patient studied, mRNA was also extracted from a comparable number of cells derived from untreated control colonies using the same technique. Eight cases were evaluated and in each case *bcr-abl* expression as detected by RT-PCR correlated with colony growth in cell culture. In cases which were inhibited by exposure to c-*myb* antisense oligomers (7/11), *bcr-abl* expression was also greatly decreased or non-detectable. These results suggested that *bcr-abl* expressing CFU might be substantially or entirely eliminated from a population of blood or marrow mononuclear cells by exposure to the antisense oligos. To explore this possibility further, re-plating experiments were carried out on samples from two patients. We hypothesized that if CFU belonging to the malignant clone were present at the end of the original 12 day culture period, but not detectable because of failure to express *bcr-abl*, they might re-express the message upon re-growth in fresh cultures. Accordingly, cells from these patients were exposed to oligomers and then plated into methylcellulose cultures formulated to favor growth of either CFU-GM or CFU-GEMM. As was found with the original specimens, untreated control cells and cells exposed to sense oligomers had RT-PCR detectable *bcr-abl* transcripts. Those cells exposed to the c-*myb* antisense oligomers had none. One of the paired dishes from these cultures was then solubilized with fresh medium, and all cells contained therein were washed, disaggregated, and re-plated into fresh methylcellulose cultures *without* re-exposing the cells to oligomers. After 14 days, CFU-GM and CFU-GEMM colony cells were again probed for *bcr-abl* expression. Control and sense treated cells had RT-PCR detectable mRNA but none was found in the antisense treated colonies. These results suggest that elimination of *bcr-abl* expressing cells and CFU was highly efficient and perhaps permanent.

2.6   Efficacy of c-myb Oligonucleotides *in vivo*

The studies described above were carried out primarily with unmodified DNA. Such molecules are subject to endo and exonuclease attack at the phosphodiester bonds and are therefore of little utility *in vivo*. Chemical modifications of oligos have sought to address this problem. For example, substitution of a sulfur for a hydrogen atom creates a class of compounds called phosphorothioates. These have a number of desirable properties including increased stability, water solubility, and ability to provide an RNaseH substrate (see also chapter 1). To examine the *in vivo* effectiveness of phosphorothioate oligos we established human leukemia/SCID mouse chimeras and then treated the animals with the nuclease-resistant phosphorothioate oligomers targeted to the c-*myb* gene. SCID mice were injected i. v. with K562 chronic myeloid leukemia cells after cyclophosphamide conditioning. K562 cells express c-*myb*, the antisense oligo target, and

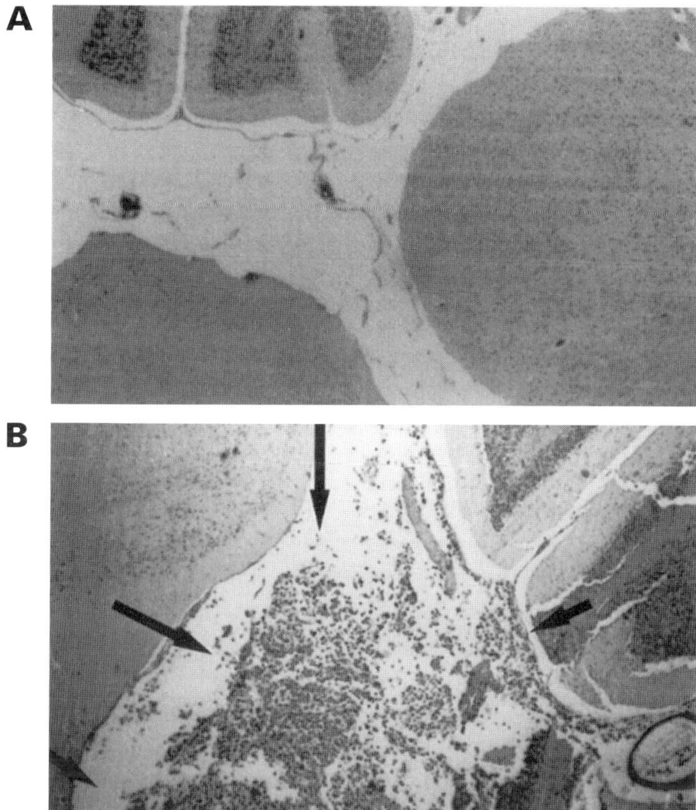

Figure 2: Composite photomicrographs (× 400) of brain stained with hematoxylin/eosin from scid-human K562 cell leukemia-bearing mice (for experimental details see section 5). Animals were treated with antisense (**A**) or sense (**B**) phosphorothioate *c-myb* oligodeoxynucleotides as described in [25]. Note extensive meningeal and subarachnoid infiltration (arrows) in **B** and the lack of involvement in **A**.

the tumor-specific *bcr-abl* oncogene which was utilized for tracking the human leukemia cells in the mouse host. After tumor cell injection, animals developed blasts in the peripheral blood within four to six weeks. After peripheral blood blast cells appeared, mean (± SD) survival of untreated mice (n = 20) was 6 ± 3 days. Dying animals had prominent central nervous system infiltration, marked infiltration of the ovaries, and scattered abdominal granulocytic sarcomas. Infusion of either sense or scrambled sequence *c-myb* phosphorothioate oligomers (24bp; codons 2–9) for three, seven, or fourteen days had no statistically significant effect on the sites of disease involvement, or animal survival in comparison to control animals. In contrast, animals treated for 7 or 14 days with *c-myb* AS oligomers survived 3.5 to 8 times longer (p < 0.001) than the various control animals (n = 60). In addition, animals receiving *c-myb* AS DNA had either rare microscopic foci or no obviously detectable CNS disease (Figure 2), and a > 50 % reduction of ovarian involvement. A 3 day infusion of *myb* AS (100 µg/d) was without effect. Infusing mice (n = 12) with AS oligomers (200 µg/day × 14 days) complementary to the c-*kit* protooncogene, which K562 cells do not express, also had no effect on disease burden or survival (n=12). These results suggested that phosphorothioate modified c-*myb* antisense DNA might be efficacious for the treatment of human leukemia *in vivo*.

3   Protocols for Studying Antisense Inhibition of c-*myb* in Cell Culture

3.1

Protocol 1: Cell Preparation

Special materials/equipment:
- Packed sheep red blood cells (SRBC)
- 0.1 % $CrCl_3$
- Anti-HPCA-1 MAb (Becton Dickinson, Mountain View, CA)
- F(ab)$_2$ goat anti-mouse affinity-purified F(ab)$_2$ (1 mg/ml; Cappel Laboratories, Cochranville, PA)
- Lysing reagent (Ortho Pharmaceutical, Raritan, NJ)

- Human bone marrow cells are obtained from normal healthy volunteers after informed consent.
- Light density mononuclear cells (MNC) are obtained by Ficoll-Hypaque density gradient centrifugation and enriched for hematopoietic progenitors by first removing adherent and phagocytic elements and then T lymphocytes as previously described [13, 5].
- MNC depleted of adherent/phagocytic and T cells (A⁻T⁻MNC) are then plated in semi-solid medium or further enriched for immature progenitors by immunorosetting with anti-HPCA-1 MAb, which recognizes the MY10 or CD34 antigen [4].
- Packed sheep red blood cells (SRBC) are incubated with an equal volume of 0.1 % $CrCl_3$ and F(ab)$_2$ goat anti-mouse affinity-purified F(ab)$_2$ for 10 min at room temperature, SRBC are then extensively washed in normal saline and resuspended at a concentration of 2 % in normal saline (200 ml of packed SRBC in a final volume of 10 ml).
- Marrow cells are incubated for 60 min at 4 °C (6×10^6 cells/ml) in the presence of 1:50 anti-HPCA-1 MAb.
- After extensive washing, antibody-treated SRBC (0.4 ml/10^6 cells) are added to cells, pelleted, and incubated for 1 h at 4 °C.
- Rosetted cells are then separated on a Ficoll gradient and SRBC are subsequently eliminated with the lysing reagent.

3.2

Protocol 2: Colony Assays

Special materials/equipment:
- Agar (Difco Laboratories Inc., Detroit, MI)
- Methylcellulose (Dow Chemical Co., Indianapolis, IN)
- β-mercaptoethanol (Sigma Chemical Co., St. Louis, MO)
- Granulocyte/macrophage colony-stimulating factor (GM-CSF)
- Granulocyte colony-stimulating factor (G-CSF)
- Interleukin-3
- Recombinant human erythropoietin alone (Amgen Biologicals, Thousand Oaks, CA)

- 2.5×10^4 A⁻T⁻MNC or 4×10^3 MY10⁺ MNC are plated in Iscove's modified Dulbecco's medium (IMDM) supplemented with 20 % fetal bovine serum (FBS)
- and 0.3 % agar or in IMDM, 30 % FBS, 10^{-4} M, β-mercaptoethanol, and 0.9 % methylcellulose in 35-mm petri dishes at 1 ml/dish.
- For myeloid colony stimulation GM-CSF (5 ng/ml), IL-3 (20 U/ml), or G-CSF (10 % of transfected Chinese hamster ovary cell supernatant) are used.
- CFU erythroid (CFU-E) colonies are obtained in the presence of erythropoietin alone (3 U/ml).
- Burst-forming units-erythroid (BFU-E) colonies are obtained in the presence of combination of IL-3 (10 U/ml), GM-CSF (5 ng/ml), and erythropoietin (3 U/ml).
- Colonies are scored after 14–16 d of culture in a humidified 5 % CO_2 incubator. CFU-E and G-CSF-stimulated granulocytic colonies are scored at 7–9 d of culture. To determine the colony size, 30 consecutive individual colonies from each plate are plucked from methylcellulose, dispersed in tissue culture medium, and then counted in a hemocytometer. Cells in smaller colonies are counted directly in the plates.

3.3

Protocol 3: Oligomer Treatment of Cells

- 2.5×10^4 A⁻T⁻MNC or 4×10^3 MY10⁺ MNC from each sample are incubated in 0.2 ml of IMDM, 20 % FBS in polypropylene tubes in the presence of optimal concentrations of the relevant hematopoietic growth factor.
- Cultures are also supplemented with 100 µg/ml of either sense or antisense oligomers. Cells are not washed before plating. Control cultures are left untreated.

- If one is working with natural (phosphodiester) DNA it is important that all serum be pre-screened for nuclease activity against the desired DNA. In our experience human serum had less nuclease activity than animal serum. If necessary, serum free conditions may also be employed.

3.4

Protocol 4: Detection of c-myb mRNA in Bone Marrow Cells by Reverse Transcriptase-polymerase Chain Reaction (RT-PCR) Analysis

Special materials/equipment:
- *Escherichia coli* ribosomal RNA
- Moloney murine leukemia virus reverse transcriptase
- Oligo(dT)
- *c-myb* mRNA sequence-specific synthetic primers (see protocol)
- *Thermus aquaticus* (Taq) polymerase

Expression of *c-myb* mRNA in A⁻T⁻MNC and MY10⁺ MNC is analyzed as follows:
- Cells seeded at 4×10^4 are collected at the end of the required time of culture and total RNA is extracted in the presence of 20 µg of *Escherichia coli* ribosomal RNA as described [24].
- RNA is then reverse-transcribed using 500 U of Moloney murine leukemia virus reverse transcriptase and 0.2 mg of oligo(dT) as primer for 1 h at 37 °C.
- The resulting cDNA fragments are amplified with 7.5 U of Taq polymerase in the presence of the primers; the 5′ oligo corresponds to the *c-myb* mRNA sequence from nucleotides 2,258–2,279 and the 3′ oligo corresponds to nucleotides 2,466–2,487. Therefore, a 230-bp fragment corresponding to a portion of the 3′ untranslated region of *c-myb* [17] will be generated by the polymerase chain reaction.
- 10 µl of the 50-µl polymerase chain reaction is separated in a 4 % Nusieve agarose gel and transferred to a nitrocellulose filter.
- The resulting blot is hybridized with a synthetic 50-base *c-myb* oligomer complementary to the amplified *c-myb* cDNA from nucleotides 2,351–2,400 [17]. The synthetic *c-myb* oligomer is end-labelled with [³²P]g-ATP and polynucleotide kinase as described [7].

3.5

Protocol 5: Detection of Myb Protein

Special materials/equipment:
- Sheep anti-human myb serum (Cambridge Research Biochemicals, Valley Stream, NY)
- GM-CSF

- IL-3
- Methanol/acetone (1:9)

- 5×10^4 MY10⁺ MNC, separated as described above, are plated in 96-well plates in 200 ml/IMDM, 20 % FBS per well in the presence of GM-CSF (5 ng/ml) plus IL-3 (20 U/ml).
- At days 0 and 21 of culture in a humidified incubator at 37 °C cells are placed on poly-L-lysine-treated slides, cytocentrifuged at 400 rpm for 5 min (Shandon II), and fixed with methanol/acetone (1:9) for 15 min at room temperature.
- Expression of Myb antigen is assessed by indirect immunofluorescence with a sheep anti-human Myb serum at a 1:40 dilution [1, 15]. Alternatively, Myb protein may be detected by Western blotting as we have recently reported.

4   Protocols for Studying Oligonucleotide Uptake and Trafficking in Hematopoietic Cells

An enhanced appreciation of uptake mechanisms and intracellular trafficking of phosphorothioate modified oligos (S-ODN) might facilitate the use of these compounds for experimental and therapeutic purposes. We addressed these issues by identifying cell surface proteins with which S-ODN specifically interact, by studying S-ODN internalization mechanisms, and by tracking internalized S-ODN through the cell utilizing immuno-chemical and ultrastructural techniques. Chemical cross-linking studies with a biotin labelled S-ODN (ᵇS-ODN), revealed the existence of five major cell surface S-ODN binding protein groups ranging in size from ~20 kD to ~143 kD. Binding to these proteins is competitively inhibited with unlabelled S-ODN, but not free biotin, suggesting specificity of the interactions. Additional experiments suggested that binding proteins likely exist as single chain structures, and that carbohydrate moieties may play a role in S-ODN binding. Uptake studies with ^{35}S labelled S-ODN revealed that endocytosis, mediated by a receptor-like mechanism, predominated at S-ODN concentrations of < 1 μM, whereas fluid-phase endocytosis prevailed at higher concentrations. Cell fractionation and ultrastructural analysis demonstrated the presence of oligos in clathrin coated pits, and in vesicular structures consistent with endosomes and lysosomes. Labelled oligos are also found in significant amounts in the nucleus, while none is associated with ribosomes, or ribosomes associated with rough endoplasmic reticulum (ER). Since nuclear uptake is not blocked by wheat germ agglutinin or concanavalin A, a nucleoporin independent, perhaps diffusion driven, import process is suggested. These data imply that antisense DNA may exert their effect in the nucleus. They also suggest rational ways to design oligos which might increase their efficiency.

While studies of the type described above have been published by several other groups, we provide the methods we have employed below. Further details may be found in Beltinger et al, J. Clin. Invest. 95:1814–1823, 1995.

Cells

K562 human leukemia cells are grown in RPMI 1640 (Gibco-BRL, Gaithersburg, MD) with 10 % bovine calf serum (Gibco-BRL) at 37 °C in 5 % CO_2.

4.1

Protocol 6: Uptake of ^{35}S Labelled S-ODN and ^{14}C-Sucrose in K562 Cells

Special materials/equipment:
- ^{14}C-sucrose (Amersham, Arlington Heights, IL)
- ^{35}S-labelled S-ODN
- Scintillation fluid (In-Flow BD, IN/US Systems, Fairfield, NJ)
- Lysis buffer: 150 mM NaCl, 50 mM TRIS pH 8.0, 0.5 % Non-idet P-40

- K562 cells (2×10^7/ml) are incubated for 6 h at 37 °C in RPMI 1640 with various concentrations of either ^{14}C-sucrose (specific activity 621 mCi/mmol) or ^{35}S-labelled S-ODN (specific activity 7×10^6 cpm/µmol, see also chapter 3, Labelling).
- After incubation the cells are washed three times in PBS.
- The washes are added to scintillation fluid and analyzed in a scintillation counter.
- The cell pellet is lysed with a lysis buffer, added to scintillation fluid and analyzed in a scintillation counter.

4.2

Protocol 7: Chemical Cross-linking of Surface Bound S-ODN

Special materials/equipment:
- Biotin (d-Biotin) labelled oligo (^{b}S-ODN)
- Bis(sulfosuccinimidyl) suberate (BS^3), 1–2 mM disuccinimidyl suberate (DSS)
  dithiobis(succinimidylpropionate) (DSP) (Pierce, Rockford, IL)
- 10 mM ammonium acetate in ice-cold PBS
- Exponentially growing K562 cells (2×10^7 cells/ml) are washed and resuspended in 0.5 ml phosphate buffered saline (PBS).
- Cells are incubated (30 minutes, 4 °C) with ^{b}S-ODN (23 nM) in the presence or absence of a 500-fold excess (25 µg) of non-biotinylated S-ODN in the presence or absence of a 2×10^6 fold excess of d-Biotin.
- Cells are washed once to remove unbound ^{b}S-ODN after which surface bound material is cross-linked (14 °C; 15 minutes) with 1–2 mM BS^3, 1–2 mM DSS, or DSP as described [8].

- Excess cross-linker is quenched by washing the cells twice with 10 mM ammonium acetate in ice-cold PBS.

4.3

Protocol 8: Biochemical Analysis of S-ODN Binding Proteins

Special materials/equipment:
- Biotin (d-Biotin labelled oligo [bS-ODN]
- Detergent buffer 1: 10mM Tris·HCl, pH 7.6, 0.6 M NaCl, 10µg/ml each of leupeptin and aprotinin, 1 mM PMSF, and 2 % NP-40
- Detergent buffer 2: 1 % NP-40 with 0.1 % BSA
- Enhanced Chemoluminescence kit with horseradish peroxidase (BRL, Gaithersburg, MD)
- ^{35}S-methionine and ^{35}S-cysteine (NEN, Boston, Mass)
- Avidin-Sepharose
- DTT

- Membrane proteins from cross-linked cells are lysed in detergent buffer followed by removal of cell nuclei and organelles by centrifugation (15 min, 14,000 g).
- Unfractionated, whole cell detergent lysates are subjected to SDS-PAGE and Western blotted to PVDF membranes (1 × 10^6 cell equivalents/lane).
- Material cross-linked to bS-ODN is detected on Western blots with horseradish peroxidase enzyme (HRP) coupled to avidin (Avidin-HRP) using the Enhanced Chemoluminescence kit as directed by the manufacturer (BRL, Gaithersburg, MD).
- bS-ODN binding proteins are directly analyzed by performing similar cross-linking experiments with metabolically labelled cells.
- K562 cells are first labelled by overnight culture with ^{35}S-methionine and ^{35}S-cysteine. Metabolically labelled cells are incubated with bS-ODN and cross-linked as described above with either DSP, a reversible, membrane impermeable agent, or with its non-cleavable cross-linker equivalent DSS.
- Cell membranes are then solubilized in NP-40 detergent and bS-ODN-binding protein complexes are affinity precipitated with avidin coupled to sepharose (Avidin-Sepharose) as described for immunoprecipitations with monoclonal antibodies [8].
- The complexes are washed (2x in detergent buffer 2; 2X in detergent buffer 2 with 0.5 % NP-40; 1x in detergent buffer 2 with 0.1 % NP-40) and eluted from Avidin-Sepharose with SDS-PAGE loading buffer.
- The proteins are then treated (100 mM DTT, 5′ at 95 °C) to cleave the reducible DSP cross-linker and release the bS-ODNs.

- Proteins are fractionated by SDS-PAGE, the gels dried and exposed to autoradiography to detect [b]S-ODN binding, affinity purified, radiolabelled cell surface proteins.
- Binding studies and Scatchard plot analysis are performed as described [8].
- K562 cells may be also treated with neuraminidase for 1 to 3 hours to determine the role of glycosylation on the ability of putative binding proteins to interact with S-ODN. Neuraminidase treatment is carried out as previously published [12].

4.4

Protocol 9: Cell Fractionation Studies

Special materials/equipment:
- ^{35}S labelled S-ODN
- Glass dounce
- Ultracentrifuge
- Hypotonic homogenizing medium: 50 mM MOPS, pH 7.4, 1mM PMSF

- K562 cells (2 × 10^7 cells/ml) are exposed to ^{35}S labelled S-ODN (2.5µM; 2.5µCi) for 1 hour at 4 °C or 37 °C. Cells are washed in tissue culture medium at 4 °C to remove free S-ODN, homogenized, and then fractionated by centrifugation.
- The washed cell pellet is diluted in 6 ml of hypotonic homogenizing medium and allowed to stand at 4 °C for 10 minutes to swell the cells.
- Cells thus treated are homogenized with 50 strokes in a glass dounce.
- The homogenized cell suspension is transferred to a polycarbonate tube, and diluted with 6 ml 0.5M sucrose in 50 mM MOPS, pH 7.4, with 1 mM PMSF.
- The sample is mixed and an aliquot removed for scintillation counting. The remaining material is centrifuged at 50 × g for 5 minutes to remove intact cells. The supernatant fraction containing the cell homogenate sub-fractions (S1) is centrifuged again at 1000 × g for 10 minutes to remove nuclei. Radioactivity associated with the isolated nuclei is quantitated using a scintillation counter.
- The resulting supernatant (S2) is centrifuged at 16,000 × g for 20 minutes to remove plasma membranes, mitochondria, endosomes, and lysosomes. Radioactivity in the pellet is counted, and the resulting supernatant (S3) is centrifuged again at 160,000 × g for 60 minutes to pellet the microsomal fraction containing free ribosomes, and ribosomes associated with rough endoplasmic reticulum. Radioactivity associated with this pellet is also counted, as are counts in the final supernatant (S4) which is the equivalent of free cytosol.

- Recovery of S-ODN at each step is compared to total cellular uptake of labelled material under each condition. This is determined by removing a small aliquot of cells immediately after incubation with the labelled S-ODN.
- The labelled cells are separated from unincorporated counts by sucrose gradient centrifugation.
- The cell pellet is resuspended in liquid scintillation cocktail in 1 N NaOH and counted in a liquid scintillation counter.
- Protein concentration is determined in each fraction by the method of Lowry et al [16].

4.5

Protocol 10: Ultrastructural Localization of Biotinylated Oligonucleotides in K562 Cells

Special materials/equipment:
- Biotin (d-Biotin labelled oligo [^{b}S-ODN]
- Bis(sulfosuccinimidyl) suberate (BS^3) (Pierce, Rockford, IL)
- Gold-labelled streptavidin (Zymed Laboratories, Inc., South San Francisco, CA)
- 0.1 M sodium cacodylate
- 2 % OsO_4 in 0.1 M sodium cacodylate buffer, pH 7.4
- 2 % uranyl acetate
- Propylene oxide
- Spurr resin (Electron Microscopy Sciences, Fort Washington, PA)
- d-biotin (Sigma Chemical Co., St. Louis, MO)
- LR White resin (Electron Microscopy Sciences, Ft. Washington, PA.
- Anti-biotin IgG (Rockland, Inc., Gilbertsville, PA; Sigma Chem. Co.)
- Gold-labelled protein A [27]
- Transmission electron microscope (e.g. JEOL 100CX)
- Solutions as indicated in the protocol

Two different techniques, pre -and post-embedding labelling, may be used to ultrastructurally localize cell associated ^{b}S-ODN. Visualization of S-ODN membrane binding protein complexes and the initial internalization process require pre-embedding labelling of bound ^{b}S-ODN because of the low concentrations of the ligand used to demonstrate receptor mediated uptake.

- K562 cells (2 × 10^7 cells/ml) suspended in RPMI 1640 with 2 % bovine calf serum are incubated in the presence or absence of 100 nM c-*myb* antisense ^{b}S-ODN at 4 °C for 30 min to saturate putative membrane receptors without allowing internalization. Cells incubated without ^{b}S-ODN are used as controls for labelling specificity.

- Cells are then washed 2x and resuspended in PBS pH 7.4. The [b]S-ODN are cross-linked to surface binding proteins using the membrane impermeable cross-linker BS^3 (20 mg/ml).
- The reaction is quenched by washing the cells 3x with 50 mM TRIS/HCL in 150 mM NaCl.
- To label the crosslinked [b]S-ODN, cells are incubated at 4 °C for one hour with 10nm gold-labelled streptavidin.
- Thereafter, the cells are warmed to 37 °C for 20 minutes to initiate uptake. The cells are diluted in PBS at 4 °C, centrifuged at 500 × g, and the resulting cell pellet fixed overnight in 2 % glutaraldehyde in PBS.
- The specimens are washed in 0.1 M sodium cacodylate and post-fixed with 2 % OsO_4 in 0.1 M sodium cacodylate buffer, pH 7.4, for 1 h at 4 °C.
- The cells are stained en bloc with 2 % uranyl acetate, dehydrated with ethanol (70–100 %) and propylene oxide (100 %) and embedded in Spurr resin.
- Thin sections are cut, collected on copper grids, and stained with saturated alcoholic uranyl acetate and bismuth subnitrate.

Because studies designed to demonstrate subsequent intracellular processing used a higher [b]S-ODN concentration and longer incubation times, post-embedding labelling is used to detect the ligand.

- K562 cells (2.5 × 10^7 cells/ml) suspended in RPMI 1640 with 2 % bovine calf serum are incubated with 10 µM c-myb antisense [b]S-ODN, or d-biotin at 37 °C (5 % CO_2) for 12 hours.
- The cells are then transferred to Dulbecco's MEM and incubated under the same conditions with 5 µM [b]S-ODN for an additional 8 h. Cells incubated with d-biotin serve as a control of labelling specificity of [b]S-ODN, as opposed to endogenous or added biotin, in these experiments.
- The cells are diluted in PBS at 4 °C, centrifuged at 500 × g, and the resulting pellet fixed in 1 % glutaraldehyde, with 0.2 % picric acid in PBS, pH 7.4, overnight at 4 °C.
- Specimens are dehydrated with ethanol (70, 80 and 90 %), and embedded in LR White resin.
- Thin sections are cut and collected on nickel grids. [b]S-ODN are detected by incubating the sections with anti-biotin IgG, normal rabbit IgG, or no primary antibody overnight at 4 °C.
- The sections are washed and incubated with gold-labelled protein A for 1 hour at room temperature. Incubations with normal IgG and no primary antibody serve as controls for the specificity of the anti-biotin and gold-labelled protein A labelling.
- After washing, grids are stained with neutralized uranyl acetate.
- All sections are observed and photographed in a transmission electron microscope.

4.6

Protocol 11: Nuclear Import of Fluorescent Oligonucleotides

Special materials/equipment:
- Transport buffer (TB): 20 mM HEPES, pH 7.3, 110 mM potassium acetate, 5 mM sodium acetate, 2mM magnesium acetate, 1 mM EGTA, 2mM DTT, 1g/ml leupeptin, 1 g pepstatin
- Digitonin (Sigma, St. Louis, MO)
- Glass cell culture slides (Lab-Tek, Nunc, Naperville, Il)
- Anti-vimentin mouse IgG (Sigma)
- Anti-PCNA mouse IgG (Boehringer-Mannheim, Indianapolis, IN)
- FITC labelled rabbit anti-mouse IgG
- FITC tagged c-*myb* antisense S-ODN
- Sheep serum
- Dilution buffers as indicated in the protocol
- Rabbit reticulocyte lysate (Promega, Madison, Wisconsin)
- Wheat germ agglutinin (WGA; Sigma, St. Louis, MO)
- Concanavalin A (Con A; Sigma)
- Solutions and fixatives as indicated in the protocol

- Proliferating cells from the melanoma cell line HS 294 T are plated at subconfluency in 96-well plates, or on glass cell culture slides.
- After 48 hours growth, cells are washed with ice-cold TB. Cell membranes are selectively permeablized by exposure to 40 µg/ml digitonin in TB for 5 minutes.
- The permeabilized cells are incubated for 30 min at 4 °C with either anti-vimentin mouse IgG or anti-PCNA mouse IgG diluted 1:160 and 1:80 in TB, respectively.
- The cells are washed 2x with ice-cold TB, and then incubated with 2 % sheep serum in TB for 10 min at 4 °C to reduce non-specific signal. Cells are then incubated for 30 min at 4 °C with FITC labelled rabbit anti-mouse IgG in a 1:128 dilution with TB and 2 % sheep serum.
- Cells are analyzed by phase contrast and epifluorescence.

- To ensure that PCNA could be detected in permeabilized nuclei, cells are also fixed in 1 % paraformaldehyde for 2 min at room temperature and then permeabilized with methanol for 10 min at –20 °C.
- The cells are then incubated with anti-PCNA antibody followed by washing with 0.1 % Triton in TB.
- The cell preparations are then incubated with the appropriate fluorescein labelled second antibody and examined with a fluorescence microscope.

- To analyze nuclear import of oligos, digitonin-permeabilized cells are incubated for 30 minutes at room temperature with transport medium (TM) containing 60 % (v/v) dialyzed rabbit reticulocyte lysate in TB, 3.3 µM FITC-tagged c-*myb* antisense S-ODN, 20 mM HEPES, pH 7.3, 110 mM potassium acetate, 5 mM sodium acetate, 2 mM DTT, 1.0 mM EGTA, 1.0 mM ATP 5 mM creatine phosphate, 20 U/ml creatine phosphokinase and 1 µg/ml each aprotinine, leupeptin and pepstatin.
- Rabbit reticulocyte lysate is prepared for use by centrifugation at 100,000 g.
- The supernatant obtained is dialyzed against transport buffer (20 mM HEPES, pH 7.3, 110 mM potassium acetate, 5 mM sodium acetate, 2 mM magnesium acetate, 1 mM EGTA, 2 mM DTT, 1 µg/ml aprotinin, 1 µg/ml leupeptin and 1 µg/ml pepstatin) in tubing (Spectrum Medical Industries, Los Angeles, CA) with a molecular weight cut-off of 12,000–14,000. Aliquots may be frozen in liquid nitrogen and stored at –80 °C.
- To investigate the effect of lectins on nuclear import of S-ODN, permeabilized cells are pre-incubated (20 minutes at room temperature) with either wheat germ agglutinin or concanavalin A (each 100 µg/ml in TM).
- Cells are then incubated for 30 minutes at room temperature with transport medium containing 3.3 µM of fluorescein tagged c-*myb* antisense S-ODN.
- After washing with TM, the cells are analyzed by phase contrast and epifluorescence.

5   *In vivo* Treatment of Human Leukemia in a SCID Mouse Model with c-*myb* Antisense Oligonucleotides

5.1

Protocol 12: Establishment of SCID Mouse-human K562 Cell Leukemia Chimera

Special materials/equipment:
- SCID mice (e.g. CB-17/ACRTAC/SCID/SDS; Taconic Laboratories, Taconic, New York)
- Cyclophosphamide
- K562 human leukemia cells (American Type Culture Collection, Rockville, Maryland)

- Six- to seven-week old female SCID mice are injected intraperitoneally with cyclophosphamide (150 mg/kg of body weight; total dose of ~3 mg) on each of two successive days.
- 24 hours after the second injection, the animals are transplanted by tail vein injection with 1x10^7 K562 human leukemia cells which have been

washed and seeded into fresh medium (RPMI + 10 % FBS) 12 hours prior to use.

- Post transplantation of the K562 cells, animals are monitored for the presence of circulating blast cells by blood sampling from the retro-orbital plexus.

5.2

Protocol 13: Oligodeoxynucleotide Administration

Special materials/equipment:
- Osmotic minipumps (Alzet, Palo Alto, CA)
- Phosphorothioate oligomers

- Beginning four weeks after leukemic cell transplantation, animals are monitored for the development of overt leukemia by retro-orbital plexus bleeding thrice weekly.
- When the peripheral blood blast cell content is at least 1–5 % of the total leucocyte count, as determined by a 200 cell differential count in a standard hemocytometer chamber, treatment is initiated. Animals are anesthetized and pre-filled constant infusion mini-osmotic pumps are inserted subcutaneously into a paraspinal pocket which is entered through a small inter-scapular incision.
- Oligomers are diluted in sterile distilled water to so that each animal received a total dose of 100 µg/day (5 mg/kg body weight; 1 M). Implanted pumps released their contents at a rate of 1 µl/h.

5.3 Statistical Analysis

- Statistical significance of survival differences among animals in the various oligomer treatment groups is determined using the Student t test for unpaired samples. Survival is determined from the time that the infusion pumps are implanted.
- In our experiments p values less than 0.05 were judged to be of statistical significance [25].

6 Use of Antisense Oligonucleotides in a Clinical Setting

Based on the type of data presented above, a favorable therapeutic index in toxicology testing, and more detailed knowledge of the pharmacokinetics of oligos, antisense oligos have begun to be evaluated in the clinic. At the University of Pennsylvania we have initiated two such trials. One of these is a pilot study to evaluate the effectiveness of a phosphorothioate modified oligo antisense to c-*myb* as a marrow purging agent in patients with chronic phase (CP) or accelerated phase (AP) CML. The other is a PhaseI/II study evaluating the toxicity of systemically infused c-*myb* antisense oligos in patients with CML in blast crisis, or refractory acute leukemia.

When designing a clinical protocol it is important that the following general format will be followed. The background for the study should be clearly and succinctly stated in an *introductory* section, which should be followed by equally clear *objectives* for the study. Patient *eligibility* for treatment on the protocol should be defined, as well as specific exclusionary criteria. For example, in the case of patients who are being transplanted with a c-*myb* purged marrow we have adopted the following rules: The growth of the patients marrow cells must be inhibited *in vitro* by exposure to the antisense c-*myb* DNA, and not by exposure to a control DNA sequence and/or both c-*myb* and *bcr/abl* gene expression must be specifically down-regulated in the patients marrow cells. Any patient with an allogeneic marrow transplant donor is NOT eligible for treatment. Next, the actual *treatment plan* should be detailed. An example, adopted from the actual clinical protocol written by Dr. Selina Luger and myself at the University of Pennsylvania Cancer Center) is given below:

6.1

Protocol 14A: Treatment Plan

6.1.1 Pre-transplant Evaluation

- Within 2 weeks preceding harvesting, a bone marrow aspirate and biopsy are performed. A marrow that is hypocellular ($< 25\,\%$ cellularity) excludes a patient from autologous bone marrow transplantation.
- Viral titers for hepatitis B, hepatitis A, herpes simplex virus, VZV and HIV are to be assayed.
- MUGA scan
- Creatinine clearance
- Chest X-ray and electrocardiogram
- Pulmonary function studies with diffusing capacity.

6.1.2. Marrow Harvesting

- A minimum of 3.5×10^8 nucleated bone marrow cells (total cells collected minus peripheral blood cell contamination)/kg of body weight must be collected. Blood removed during the procedure will be replaced by transfusion of *irradiated* blood, administered as packed cells.
- Marrow is extracted by multiple bone marrow aspirates into syringes treated with RPMI solution with 10 % sterile Heparin.
- An aliquot of processed bone marrow will be stained with supravital stain to assess percent viability prior to initiation of chemotherapy treatment. If insufficient numbers of cells are obtained, a second harvest may be performed.

6.1.3. Marrow Processing

- Marrow is filtered through stainless steel mesh filters (0.33 mm and 0.2 mm pores) to remove particulate matter, and the buffy coat is extracted. All manipulations of the bone marrow are performed in a closed system to avoid bacterial contamination. Acceptable procedures to obtain the buffy coat include centrifugation in blood transfer packs in a blood bank centrifuge, or use of the Cobe 2991 cell washer or Haemonetics cell separator.
- Heparinized plasma from the marrow is saved for use in the processing and freezing, and should be irradiated or filtered to remove any viable leukemic cells, and tested for nuclease activity.
- The buffy coat procedure selected should retain 80 % of the mononuclear cells (lymphocytes and monocytes). Of the buffy coat cells, at least 3×10^8 cells/kg will be selected for CD34$^+$ cells and treated with c-*myb* (LR-3001) antisense oligo. The remaining 0.5×10^8 cells/kg are cryopreserved and saved as backup.

6.1.4 Oligonucleotide Treatment

- Endotoxin-free and sterilized c-*myb* antisense DNA (lyophilized sodium salt form) is dissolved in sterile tissue culture medium (TC199) and added to the marrow suspension through a sterilizing filter to achieve a final concentration of 100 µg/ml.
- The cell suspension (2×10^6 cells/ml TC199 with 2 % autologous serum) is then incubated for 18 hours at 37 °C.
- At the end of 18 hours, an additional bolus (50 µg/ml) of c-*myb* is added to the cell suspension to bring the final oligos concentration to 150 µg/ml.
- The cell suspension is incubated for an additional 6 hours at 37 °C after which it is then cooled to 4 °C.
- The cells are then collected by centrifugation at $2200 \times g$ for 10 minutes.
- The supernatant containing the bulk of the unincorporated oligomer is removed. The cell pellet is then diluted to a final volume of 40–60 ml with an equal volume of freezing solution containing 10 % irradiated autologous plasma, 15 % DMSO, and 75 % TC 199. The cells are cryopreserved in 5 ml aliquots using a programmable freezer. The frozen cells are stored in the vapor or liquid phase of liquid nitrogen. Samples for CD34+ assay, viability, culture, sterility and PCR will be obtained after processing. Post-purge, a minimum of 0.5×10^6 CD34$^+$/kg must remain. If too few cells are recovered, a repeat bone marrow harvest will be performed.

6.1.5 Treatment of Patients

> • Patients receive myeloablative therapy with myleran (4 mg/kg/day × 4)
> and cyclophosphamide (60 mg/kg/day × 2), followed by re-infusion of
> S-ODN purged MNC. G-CSF is administered concurrently.

Details of the intercurrent management plans during the transplant period should also be provided, along with detailed plans as to how the *effect* of the transplant will be measured. Criteria which need to be met in order to decide if the patient has had a response should be listed. Categories for response include complete, partial, stable disease, and no response.

The studies which will be performed on patients treated on the protocol need to be described, and it must be stated explicitly when the studies need to be carried out. An example of this, again adapted from the U Penn protocol is given below.

6.2

Protocol 14 B: Study Parameters

> • All scans and x-rays should be done *6 weeks* before registration.
> • CBC with differential should be done *48 hours* before registration.
> • All chemistries should be done *2 weeks* before registration. If abnormal, they must be repeated within 48 hours prior to registration.

	Prior	ABMT- Daily	ABMT- Weekly	Follow-up Monthly
Physical exam	X	X		X
Weight	X	X		X
Hemoglobin, Hematocrit, WBC, Differential, Platelets	X	X		X
Elecrolytes, BUN, Creat.	X	X		X
Bilirubin, SGOT/PT, Uric Acid, LDH, Alk.Phos.	X		X	
Chest X-ray	X		prn	
Electrocardiogram	X		prn	
MUGA Scan	X			
PFT's/DLCO	X			
Creatinine Clearance	X			
Bone Marrow Asp/Biopsy	X			X[1]
Adverse Reactions		X[2]		X[2]

[1] Every 2–3 months for 2 years unless evidence of relapse. Detection of minimal residual disease will also be performed.

[2] Patients will be monitored daily during therapy and monthly thereafter for any adverse reactions which may be treatment related. Adverse reactions will be reported to the IRB and FDA.

The nature of any companion laboratory tests, what they are and when they will be performed should also be "spelled out." In our study, for example, we periodically study *bcr/abl* gene expression and look for the Philadelphia chromosome in all patients that we have treated.

Finally, it should be obvious that a mechanism for identifying, following, and reporting to the appropriate authorities potential drug toxicity must also be provided.

The general outline of the purging protocol is shown in Figure 3.

As noted earlier, we have employed similar considerations in the design of a Phase I study to examine the toxicity of infused *c-myb* antisense DNA. At doses of up to 2 mg/kg/day × 7 days the material appears to be well tolerated.

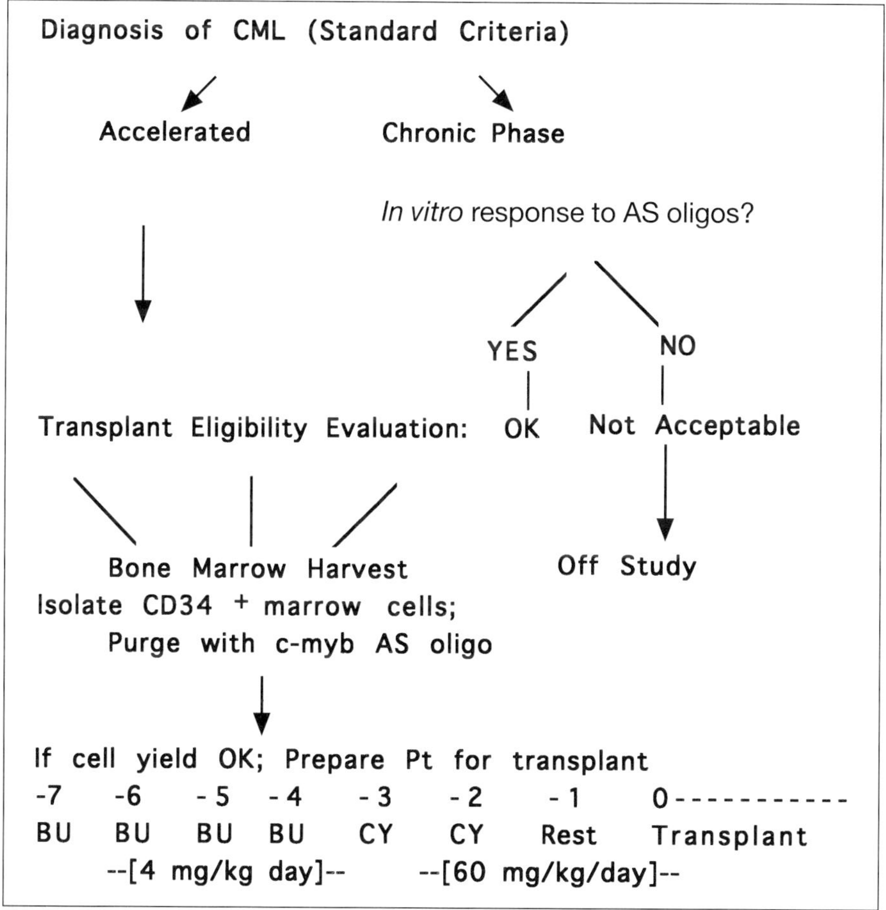

Figure 3: Purging with c-myb AS oligo: Evaluation decision tree and treatment protocol outline. BU = Busulfan (Myleran®), CY = Cyclophosphamid

7   Conclusions

The ability to block gene function with antisense oligos has become an important tool in many research laboratories. Since activation and aberrant expression of protooncogenes appears to be an important mechanism in malignant transformation, targeted disruption of these genes and other molecular targets with oligos could have significant therapeutic utility as well. In this regard, the potential therapeutic usefulness of oligos has been demonstrated in many systems and against a number of different targets including viruses, oncogenes, protooncogenes, and an increasing array of cellular genes. These studies in aggregate suggest that synthetic oligomers have the potential to become an important new therapeutic agent for the treatment of human cancer. Nevertheless, it is clear that considerable optimization will be required before antisense oligos will emerge as an effective agent for treating human disease. Progress will need to occur on several fronts. These include issues related to the chemistry of the molecules employed. For example, how chemical modification impact on uptake, stability, and hybridization efficiency of the synthetic DNA molecule. A clearer understanding of the mechanism of antisense mediated inhibition, including where such inhibition takes place, will also be required. Finally, cellular "defense" mechanism such as increasing transcription of the targeted message, may also be factors to consider in planning effective treatment strategies with these agents. It is also clear that choice of target is an important issue. Nevertheless, while many issues remain to be resolved, we remain optimistic that this approach will one day prove useful for the treatment of patients with a variety of hematologic malignancies.

Acknowledgment

The work presented in this chapter was supported by grants from the NIH and the Leukemia Society of America

References

1. Anfossi, G., A. M. Gewirtz, and B. Calabretta: An oligomer complementary to c-myb encoded mRNA inhibits proliferation of human myeloid leukemia cell lines. Proc. Natl. Acad. Sci. USA 86 (1989) 3379–3383.

2 Barletta, C., P. G. Pelicci, L.C. Kenyon, S. D. Smith, R. Dalla-Favera: Relationship between the c-myb locus and the 6q- chromosomal aberration in leukemias and lymphomas. Science 235 (1987) 1064–1067.

3. Biedenkapp, H., U. Borgmeyer, A. E. Sippel and K. H. Klempnauer: Viral myb oncogene encodes a sequence-specific DNA-binding activity. Nature 335 (1988) 835–837.

4. Burk, O., S. Mink, M. Ringwald, and K. H. Klempnauer: Synergistic activation of the chicken mim-1 gene by v-myb and C/EBP transcription factors. EMBO J 12 (1993) 2027–2038.

5. Caracciolo, D., N. Shirsat, G. G. Wong, B. Lange, S. Clark, and G. Rovera:

Recombinant human M-CSF requires subliminal concentration of GM-CSF for optimal stimulation of human mcrophage colony formation in vitro. J.Exp. Med. 166 (1987) 1851–1860.

6. CLEVENGER C. V., W. NGO, D. SOKOL, S. M. LUGER, and A.M. GEWIRTZ: Vav is necessary for Prolactin-Stimulated Proliferation and is Translocated into the Nucleus of a T-cell line. J. Biol. Chem. 270 (1995) 13246–13253.

7. COGSWELL, J. P., P.C. COGSWELL, M. KUEHL, A. M. CUDDIHY, T. M. BENDER, U. ENGELKE, K. B. MARCU and J.P.-Y. TING: Mechanism of c-myb regulation by c-myb in different cell lineages. Mol. Cell Biol. 13 (1993) 2858–2869.

8. COLIGAN, J. E., A. M. KRUISBEEK, D. H. MARGULIES, E. M. SHEVACH, and W. STROBER. (eds.) In: Current Protocols in Immunonology. Greene Publishing and Wiley-Interscience, New York (1993) 6.1.7–6.1.10.

9. DINI, P. W., J. T ILTMAN, J. S. LIPSICK: Mutations in the DNA-binding and transcriptional activation domains of v-myb cooperate in transformation. J. Virol. 69 (1995) 2515–2524.

10. GEWIRTZ, A. M., G. ANFOSSI, D. VENTURELLI, S. VALPREDA, R. SIMS, and B. CALABRETTA: G1/S transition in normal human T-lymphocytes requires the nuclear protein encoded by c-myb. Science (Wash,. DC) 245 (1989) 180–183.

11. GEWIRTZ, A. M. and B. CALABRETTA: A c-myb antisense oligodeoxynucleotide inhibits normal human hematopoiesis in vitro. Science 245 (1988) 1303–1306.

12. GEWIRTZ A. M., W .Y. XU, and K. F. MANGAN: Role of Natural Killer Cells, in Comparison to T lymphocytes and Monocytes, in the Regulation of Normal Human Megakaryocytopoiesis In Vitro. J. Immunol. 139 (1987) 2915–2925.

13. KANEI-ISHII, C., E. M. MACMILLAN, T. NOMURA, A. SARAI, R. G.RAMSAY, S. AIMOTO, S. ISHI, and T .J. GONDA: Transactivation and transformation of MYB are negatively regulated by a leucine-zipper structure. Proc. Natl. Acad. Sci. USA 89 (1992) 3088–3092.

14. KITAJIMA, I., T. SHINOHARA, J. BILAKOVICS, D. A. BROWN, X. XU, and M. NERENBERG: Ablation of transplanted HTLV-I Tax-transformed tumors in mice by antisense inhibition of NF-kappa B. Science 259 (1993) 1523–1523.

15. KU, D. H., S. C. WEN, A. ENGELHARD, N. C. NICOLAIDES, K. E. LIPSON, T. A. MARINO and B. CALABRETTA: c-myb transactivates cdc2 expression bia MYB binding sites in the 5'-flanking region of the human cdc2 gene. J. Biol. Chem. 268 (1993) 2255–2259.

16. LOWRY, O. H., N. J. ROSEBROUGH, A. L. FARR, and R. J. RANDALL: Protein measurement with the Folin phenol reagent. J. Biol. Chem. 193 (1951) 265–267.

17. MAJELLO, B., L. C. KENYON, and R. DALLA-FAVERA: Human c-myb proto-oncogene: nucleotide sequence of cDNA and organization of the genomic locus. Proc. Natl. Acad. Sci.USA 83 (1986) 9636–9640.

18. MELOTTI, P. and B. CALABRETTA: ets-2 and c-myb act independently in regulating expression of the hematopietic stem cell antigen CD34. J. Biol. Chem. 269 (1994) 25303–25309.

19. NAKAYAMA, K., R. YAMAMOTO, S. ISHII and H. NAKAUCHI: Binding of c-Myb to the core sequence of the CD4 promoter. Int. Immunol. 5 (1993) 817–824.

20. NESS, S. A., A. MARKNELL, and T. GRAF: The v-myb oncogene product binds to and activates the promyelocytic-specific mim-1 gene. Cell 59 (1989) 1115–1125.

21. NOMURA, T., N. SAKAI, A. SARAI, T. SUDO, C. KANEI-ISHII, R. G. RAMSAY, D. FA-

VIER, T.J. GONDA and S. ISHII S: Negative autoregulation of c-myb activity by homodimer formation through the leucine zipper. J. Biol. Chem. 268 21914–21923.

22. NOMURA, N., M. TAKAHASHI, M. MATSU, S. ISHII, T. DATE, S. SASAMOTO, and R. ISHIZAKI: Isolation of human cDNA clones of myb-ralted genes, A-myb and B-myb. Nucl. Acids Res. 16 (1988) 11075–11089.

23. PRESS, R. D., E. P. REDDY and D. L. EVERT: Overexpression of c-terminally but not n-terminally truncated Myb induces fibrosarcomas: a novel nonhematopoietic target cell for the myb oncogene. Mol. Cell Biol. 14 (1994) 2278–2290.

24. RAPPOLEE, D. A., D. MARK, M. J. BENDA, and Z. WERB: Wound macrophages express TGF- and other growth factors in vitro: analysis by mRNA phenotyping. Science (Wash., DC) 239 (1988) 487–491.

25. RATAJCZAK, M. Z., J. A. KANT, N. HIJIYA, J. ZHANG, S. M. LUGER, G. ZON, and A. M. GEWIRTZ: In Vivo Treatment of Human Leukemia In a SCID Mouse Model With c-Myb Antisense Oligodeoxynucleotides. Proc. Natl. Acad. Sci. USA 89 (1992) 11823–11827.

26. RATAJCZAK, M. Z., S. M. LUGER, K. DERIEL, J. ABRAHM, B. CALABRETTA and A. M. GEWIRTZ: Role of the KIT protooncogene in normal and malignant human hematopoiesis. Proc. Natl. Acad. Sci. USA 89 (1992) 1710–1714.

27. ROMANO, E. L., and M. ROMANO: Staphylococcal protein A bound to colloidal gold: a useful reagent to label antigen-antibody sites in electron microscopy. Immunocytochemistry 14 (1977) 711–715.

28. SAKURA, H., C. KANEI-ISHII, T. NAGASE, H. NAKAGOSHI, T. J. GONDA and S. ISHII: Delineation of three functional domains of the transcription activator encoded by the c-myb protoonocogene. 1989. Proc. Natl. Acad. Sci. USA 86 (1989) 5758–5762.

29. SIMONS, M., E. R. EDELMAN, J. L. DEKEYSER, R. LANGER, and R. D. ROSENBERG: Antisense c-myb oligonucleotides inhibit intimal arterial smooth muscle cell proliferation in vivo. Nature 359 (1992) 67–70.

30. TRAVALI, S. , K. REISS, A. FERBER, S. PETRALIA, W. MERCER, B. CALABRETTA and R. BASERGA: Constitutively expressed c-myb abrogates the requirment for insulin like growth factor in 3T3 fibroblasts. Mol. Cell Biol. 11 (1991) 731–736.

31. WEBER, B. L., E. H. WESTIN and M. F. CLARKE: Differentiation of mouse erythroleukemia cells enhanced by alternatively spliced c-myb mRNA. Science 249 (1990) 1291–1293.

32. YU, H., B. BAUER, G. K. LIPKE, R.L. PHILLIPS and G. VAN ZANT: Apoptosis and Hematopiesis in Murine Fetal Liver. Blood 81 (1993) 373–384.

Additional Suggested Readings

BAYEVER, E., P. IVERSEN, L. SMITH, J. SPINOLO, G. ZON: Guest editorial: Systemic human antisense therapy begins. Antisense Research and Development 2 (1992) 109.

GEWIRTZ, A. M.: Therapeutic applications of antisense DNA in the treatment of human leukemia. Ann. N.Y. Acad. Sci. 660 (1992) 178.

GEWIRTZ, A. M. and B. CALABRETTA: A c-myb antisense oligodeoxynucleotide inhibits normal human hematopoiesis in vitro. Science 245 (1988) 1303.

HIJIYA, N., J. ZHANG, M. Z. RATAJCZAK, K. DERIEL, M. HERLYN, A. M. GEWIRTZ: The biologic and therapeutic significance of c-Myb expression in human melanoma. Proc. Natl. Acad. Sci. USA 91 (1993) 4499.

MILLIGAN, J. F., M. D. MATTEUCCI, J. C. MARTIN: Current concepts in antisense drug design. J. Med. Chem. 36: 1993.

RATAJCZAK, M .Z., J. A. KANT, S. M. LUGER, N. HIJIYA, J. ZHANG, G. ZON, A. M. GEWIRTZ: In vivo treatment of human leukemia in a scid mouse model with c-myb antisense oligodeoxynucleotides. Proc. Natl. Acad. Sci. USA 89 (1993) 11823.

STEIN, C. A. and Y. C. CHENG: Antisense oligodeoxynucleotides as therapeutic agents – is the bullet really magical? Science 261 (1993) 1004.

WAGNER, R. F.: Gene inhibition using antisense oligodeoxynucleotides. Nature 372 (1994) 333.

III. Perspectives for Drug Development

14 Pharmacology and Drug Safety

Pharmacology and Toxicology Studies in Oligonucleotide Drug Development – Supporting Clinical Evaluation of Antiviral Agents

Lauren E. Black, Ph. D.
Divison of Antiviral Drug Products, FDA, Rockville, USA

1   Introduction

Oligodeoxynucleotides (oligos), mainly modifications of the natural phosphodiester structure, have been under development as drug candidates since 1991. In general, clinical trials are being sponsored by venture sponsored pharmaceutical companies and are focusing on a subclass of oligos referred to as phosphorothioates. These oligos have a sulfur substituted for oxygen in the backbone of the deoxyribonucleotide (DNA) chain. Phosphorothioates are very nuclease resistant and possess increased stability in serum and tissues relative to "natural" phosphodiesters; other structural "cousins" are in the pipeline as second generation drug candidates [22].

Just as for somatic cell "gene therapy", momentum for the development of oligos as drugs has been derived from molecular biology and cellular studies, especially in the case of "antisense" drugs, since sequencing genes simultaneously reveals the target sequence and structure of the proposed drug.

Due to the wealth of information now available on gene sequences, there has been a burst of exploration of the capacity for oligos to inhibit protein expression. These efforts have been directed mainly at downregulating proteins (ranging from growth factors to neurotransmitter receptors and enzymes critical for viral replication) whose independent overexpression critically contributes to human disease [2, 23]. Other support for development in this field has been fueled by promising results from animal studies as well as preliminary clinical trials. For instance, an oligo may have delayed cytomegalovirus (CMV) retinitis progression in a limited number of AIDS patients with advanced disease [17], and may have indicated that clinical efficacy is possible in other drugs of this class.

Thus, oligos represent a family of novel structures which may offer new therapeutic options. However, the potential limitations to therapy, or "safety issues", have been incompletely elucidated by early, cautious clinical trials and short-term animal studies that have been conducted to date. Various oligos now in clinical trials have been noted in animal studies to cause both acute and chronic toxicity to the liver, kidney, and blood, despite references in a number of published articles to the contrary. These early trials have offered critical information for determining starting and maximum dose, administration rates, trial duration, and patient enrollment. This chapter sets out some of the current safety concerns which have arisen in recent Food and Drug Administration (FDA) reviews of Investigational New Drug Applications (INDs), and describes some pharmacologic and toxicologic methods that have been successful in obtaining regulatory approval for the clinical trials[1]. The information presented here is based on the types of data viewed in INDs; current findings from published literature are mentioned where these are supported by more extensive, proprietary toxicity data.

2   Oligonucleotide Action and Drug Discovery

One hypothesis explaining the observed activity of oligos has been based largely on cell-free experimental observations in which oligo drugs (antisense) formed double helices by Watson-Crick base pairing with complementary ribonucleic acid (RNA; sense) sequences. In cultured cells and *in vivo*, incubation of cells with oligos has resulted in reduced cellular expression of target proteins. Two main mechanisms have been proposed to explain this observation, and have been supported by experimental findings. Binding and double helix formation of the DNA oligo to mRNA may hinder ribosomal translation of the message, limiting protein synthesis. Also, an important role may be served by activation of RNase H [10], a cytoplasmic enzyme which recognizes and cleaves DNA/RNA hybrid helices, resulting in decreased message quantity or half-life. Viewed from this mechanistic perspective, oligos have often been referred to as "antisense drugs" and several investigational drug structures have been based on the antisense/sense model.

However, a number of other cellular and systemic effects of oligos may be attributable to alternative mechanisms, such as non-sequence dependent effects [21]. These effects may occur by virtue of physicochemical properties, and may be expected to be occur in association with any nucleotide sequence. Alternatively, effects may be sequence-dependent, i.e. due to short sequences which form unique structural motifs, and confer target selectivity by virtue of three-dimensional structure [15, 25, 26]. Therefore, various properties of oligos offer focal points for design optimization: these include direct binding and affinity for protein targets (such as specific receptor or enzyme inhibition discovered in a broad screen of random sequences), ability to activate RNase H, or demonstration of favorable body distribution patterns or kinetics (that may contribute to prolonging active levels of drug in the diseased tissue). Phosphoro-

---

[1] Additional advisement regarding oligonucleotide drug chemistry standards have been published [13] which represents a consensus position among a number of FDA reviewing divisions. Persons, whether interested individuals or sponsoring firms, persuing drug development are encouraged to seek out information from the specific FDA reviewing divisions which regulate the diseases of interest and to request available Pre-IND points to consider documents which pertain to chemistry, pharmacology/toxicology, microbiology (in the case of infectious diseases), and clinical sections of an IND. These working standards may be of some assistance in clarifying the "state of art" while more formal guidelines are finalized through the International Conference on Harmonization (ICH) negotiations.

thioates containing a binding motif for transcription factors may also regulate transcription of DNA through direct interaction [11], offering other pathways for affecting gene expression than previously appreciated. Careful selection of oligo sequence is mandatory to allow for specific suppression of translation of transcription factors of the mRNA level [19, 20].

Because oligos are capable of affecting biological systems through multiple mechanisms, drug candidates may be selected based on any of, or several of, these properties which offer opportunities for therapeutic intervention. Therefore, drug development and regulatory strategies for oligos are and will continue to be similar to those used for other drug structural classes, especially regarding the exploration of class-related or unique toxicities.

3   Drug Development of Antiviral Oligonucleotides

The identification of a mechanism for oligo drug action is very important from a standpoint of drug *discovery*. For instance, the mechanism of a drug may critically determine its efficacy against a virus in the face of resistance development, and additionally determine which drugs are proposed for combination therapy use, i.e. to prevent resistance development. However, from the viewpoint of U.S. regulatory standards, drug *development* (clinical investigation) of new molecular entities does not require proof of mechanism, only setting out a reasonable rationale for testing a given drug for a given disease indication [6]. This rationale is explained in the investigational plan submitted with an IND, and is adequately supported by results of in vitro studies. The rationale for clinical trials of oligo drugs to inhibit viral replication have been adequately supported, for example, by *in vitro* studies documenting reduced numbers of viral particles, reduced spread of infection to uninfected cells, or prolonged survival of cells in culture. Mechanistic studies and demonstrations of activity in animal models of disease (where possible), as well as journal publications (such as those relating the target molecule to pathophysiologic changes) are also supportive, but are rarely specified as a requirement prior to conduct of a clinical trial (a hold issue).

But while regulatory codes do not require *in vivo* data to support rationale, it may be in the long view misleading to rely on purely *in vitro* data to identify drug candidates ("leads"). Many INDs have been submitted for molecules which have only been shown to permeate a (single) biological membrane, possess activity against a tested experimental endpoint, and pass a gross screen for cytotoxicity over the course of several days; in other words, meeting the criteria for a useful, and potentially selective, *research tool.* Successfully meeting these criteria establishes only the possibility that a given lead compound may perform as sucessfully *in vivo.*

Drugs must be bioavailable (i.e. injected or orally absorbed) and overcome a number of disposition barriers and methods of elimination to attain therapeutic levels in a patient for an adequate period of time. Additionally, sustained levels (for weeks to years) may be required to treat serious diseases; a concept widely ascribed to in Western pharmacotherapeutic practices. This continual assault on the organs engaged in metabolism and clearance favors the likelihood of toxicities[2], and tolerance development, as well as the predominance of non-specific activities. Consequently, patient intolerance or absence of therapeutic benefit has resulted in sponsors terminating INDs; approximately 85 % of all INDs for new molecular entities have been terminated due to limitations discovered mainly in Phase 2 trials [8].

4 Clinical Risk Assessment

Since investigational drugs, including oligos, may be clinically toxic while having unproven or no activity, FDA regulations require that early IND clinical studies stress safety over mechanistic evaluation. The goal is to protect patients from unacceptable risk in the face of uncertain therapeutic benefit. In viewing the data gathered on an investigational compound, FDA reviewers balance risks due to toxicity, lack of efficacy, and disease progression; a "risk-benefit" analysis. Consideration of the economic benefit to industry (cost-benefit balancing) is not an FDA function.

The aspects of the clinical study design that are most critically viewed for safety include the intended patient population (healthy, early or advanced disease), the disease indication (life-threatening or not), doses (starting and maximum), dose escalation and frequency, and duration of the dosing period; as well as the kinetic and disposition issues which may arise from formulation (i.e. improving bioavailability) or delivery method (i.e. route or rate of administration) adjustments. These latter issues are especially important for drugs early in development, when adjustments may make large differences in maximum plasma levels (Cmax) or total exposure (i.e. Area Under the Curves, or "AUCs"). Toxicities seen in animal studies frequently can be ascribed to AUC/duration issues, or to more acute, Cmax-dependent events, when pharmacokinetic data are available in the same species. These non-clinical data are relied upon in order to gauge the clinical risk factors associated with increased blood levels.

---

[2] An example of non-specific effects may be the case for some oligo drugs: if the drug has non-selective inhibitory effects on nucleic acid polymerases, it may cause interference in cell growth regulation and disrupt tissues which require continual or rapid cell turnover (gastrointestinal, immune, reproductive, and hematopoietic systems) to remain healthy.

5   Non-clinical Studies – Evaluating Risk and Supporting Clinical Trials

A number of differing types of information submitted with an IND are utilized to evaluate the safety of the trial. Chemistry information is required to document the results (quality and purity) and reproducibility of the drug manufacturing processes. Microbiologic activity data are submitted for drugs indicated for infectious diseases; these data can bear on safety issues when a drug is found to activate viral replication, or result in accelerated resistance development, and are often used by drug sponsors to identify target, "active" blood levels in patients.

However, mainly *in vivo* pharmacologic and toxicologic data are submitted for safety assessment, especially for gauging a safe starting dose[3]. Pharmacology data are utilized in determining whether a clinical protocol is allowed to proceed, is required to be modified, or must be placed "on hold" until further information is submitted to elucidate a safety concern. In the case of an initial IND submission, most safety data are derived from only animal studies. However, single dose studies, conducted in healthy volunteers, can be submitted with the IND in the section entitled "previous human experience"; these studies can complement animal studies and support IND approval where initial IND trials are proposed using short-term multiple dosing. The manufacturer or IND sponsor (not always the identical firm or individual) is responsible for conducting studies, compiling and submitting the IND data, including further preclinical work, patient adverse drug reaction reports and the clinical trial results specified by the protocol. FDA staff receive complete copies of the submitted data, and draft scientific reviews based on published scientific literature and data from similar products studied under other INDs.

The criteria for evaluating IND data are outlined in the Code of the Federal Register (CFR) in Title 21, Part 312 [6], which states that the data submitted to the IND file should be adequate to support the safety of the (accompanying) clinical investigation. Therefore, non-clinical safety, or "toxicology", studies submitted are viewed as particularly important, especially early in the IND while there is a paucity of clinical data on a new molecular entity. These studies are designed to evaluate a broad spectrum of clinical pathology and histopathologic endpoints under conditions which emulate the clinical administration conditions. Studies are conducted following a period of sustained supra-clinical exposure to the drug, and evaluated by comparing responses in drug-treated with vehicle-treated control

---

[3] Even if a manufacturing process results in an "impure" product, clinical trials can proceed so long as the impurities are characterized and toxicology tests are performed with the material produced and intended for clinical use. Subsequent removal of impurities is rarely regarded as being as difficult an issue as the converse, the appearance of new impurities.

groups. Doses chosen for toxicology studies generally match or exceed the doses intended for human use, so that upper dose groups demonstrate severe toxicities, and hopefully, lower dose groups identify a No Effect Level (NOEL).

The adequacy of the toxicology data is decided during an independent evaluation conducted by FDA reviewing pharmacologists. One step in evaluating the toxicology study data involves determining the NOELs, No Adverse Effect Levels (NOAELs), Minimum Efficacious Levels (MELs; derived from activity models), and Minimum Lethal Doses (MLDs) associated with the various effects seen. Doses are most often expressed in milligrams per kilogram (mg/kg), or "nominal dose", associated with the most sensitive indicators of activity or toxicity. These doses are corrected for body surface area using correction factors[4]. For an initial IND submission, this evaluation must be complete and the decision relayed to the sponsor within 30 days of IND receipt. An absence of response from FDA results in the sponsor proceeding with the clinical trial as proposed.

Due to the rigor with which toxicity studies are conducted (many monitored parameters, carefully standardized and documented procedures according to Good Laboratory Practices or GLPs), and their critical role in clinical risk assessment, these studies are normally carried out at specialized animal facilities in large pharmaceutical houses or contract toxicology firms. Quality assured, GLP toxicology studies are designed to far outstrip, in sensitivity and scope, the acute observational study in which several animals are injected and monitored for clinical signs of agonal (life-threatening) responses (such as respiratory distress, bloody urine or feces, temperature drops, or changes in seizure/spasm, activity level, posture or gait).

Acute studies are generally only sensitive enough to identify severe, acute effects, particularly if a compound has pharmacologic activity on central or peripheral nervous, cardiovascular, or respiratory systems. In this context, the absence of observable acute effects should not be interpreted by the investigator as meaning a compound is "non-toxic", a common comment found in published toxicity studies performed in academic laboratories. Chronic effects, not acute ones, are likely to be the most limiting factor to therapeutic drug use as the most serious or life-threatening diseases (those most often earmarked for development effort by venture capital-

---

[4] The NOEL is the highest nominal dose which causes no adverse effects. The human dose which is equivalent to this would be the NOEL in the most sensitive species, divided by the following correction factors depending on the species utilized: rats, 6.3; mice, 12.3; rabbits, 3.1; monkeys, 3.1; dogs, 1.8. Thus the human proposed safe dose would be smaller than that given to the animal. This method is utilized for risk assessment for most metabolized drugs early in development, on the assumption that metabolic rates are better correlated with surface area than other physiologic parameters.

funded drug sponsors) may only respond to aggressive, chronic treatment. For instance, it is not at all uncommon for the minimum lethal dose in rodents to decrease 10-fold, when comparing results following a single injection with results during 2 weeks of daily injections. While repeat-dose studies are much more revealing of a drug's potential for toxicity, they are often very expensive (especially since they utilize milligram quantities of oligos) and are therefore rarely, if ever, conducted in academic laboratories funded by federal grants.

6   Toxicology Studies of Oligonucleotides and Safety Evaluation

Consideration of clinical trials for oligo agents has been unique in some respects. Unlike many other classes of drugs under development, oligos had not been evaluated for pharmacologic and toxicologic effects in animals to any significant extent prior to the first IND application to the FDA. Only a few papers published in the late 80's and early 90's touched on acute effects of phosphorothioates in animals [1], and none reported on drug safety in more depth than observations of readily apparent toxic effects following acute dosing.

In fact, at this writing, there are still only a few published articles reporting data on toxicities or drug disposition following repeated doses [9, 12, 18]. The absence of a published scientific database of in vivo physiologic or pharmacologic data has placed an extra burden on oligo drug sponsors to independently generate their own, proprietary databases for drug safety, which must be very complete in order to independently elucidate the clinical dosing "safe range" for their product. These proprietary safety studies are infrequently discussed at meetings during the IND and are more rarely published prior to the drug's approval. So these data from multiple drug development programs, sometimes generated in studies of very similar structures are not freely available.

These proprietary data help FDA reviewers note class-related (sequence-independent) toxicities, but do not assist the multiple drug sponsors who may be unaware that they are duplicating each other's efforts, or all working on a structural class bound to have a relatively low therapeutic index.

7   Toxicity Studies for Drug Regulatory Approval

Most toxicity studies assessing drug-induced toxicity adhere to standard protocols and evaluate the health of an animal following a period of time in which animals are dosed in a manner similar to that planned for clinical study, often at 1-day intervals. Under special circumstances, longer dose intervals (the interval between doses) are utilized, such as every-other-day or weekly. These are normally employed only if the drug is thought to have a very long half-life at the site of activity, or accumulate in a body compartment to a sufficient degree to necessitate infrequent clinical dosing (noting that accumulation may not necessarily cause toxicities). Dose interval is usually constant, is similar for studies conducted in multiple species, and matches, or is shorter than, the interval proposed for clinical use.

While toxicology studies are most frequently conducted using daily dosing, sometimes (like for short half-life drugs) twice- or thrice-daily dosing is justified; or (for long half-life drugs) every-other-day may be appropriate. However, infrequent dosing in preclinical studies can be problematic. It is a common pattern in clinical development programs of long half-life drugs (such as nuclease-stabilized oligos) for the sponsor to initiate toxicity studies and Phase 2 clinical trials with infrequent dosing. But, when no early clinical activity is seen, protocols are submitted to the IND for higher doses and shorter dose intervals, to maintain higher plasma levels. Sponsors are best advised to ensure that early toxicity studies include a daily dosing group to ensure that this possibility is covered. Otherwise such a change in clinical investigation may be held up by new toxicity studies.

Typical toxicity study designs employ standardized study durations that fall into 2 main categories: acute (1–3 days with immediate sacrifice or 1-day followed by a 14-day, drug-free observation period and sacrifice); and subchronic, meaning 2 weeks, 1, 3, 6 or 12 months. Chronic assays for carcinogenicity are normally conducted with daily dosing for 2 years. Parameters studied in toxicology protocols include thorough evaluation of histopathologic (macroscopic and microscopic), hematologic, and serum chemical changes; as well clinical observation, body weight, food consumption, and urinalysis measures. Tissues which have often proved to be sites of oligo (or nucleotide) toxicity include (among others) spleen/lymph node, heart, bone marrow, gut, testes/ovaries, liver, pancreas, nerve/brain, lung, and kidney.

Other aspects of toxicity study design follow the predicted clinical use of the drug. The aim of the study should be to "match and exceed" clinical exposure, using preclinical pharmacokinetic studies performed to ensure that doses used in the toxicity studies accomplish this aim. Additionally, when exceeding clinical exposure, dosing should be adequate to cause marked multi-system toxicities or mortality in the high-dose group; while the mid- or low-dose (i.e. 2 to 10-fold lower) group might indicate the MELs or NOELs. At least 2 species (a rodent and non-rodent) should be used for studies throughout development. Study designs to meet regulatory guidelines are often standardized; preclinical programs routinely include evaluation of subchronic, reproductive, and genetic toxicity; as well as of carcinogenic potential.

However, drug development programs can be specifically tailored based on early clinical and preclinical findings, through negotiation with the FDA reviewers. Care should be taken in agreements with a specific division when a traditional program is cut down to meet the minimum needs of a specific (i.e. life-threatening) clinical indication; drugs are often resubmitted under new INDs for other (less risky) diseases, reviewed by a separate, autonomous FDA team. While the new molecular entity evaluated under each of several INDs may be the same, the risk-benefit analysis conducted separately for each different circumstance (disease type or stage, or patient population such as adult versus pediatric) can alter the FDA requirements regarding the nature or timing of long-term preclinical studies.

Sometimes circumstances (especially a formulation, population, or route of administration change such as going from oral to topical exposure) necessitate extra, "special toxicity studies", which are needed to address specific new clinical risks. Examples include a possible need for phototoxicity studies for a drug used orally in one disease and topically in another; occasionally, extended studies in young, developing animals may be needed before pediatric diseases can be studied.

Timing of preclinical study report submission, in comparison to planned start dates for clinical studies, is an issue that is best discussed in detail with each reviewing division. In general, to support the *initial* Phase 1 clinical trial of an oligo drug, the following types of non-clinical toxicity studies are needed: (1) single-dose toxicity studies that assess body weight, clinical signs, and necropsy results following a 14-day observation period; (2) 2-week, repeated-dose toxicity studies; and (3) *in vitro* and *in vivo* studies assessing genetic toxicity in both mammalian and bacterial systems. To support Phase I trials longer than one week in duration, Phase II and Phase III clinical trials, the requirements for non-clinical studies are mainly determined by the projected duration of clinical trials, the types of toxicities noted in animals and Phase 1 patients, and the maximum proposed therapeutic dose.

8   Special Considerations for Oligonucleotides

The issue of the relationship of a drug's observed activities and its mechanism is one which may be critical to determining the eventual clinical success of a drug. For instance, an antiviral drug may be especially successful if it, by virtue of its mechanism, is able (1) to subvert a viruses' attempts to develop resistance, or (2) to force, perhaps in combination with another agent, a lethal mutation on the part of the virus. The regulatory importance of mechanism from a safety perspective is highlighted by the following discussion of targets. It is recognized that *in vitro* studies, but not *in vivo* studies, can be conducted practically with a number of sequence controls to aid in mechanistic characterization, and that it may be difficult to accurately attribute activities to a specific mechanism. Such attribution is not required to conduct a clinical trial, but a thorough understanding of the mechanism may assist the developer in designing clinical trials which may optimally detect the clinical activity at an early stage of Phase 2 development. Achieving this goal is often crucial for small firms to attract the investment funding needed for larger, Phase 2/3 trials. Sponsors are encouraged to submit to the IND activity (i.e. *in vitro*) studies which utilize control oligo sequences to assist in defining the mechanism of action for their drug [14]. However, defining a proposed drug's mechanism is challenging, and essential contributions to the drug's activity from secondary mechanisms (i.e. *in vivo*) can rarely be ruled out. Therefore, statements regarding the drug mechanism in the IND (Investigator's Brochure and in the patient informed consent form) should be phrased to reflect a conservative interpretation of the data.

8.1 Determinants of Activity and Toxicity

In contrast to activity study goals, safety evaluation of toxicity studies is often essentially phenomenologic, emphasizing dose-effect relationships without much speculation on mechanism. However, there are situations in which toxicity mechanisms may need clarification, as may be the case in broad oligo drug development programs with the goal of identifying new leads which lack a given toxic property. Toxicity measures for these development purposes need not be broad, but rather focused to screen for a specific property (i.e. complement activation, anemia, or renal cortical cell toxicity). In this case, conducting some short-term (i.e. 1–2 week) toxicity studies which compare different sequences and backbone structures may be helpful.

In order to identify the structural or sequence determinants of toxicity, a variety of control oligos (i.e. inverse, scrambled, or truncated analogs) would have to be tested in repeat-dosing animal studies or in alternative systems such as primary or cell-line cultures derived from organs manifesting toxic responses in more standard assays. In theory, toxicities, or activities for that matter, may be due to *sequence-dependent* effects. These effects may be either *antisense* (complementary binding to a mRNA) effects, or sequence-dependent, but *non-antisense*, effects (like motifs contributing to specific protein-binding). Alternatively, some oligo effects may be completely unrelated to sequence, and are due to other chemical properties of the drug such as the oligo backbone chemistry or chemical adducts.

Regulatory safety study expectations will ordinarily be limited to traditional toxicology studies. These studies should be adequate to detect *non-sequence-related* toxicities such as those stemming from biochemical properties of the backbone (e.g. linkage chemistry). Toxicities seen so far in preclinical safety studies have likely been due to this mechanism, and are therefore, equally relevant in animals and humans. For drug discovery purposes, the drug sponsor may elect to clarify the role of backbone chemistry in the adverse effects. To do so, the same nucleotide sequence could be synthesized yielding several drug analogs using different nuclease-resistant linkage chemistries for each one; these analogs then could be compared with the lead investigational drug in animal screening assays.

To evaluate an oligo for *sequence-dependent, antisense effects* (activity or toxicity), the target gene sequence must be the same in humans and in the test species. If there is a gene in a convenient test animal serving an equivalent purpose to that in humans, but which is not entirely identical in sequence, then one approach taken by several drug sponsors is to make an experimental oligo drug analog that is complementary to the sequence found in the test species. In this manner, adverse responses to antisense effects (i.e. downregulation of a protein which serves both pathophysiologic and homeostatic roles) might be evaluated in an animal lacking the target human sequence. In a regulatory context, the value of comparing such a drug analog side-by-side with the human drug in a short-term toxicity study (i.e. 2 high-dose groups) would be to determine if the effects were very similar. If they were, the dose-limiting toxicities would be ascribed to non-sequence-dependent effects. In this case animal studies would be viewed as appropriate for modeling potential human toxicities.

For antisense molecules directed at nucleic acid targets in infectious agents, *sequence-dependent, non-specific* activities may result from (1) conformation-related binding to nucleic acids with a complementary structure (as opposed to sequence); (2) binding to macromolecules other than nucleic acid targets; or (3) partial binding to short sequences in the human genome which have partial sequence complementarity "by chance". While no animal studies would be suitable for detecting these sequence-dependent, but unintended activities, studies of responses in human cell lines might be valuable in ruling out or identifying such effects. Chance complementarity may be reduced by designing oligos to avoid complementary to known, common human sequences through the use of genetic data banks.

8.2   Special Toxicity

The conduct of special toxicity or pharmacodynamic studies to examine immediate or delayed toxicity in specific end-organs by antisense therapeutics should be guided by non-clinical toxicity findings in 14- and 28-day studies, mechanistic considerations, and clinical findings [3]. In addition to special toxicity studies, studies may be needed to characterize toxicities specific to the route of administration. For instance; topical, intradermal, or intramuscular administration may warrant local irritancy or hypersensitivity testing; intravenous administration may warrant cardiovascular testing [4]. Additionally, assessments of immunotoxicity may reveal mitogenic or inflammatory effects since enlarged spleens, B-cell stimulation, lymphocytic organ infiltrations, and elevated immunoglobulin levels have been noted in rodents treated with phosphorothioate or phosphodiester oligos [5, 16]. In the light of these findings, the following measures, when included in toxicity studies or programs could offer some clues regarding the doses and clinical relevance of these effects: (1) flow cytometry counts of blood cell sub-populations; (2) histopathological evaluations of bone marrow and organs of the immune system; and (3) studies of immunogenicity (including serum measures of specific antibody production). Additional *in vitro* assays of cytotoxic effects may provide useful insights; parameters evaluated could include nucleic acid synthesis and degradation mechanisms, cell growth, and cell growth in response to known stimuli.

Special case toxicities may occur which are unique to novel drug delivery systems [24] due to solid materials, vehicles, or unique absorption, distribution, or metabolic profiles (for instance, depots, where high concentrations of drugs are held in one place for an extended time period). To help model these conditions, toxicity studies should be conducted using a formulation and delivery method similar to that proposed for clinical use; bearing in mind that route, formulation, and rate of delivery can dramatically alter drug kinetics and disposition, and result in different patterns of toxic responses. Studies using novel drug vehicles, or ones known to have unique biologic effects, should include untreated as well as vehicle-treated control groups. In cases where formulations change, completely new toxicity studies are rarely needed; often pharmacokinetic evaluations (Area Under the Curve) and "bridging" groups (2 high-dose groups in a new toxicity study, one with the new, the other with the old formulation) can be used to reapply old data to new risk assessments.

9   Non-Clinical Pharmacokinetics and Disposition Studies

Central differences in drug response between research animals and humans are often found to be due to metabolic or drug kinetic differences. These factors also often contribute to differing responses seen in young, adult, and geriatric human populations, or differing responses between genders. Drug kinetic and metabolic studies find a somewhat different utility in preclinical development. Performing paired studies of pharmacokinetics and toxicity can facilitate analyses of correlations among structure, kinetics, non-clinical toxicities, and clinical adverse reactions. These data would provide additional interpretive power to the preliminary toxicity results and may permit less conservative clinical risk assessments. These data also would be useful in refining the clinical dosing plan, especially regarding dose scheduling and route of administration. If possible, sponsors should aim to develop and validate a method for measuring the parent drug in plasma before engaging in GLP studies in either rats or monkeys. Obtaining estimates of the half-life $(t_{1/2\,beta})$[5], area under the curve (AUC), volume of distribution (Vd), and plasma protein binding following acute and repeated dosing in animals may be advised. Due to the

---

[5] The $t_{1/2\,beta}$ is the half-life characterizing the 2nd, or elimination, phase of a two-compartment model of drug levels in plasma.

difficulty in obtaining pharmacokinetic data in a limited number of mice, extended GLP toxicity and disposition studies in mice may not be convenient, especially if a route of administration is used other than that proposed for the clinical use. Mice are typically used for 2-year carcinogenicity assessment, following a 3-month study, dose-ranging whether or not mice were used in safety, subchronic studies.

Many alternative backbone structures have been proposed for oligo drugs in order to prevent immediate hydrolysis of the drug by serum and cellular nucleases. However, these "nuclease-stable" linkages may not completely prevent hydrolysis by endogenous nucleases. Therefore, consideration should be paid to the toxicities of the degradation products of this cleavage.

The safety implications of oligo degradation products or metabolites are currently unknown but may be elucidated by comparison of data from safety studies with data from detailed pharmacokinetic and metabolic studies. Some natural and synthetic nucleoside analogs are known to be toxic to tissues with high cell turnover as well as those with very active metabolisms. Therefore, characterizing the *in vivo* metabolites of the parent drug could be studied early in development to help address this concern. *In vitro* metabolic evaluation in systems such as human hepatocyte preparations (as well as hepatic preparations derived from the animal species used in non-clinical toxicity studies) also may yield some useful insights. The placement of the radiolabel in the drug could be critical to obtaining a complete picture of the drug's metabolism. Metabolic enzymes may hydrolyze bases from the phosphoribose backbone; ^{14}C-labelling of the bases may be helpful in assessing the significance of this metabolic route. Studies using ^{35}S-radiolabelled phosphorothioates may clarify whether backbone cleavage can result in the production of monothionucleotides, whether they may accumulate, or if they are substrates for nucleic acid polymerases and are incorporated into the genome. A discussion of this topic can be found in chapter 3, Labelling.

Assays for drug levels that employ chromatographic methods or radiolabelled tracers should be sensitive enough to provide an accurate assessment of drug exposure through measurements made on biological samples. Problems caused by using insensitive assays include calculating too short a $t_{1/2}$ due to a misattribution of a rapid distribution rate from plasma as a single plasma clearance rate. In fact, if a more sensitive assay were used, the $t_{1/2\,beta}$ might be found to be prolonged. The degree of regulatory clinical concerns in this case could be underestimated by the sponsor, because they might dismiss the potential for target organ accumulation with daily dosing. Early development of assay methods (monoclonal antibody, gel electrophoresis, high performance liquid chromatography, and mass spectrometry systems) which detect "cold" drug and metabolites in biologic sample following *in vivo* dosing are highly encouraged because of their utility in toxicology and clinical studies. Due to the concern for long $t_{1/2\,beta}$s seen with several phosphorothioates, regulatory recommendations have often included evaluating intact drug levels in critical target organs (i.e. liver and kidney) at the termination of repeat-dose toxicity or disposition studies.

10   Special Considerations for *ex vivo* Uses

Antisense drugs may be used for immunosuppressant therapies; these include inhibiting solid organ graft rejection, an indication regulated by the Antiviral Division since 1991. (Other immunosuppressant drugs, such as those for multiple sclerosis, colitis, psoriasis, and rheumatoid arthritis are each regulated in separate divisions.) Two main approaches to treating this indication have been suggested: 1) systemic treatment of the patient to suppress host immune response to the engrafted organ, and 2) *ex vivo* pretreatment of the donated organ to minimize host recognition of the foreign tissue. In this latter case, which would employ a single *ex vivo* treatment of the organ prior to transplantation, chronic toxicity data in animals would not be required. Instead, several types of special pharmacology and toxicology evaluation would be in order, prior to the initiation of Phase I clinical trials. Activity studies often can be designed to include all the necessary toxicologic endpoints (such as histopathologic assessments), and to correlate tissue levels with adverse responses. Transplantation models (treating the organ ex vivo, then engrafting and monitoring the graft's survival time) could be designed to simultaneously track pharmacokinetics and disposition in the host. This would be valuable in estimating the patients' systemic exposure to the oligo antisense drug following the *ex vivo* treatment.

This system would lend itself to the evaluation of sequence-specific and -non-specific effects if the sponsor wished to identify the compound as an antisense agent and if a test species, which was known to possess the drug's target sequence (see Section 8), can be identified. If the human drug was intended to bind to a specific gene, not present in a suitable animal transplantation model (usually mouse, rat, beagle dog, or cynomolgus monkey), then an analog to the animal sequence could be tested. In this case, it would be interesting to also test the human drug side-by-side with the animal analog – if both worked, the drug might be interpreted as having a non-antisense effect. The molecular mechanism of action is more in the sponsor's interest, though, and would have little influence on reviewers' view of the drug. Finally, additional insights might be gained from *in vitro* toxicity studies, noting adverse effects of the compound on the tissue intended for engraftment by utilizing cells (primary or cultured lines) or tissues (slices) from the donor species.

11   Updated Recommendations – Intravenous Infusion

Recent findings in animal experiments have indicated that oligo drugs (specifically phosphorothioates) are associated with some sequence-independent, class-related adverse effects which are important determinants of initial clinical trial design. These findings were apparent in 1993, by the time several INDs for systemically administered phosphorothioates had been reviewed by the FDA. For instance, common effects seen for these drugs in rats dosed for 14 days were: injection site inflammation, increased spleen weight, and lymphocytosis in organs (2–3 mg/kg), as well as increased clotting time, liver, kidney, and hemato-toxicities and decreased body weight (20–30 mg/kg). Since similar findings were apparent at similar doses for different phosphorothioate sequences, it seemed clear that non-sequence related properties of the oligos, like the backbone or charge, were causing the responses. Since these results were not being published or discussed at meetings, regulatory guidance was published [3] to convey that safety concerns were occurring close to clinical dosing ranges for these drugs. (As discussed previously, 30 mg/kg converts to approximately a 5 mg/kg human dose when a rat dose is corrected for differences in body surface area.)

Multiple INDs proposing the use of intravenous administration were not available for review until 1994. In addition to proprietary data archived with the FDA, two reports in the literature now show that hemodynamic toxicity and mortalities have occurred in monkeys following rapid intravenous infusions (i.e. over 10–15 minutes) of phosphorothioates [7, 9]. These effects have been corroborated by 3 other data sets from sponsors using phosphorothioates of 20–26 nucleotides in length and of varying sequence. The effects of main concern were broad fluctuations in mean arterial pressure and cardiac output, occurring as early as 10 minutes following initiation of infusion, and lasting over 2 hours. Doses associated with these effects were 5 to 20 mg/kg, with the higher doses being associated with mortality within 24 hours. Mortalities were not always associated with fast infusion.

Similar cardiovascular responses were not, and have not been, reported in species other than monkeys; which raised another concern. In particular, these effects were not noted in rats or mice, often used for safety pharmacologic (i.e. cardiovascular and receptor-mediated effects) screening by small firms. Again, the effect seemed to be sequence-independent (although CpG motifs are present in all oligos currently in clinical trials; [15]) and remains largely unpublished. New sponsors working on phosphorothioates might, understandably, not know of the findings, and submit INDs without adequate understanding of the clinical risks. If this is the case, the FDA typically prevents, or "holds", sponsors from initiating clinical trials, a difficult situation for all concerned. To prevent this from occurring, guidance was published [4] to call attention to the need for pharmacologic, specifically hemodynamic screening for these effects prior to IND submission.

The possibility of modeling this response has been investigated, but not published, and it is difficult to know how rigorously this path is being pursued. But even if the mean arterial pressure responses are not duplicated in species other than rhesus and cynomolgus monkeys, this is *not* a reason to dismiss monkey responses when assessing clinical risk, as the phylogenetic similarities between monkeys (especially Old World monkeys) and humans are extensive. Proposed biochemical or cellular mechanisms associated with the cardiovascular changes include direct binding of these polyanionic drugs to various serum and cellular proteins, and subsequent release of anaphylatoxins (such as complement fragments), eicosanoids, leukotrienes, and/or cytokines. If complement activation ends up being identified conclusively as the causative factor, this may explain the apparent species-specificity of these responses to date.

Based on the monkey results, sponsors considering intravenous administration of charged-backbone oligos should use primates to evaluate drug safety by an integrated analysis of acute toxicity, pharmacokinetics, and pharmacodynamics. Sponsors could accomplish these aims by performing some preliminary studies in primates at the initiation of drug safety testing. During studies utilizing short (15 min or less) infusions, measures should include at least drug levels, clinical signs, central mean arterial pressure changes, blood cell counts, electrocardiograms, and alterations in activated partial thromboplastin time (aPTT). Animals should be re-tested for these parameters 24 h following the acute dose to evaluate normalization of the affected parameters. Measurements of drug levels should aim to identify the best correlate of toxic responses; in other words, to identify a blood threshold level, a time above threshold, and/or a rate of infusion correlating with toxic responses. This would provide a method for monitoring clinical blood levels or infusion rates to ensure that clinical toxicities will be avoided. If cardiovascular (and/or other hemodynamic or hematologic) changes were evident, subsequent experiments would be designed to identify the lowest adverse effect level (LOAEL) and no effect level (NOEL) and would form an important basis for calculations of safe human doses.

The accuracy of interspecies extrapolation of safe doses in primates to safe dose estimates in humans could be improved by correlating these effects with results of *in vitro* mechanistic studies comparing biochemical changes in human and primate blood. Initial clinical protocols (doses, patient inclusion and exclusion criteria, monitoring, follow-up) then could be based on the findings seen in all animal studies and could evaluate pharmacokinetics and safety following an acute dose administration.

Until more is known about the structure-activity relationships of these drugs, the similarity of the oligo cardiovascular responses to those seen with other polyanionic or polycationic compounds suggest that these concerns regarding hemodynamics pertain to all charged-backbone oligo drugs, not just phosphorothioates. Hopefully, a rational basis will be found for limiting *in vivo* screening to drugs with specific structural determinants. Also, when the mechanisms underlying the cardiovascular toxicity have been identified, an *in vitro* screening assay might be validated which could potentially replace *in vivo* screening, reducing the demand for research primates. Drug sponsors therefore are encouraged to contact the FDA reviewing division, and to develop new animal models for safety assessment, and to openly discuss and publish results found in pharmacodynamic and mechanistic studies.

12   Updated Recommendations – Protein Targets

IND applications for oligo drugs that act on diverse cellular and serum proteins through conformational, or "non-antisense", mechanisms are already being submitted. Although unique recommendations to address safety concerns for these cases are unlikely, the following example provides one case where a scientific prediction of possible non-selective effects suggests some practical studies. The intended effect of some proposed oligo drugs is to inhibit *viral* enzymes which employ RNA and DNA as substrates (for example, reverse transcriptase). In this case, to evaluate the specificity of the effects for the target enzyme, a drug sponsor could consider performing *in vitro* evaluations of effects of the drug product on *human* enzymes which employ RNA and DNA as substrates, such as nucleic acid polymerases. Assays monitoring normal cellular growth rates and viability, as well as these endpoints following mitogenic stimuli, may address the questions regarding specificity. Additionally, cell-free enzymatic activity studies could be used as screens for non-specific inhibitory effects using panels of purified human enzymes.

13   Conclusions

Pharmacology safety concerns have been raised for phosphorothioates which have appeared to be independent of sequence, both in the context of acute responses in monkeys and subchronic responses in rodents. Toxic responses in animals have occurred at plasma levels and doses 1- to 10-fold above the projected active human doses. These concerns have not prevented clinical investigation of oligo drugs for life-threatening diseases; however, some of these effects (anticoagulant effects and potential hemodynamic toxicity) may be viewed in a risk-benefit context as unsuitable for less serious disease, and/or chronic therapy. For trials currently in progress, attention has been focused (1) on screening for effects which may prove to be dose-limiting, (2) on avoiding situations which may lead to serious adverse reactions, and (3) on studying patient populations with few therapeutic options.

In most situations, careful pursuit of clinical investigation is enabled by thorough characterization of the dose (or exposure)-response curves for drug effects, allowing an estimate to be made of a human NOEL as a clinical starting dose. A less efficient (and more costly) alternative is to proceed with less non-clinical information and more clinical caution. This path is often necessary when target organs of toxicity have not been identified in animal studies. This can occur when doses chosen for evaluation in toxicity studies are too low and the highest dose tested reveals no toxic effects. Since the minimum toxic dose and the dose response curve for toxicity are unknown in this case, only a conservative human dose extrapolation (incorporating higher safety factors) and clinical dose escalation may be initiated.

This chapter has highlighted the role of pharmacology and toxicology studies in identifying safety concerns for oligo drugs reviewed in the Division of Antiviral Drug Product of the U.S. FDA. While several phosphorothioates are currently being investigated in clinical trials with few adverse findings, the doses under investigation are low, and there are a number of cautionary steps in place to ensure that few adverse reactions occur; steps which might be viewed as impractical for chronic therapy or large patient numbers. It is, therefore, not yet known whether the effects seen in animal studies will, or can, occur in humans. Careful non-clinical safety studies may provide the data to quantitatively determine a margin of safety and facilitate approvals for initial clinical trials of these oligos, and clinical evaluation will reveal the potential therapeutic utility of these compounds. Prior to and throughout IND investigation of oligos, sponsors are encouraged to correspond with their Reviewing Division at the FDA, and to publish the results of their pharmacodynamic, pharmacokinetic, and mechanistic studies.

References

1. AGRAWAL, S.: Antisense Oligonucleotides: a possible approach for chemotherapy of AIDS. In: Prospects for antisense nucleic acid therapy of cancer and AIDS. Wiley-Liss, Inc., NY, (1991) 143–158.
2. AGRAWAL, S. , T. IKEUCHI, D. SUN, P.S. SARIN, A. KONOPKE, T. MAIZEL, P. C. ZAMECNIK: Inhibition of human immunodeficiency virus in early infected and chronically infected cells by antisense oligodeoxynucleotide and its phosphorothioate analogue. Proc. Natl. Acad. Sci. USA 86 (1989) 7790–7794.
3. BLACK, L. E., J. J. DeGEORGE, J. A. CAVAGNARO, A. JORDAN, C. H. AHN: Regulatory considerations for evaluating the pharmacology and toxicology of antisense drugs. Antisense Res. Dev. 3 (1993) 399–404.
4. BLACK, L. E., J. G. FARRELLY, J. A. CAVAGNARO, C. H. AHN, J. J. DeGEORGE, A. S. TAYLOR, A. F. DeFELICE, A. JORDAN: Regulatory considerations for oligonucleo-

tide drugs: Updated recommendations for pharmacology and toxicology studies. Antisense Res. Dev. 4 (1994) 299–301.

5. BRANDA, R. F., A. L. MOORE, L. MATHEWS, J. J. MCCORMACK, G. ZON: Immune stimulation by an antisense oligomer complementary to the rev gene of HIV-!. Biochem. Pharmacol. (1993) 2037–2043.

6. Code of Federal Regulations (CFR), 1994. 21 CFR 312.20 – .160. Contains the codified U.S. Government rules regarding INDs. Published by the Office of the Federal Register, and available from: U.S. Government Printing Office, Superintendent of Documents, Mailstop: SSOP, Washington D.C. 20402–9328.

7. CORNISH, K. G., P. IVERSEN, L. SMITH, M. ARNESON, E. BAYEVER: Cardiovascular effects of a phosphorothioate oligonucleotide with sequence antisense to p53 in the conscious rhesus monkey. Pharmacol. Communications 3(3) (1993) 239–247.

8. Department of Health and Human Services (DHHS) Publication No. 90–3168. From test tube to patient: new drug development in the United States, an *FDA Consumer* special report (revised 3/90). Available on request from: U.S. FDA, HFI-40, 5600 Fishers Lane, Rockville MD, 20857.

9. GALBRAITH, W. M., W. C. HOBSON, P. C. GICLAS, P. J. SCHECHTER, S. AGRAWAL: Complement activation and hemodynamic changes following intravenous administration of phosphorothioates oligonucleotides in the monkey. Antisense Res. Dev. 4(3) (1994) 201–207.

10. GAO, W. Y., F. S. HAN, C. STORM, W. EGAN, Y. C. CHENG: Phosphorothioate oligonucleotides are inhibitors of human DNA polymerases and RNase H: implications for antisense technology. Molecular Pharmacol. 41 (1992) 223–229.

11. HIGGINS, K. A., J. R. PEREZ, T. A. COLEMAN, K. DORSHKIND, W A. MCCOMAS, U. M. SARMIENTO, C. A. ROSEN, R. NARAYANAN: Antisense inhibition of the p65 subunit of NF-kB blocks tumorigenicity and causes tumor regression. Proc. Natl. Acad. Sci. USA 90 (1993) 9901–9905.

12. IVERSEN, P. L., J. MATA, W. G. TRACEWELL, G. ZON: Pharmacokinetics of an antisense phosphorothioate oligodeoxynucleotide against rev from human immunodeficiency virus type 1 in the adult male rat following single injection and continuous infusion. Antisense Res. Dev. 4 (1994) 43–52.

13. KAMBHAMPATI, R. V. B., Y. Y. CHIU, C.W. CHEN, J. J. BLUMENSTEIN: Regulatory concerns for the chemistry, manufacturing, and controls of oligonucleotide therapeutics for use in clinical studies. Antisense Res. Dev. 3 (1993) 405–410.

14. KRIEG, A. M., C. A. STEIN: Problems in interpretation of data derived from *in vitro* and *in vivo* use of antisense oligodeoxynucleotides. Antisense Res. Dev. 4(2) (1994) 67–69.

15. KRIEG, A. M., A. K. YI, S. MATSON, T. WALDSCHMIDT, G. A. BISHOP, R. TEASDALE, G. A. KORETZKY, D. KLINMAN: CpG motifs in bacterial DNA trigger direct B-cell activation. Nature 374 (1995) 546–549.

16. MOJCIK, C. F., M. F. GOURLEY, D. M. KLINMAN, A. M. KRIEG, F. GMELIG-MEYERS, A. D. STEINBERG: Administration of a phosphorothioate oligonucleotide, antisense to murine endogenous retroviral MCF env, causes immune effects in vivo in a sequence-specific manner. Clinical Immunol. Immunopath. 67 (1993) 130–136.

17 PALESTINE, A. G., H. L. CANTRILL, A. P. LUCKIE, E. AI: Intravitreal treatment of CMV retinitis with an antisense oligonucleotide, Isis 2922. Abstracts of the 10th International Conference on AIDS, Yokohama, Japan (1994).

18. SARMIENTO, U. M, J. R. PEREZ, J. M. BECKER, R. NARAYANAN: In vivo toxicologic effects of rel A antisense phosphorothioates in CD-1 mice. Antisense Res. Dev. 4 (1994) 99–107.

19. PEREZ, J. R., Y. LI, C. A. STEIN, S. MAJUMDER, A. Van OORSCHOT, R. NARAYANAN: Sequence-independent induction of Sp1 transcription factor activity by phosphorothioate oligodeoxynucleotides. Proc. Natl. Acad. Sci. USA 91 (1994) 5957–5961.

20. SCHLINGENSIEPEN, K. H., F. WOLLNIK, M. KUNST, R. SCHLINGENSIEPEN, T. HERDEGEN, W. BRYSCH: The role of Jun transcription factor expression and phosphorylation in neuronal differentiation, neuronal cell death, and in plastic adaptation in vivo. Cell. Mol. Neurobiol. (1995) in press.

21. STEIN, C. A., Y. C. CHENG: Antisense oligonucleotides as therapeutic agents, is the bullet really magical? Science 261 (1993) 1004–1012.

22. TONKINSON, J. L., M. GUVAKOVA, Z. KHALED, J. LEE, L. YAKUBOV, W. S. MARSHALL, M. H. CARUTHERS, C. A. STEIN: Cellular Pharmacology and protein binding of phosphoromonothioate and phosphorodithioate Oligodeoxynucleotides: a comparative study. Anisense Res. Dev. 4 (1994) 269–278.

23. WAHLESTEDT, C.: Antisense oligonucleotide strategies in neuropharmacology. Trends Pharmacol. Sci. 15 (1994) 42–46.

24. WICKSTROM, E.: Strategies for administering targeted therapeutic oligodeoxynucleotides. Trends in Biotech. 10 (1992) 281–287.

25. YAMAMOTO, T., S. YAMAMOTO, T. KATAOKA, T. TOKUNAGA: Ability of oligonucleotides with certain palindromes to induce interferon production and augment natural killer cell activity is associated with their base length. Antisense Res. Dev. 4 (1994) 119–122.

26. YASWEN, P., M. STAMPFER, K. GHOSH: Effects of sequence of thioated oligonucleotides on cultured human mammary epithelial cells. Antisense Res. Dev. 3 (1992) 67–77.

Index